Hardware Acceleration of Computational Holography

Tomoyoshi Shimobaba · Tomoyoshi Ito
Editors

Hardware Acceleration of Computational Holography

Springer

Editors
Tomoyoshi Shimobaba
Chiba University
Chiba, Japan

Tomoyoshi Ito
Chiba University
Chiba, Japan

ISBN 978-981-99-1937-6 ISBN 978-981-99-1938-3 (eBook)
https://doi.org/10.1007/978-981-99-1938-3

This Springer imprint is published by the registered company Springer Nature Singapore Pte Ltd.
The registered company address is: 152 Beach Road, #21-01/04 Gateway East, Singapore 189721, Singapore

Preface

Invented by the Hungarian physicist Dennis Gabor in 1947 to improve the performance of electron microscopes, holography uses interference and diffraction of light to record three-dimensional information on two-dimensional recording media. The resulting records are referred to as holograms, and the recorded three-dimensional image is faithfully reproduced when such holograms are irradiated with visible light. Hence, holography has been described as the ultimate 3D imaging technology. In the 1960s, after lasers with good coherence were developed, E. Leith and J. Upatnieks showed that off-axis holography could be used to obtain 3D images that could be mistaken for real objects. This led to holography attracting attention as a promising 3D imaging technology.

Conventional holography was developed as an analog method involving recording holograms on photosensitive materials. However, research began to be conducted on recording holograms electronically instead of on photosensitive materials, as well as on calculating holograms with computational hardware. For example, works by B. R. Brown and A. W. Lohmann (Appl. Opt. **5**, 967–969 (1966)) and J. W. Goodman (Appl. Phys. Lett. **11**, 77 (1967)) are representative pioneering studies on computer-generated and digital holography.

In the 1990s, S. A. Benton of MIT showed that an electronic holographic display could be realized using an acousto-optic modulator (Proc. SPIE **1212**, Practical Holography IV, (1 May 1990)) and N. Hashimoto of Citizen reported a holographic LCD device (Proc. SPIE **1461**, Practical Holography V, (1 July 1991)).

The theory of holography was intensively studied until the 1970s, and the theoretical foundations of the subject were extensively elaborated. Research on computational holography began around 1970, and has now been adapted to a wide range of applications with the rapid development of computer hardware. Typical applications include holographic displays, digital holography, computer-generated holograms (CGH), holographic memory, and optical cryptography.

Digital holography techniques have been developed to capture holograms with resolutions greater than one gigapixel, and holographic displays ultimately need to compute holograms with resolutions that exceed terapixels to generate highly realistic images. Hence, accelerating the computation of holographic images is an important issue and is expected to require the development of new algorithms and the adoption of specialized hardware. The hardware used to control holographic displays includes multi-core CPUs, graphics processing units (GPUs), and field programmable gate arrays (FPGAs), which are integrated circuits to that can be freely configured by users. Owing to the evolution of the associated development environment, the difficulties associated with designing FPGAs have subsided in recent years.

This book is a guide to computational holography and its acceleration. Most of the chapters were written by young researchers who are expected to play an active role in this field in the future. The book is divided into four parts.

Part I consists of Chaps. 1–3, which explain the basic mechanics of light in terms of wave optics along with the basics of holography and CGH. This introduction to the subject was written by Prof. Takashi Kakue (Chiba Univ. Japan) and Dr. Yasuyuki Ichihashi (NICT, Japan).

Computational holography requires high-speed computation that makes full use of CPU, GPU, and FPGA hardware. In Chaps. 4 through 7 of Part II, the features and usage of these types of hardware are explained by Takashige Sugie (formerly of Chiba Univ. Japan), Minoru Oikawa (Kochi University Japan), Takashi Nishitsuji (Tokyo Metropolitan Univ. Japan), and Yota Yamamoto (Tokyo Univ. of Science, Japan).

Part III consists of Chaps. 8 through 18. Specific examples of implementations using C++, MATLAB, and Python are provided for diffraction and hologram calculations, which are important in computational holography. Mr. Soma Fujimori (Master's student, Chiba University Japan) explains CPU and GPU implementations of diffraction calculations. CGH algorithms include point-cloud, polygon, layer, and light-field methods. Specific methods to implement each of these approaches are provided by Prof. Takashi Nishitsuji, Mr. Fan Wang (Ph.D. student, Chiba University Japan), Dr. Yasuyuki Ichihashi, Mr. Harutaka Shiomi (Ph.D. student, Chiba Univ. Japan), and Prof. David Blinder (Vrije Universiteit Brussel and Imec, Belgium). Visual quality assessments for holography are provided by Dr. Tobias Birnbaum (Vrije Universiteit Brussel and Imec, Belgium). An overview written by Dr. Tatsuki Tahara (NICT, Japan) from the perspective of the high-speed reproduction of holograms acquired by digital holography is also included, along with an overview of parallel computing using PC and GPU clusters by Prof. Naoki Takada (Kochi Univ., Japan). A specific implementation of compressed holography is provided by Dr. Yutaka Endo (Kanazawa Univ. Japan) to demonstrate the advantages of compressed sensing in reducing the noise inherent in digital holography by optimizing the sparsity of the signal.

Part IV comprises Chaps. 19 and 20. Here, Dr. Yota Yamamoto describes an approach to implement point-cloud hologram calculations on FPGA hardware, and Prof. Nobuyuki Masuda (Tokyo Univ. of Science, Japan) and Dr. Ikuo Hoshi (NICT, Japan) explain the implementation of diffraction calculations used in digital holography on this hardware.

Many excellent textbooks and commentaries on computational holography are available, such as "Introduction to Modern Digital Holography" (Cambridge University Press, 2014) by T.-C. Poon and J.-P. Liu; "Analog and Digital Holography with

MATLAB" (SPIE, 2015) by G. T. Nehmetallah, R. Aylo, and L. William; and "Introduction to Computer Holography" (Springer, 2020) by K. Matsushima. Compared with these works, one unique aspect of this book is that specific implementations of methods to accelerate computational holography are provided from both algorithmic and hardware-centered perspectives. This book contains many sample source codes, which can be downloaded from the book's website.

We hope that this book will be helpful for students and researchers working on computational holography in the future, as well as for those who have been actively engaged in this field.

The editors hereby acknowledge the support of the Japan Society for the Promotion of Science (22H03607).

Chiba, Japan
December 2022

Tomoyoshi Shimobaba
Tomoyoshi Ito

Contents

Part I Introduction to Holography

1 **Light Wave, Diffraction, and Holography** 3
Takashi Kakue

2 **Computer-Generated Hologram** 25
Yasuyuki Ichihashi

3 **Basics of Digital Holography** 31
Takashi Kakue

Part II Introduction to Hardware

4 **Basic Knowledge of CPU Architecture for High Performance** 47
Takashige Sugie

5 **Basics of CUDA** 67
Minoru Oikawa

6 **Basics of OpenCL** 83
Takashi Nishitsuji

7 **Basics of Field-Programmable Gate Array** 97
Yota Yamamoto

Part III Acceleration and Advanced Techniques in Computational Holography

8 **CPU and GPU Implementations of Diffraction Calculations** 119
Soma Fujimori

9 **Acceleration of CGH Computing from Point Cloud for CPU** 137
Takashige Sugie

10 Computer-Generated Hologram Calculation Employing the Graphics Processing Unit 169
Takashi Nishitsuji

11 Computer-Generated Hologram: Multiview Image Approach 185
Yasuyuki Ichihashi

12 Hologram Calculation Using Layer Methods 193
Harutaka Shiomi

13 Polygon-Based Hologram Calculation Methods 207
Fan Wang

14 Real-Time Electroholography Based on Multi-GPU Cluster 227
Naoki Takada

15 GPU Acceleration of Compressive Holography 241
Yutaka Endo

16 Sparse CGH and the Acceleration of Phase-Added Stereograms 253
David Blinder

17 Efficient and Correct Numerical Reconstructions 271
Tobias Birnbaum

18 Digital Holography Techniques and Systems for Acceleration of Measurement 303
Tatsuki Tahara

Part IV FPGA-Based Acceleration for Computational Holography

19 FPGA Accelerator for Computer-Generated Hologram 329
Yota Yamamoto

20 Special-Purpose Computer for Digital Holography 343
Nobuyuki Masuda and Ikuo Hoshi

Index 365

Contributors

Tobias Birnbaum Vrije Universiteit Brussels, Brussels, Belgium; IMEC, Brussels, Belgium

David Blinder Vrije Universiteit Brussels, Brussels, Belgium; imec, Brussels, Belgium

Yutaka Endo Institute of Science and Engineering, Kanazawa University, Kanazawa, Ishikawa, Japan

Soma Fujimori Graduate School of Science and Engineering, Chiba University, Chiba, Japan

Ikuo Hoshi Radio Research Institute, National Institute of Information and Communications Technology (NICT), Koganei, Tokyo, Japan

Yasuyuki Ichihashi Radio Research Institute, National Institute of Information and Communications Technology (NICT), Koganei, Tokyo, Japan

Takashi Kakue Graduate School of Engineering, Chiba University, Chiba, Japan

Nobuyuki Masuda Department of Applied Electronics, Tokyo University of Science, Katsushika-ku, Tokyo, Japan

Takashi Nishitsuji Faculty of Systems Design, Tokyo Metropolitan University, Hino-shi, Tokyo, Japan

Minoru Oikawa Kochi University, Kochi, Japan

Harutaka Shiomi Graduate School of Engineering, Chiba University, Chiba, Japan

Takashige Sugie Chiba University, Inege-ku, Chiba, Japan

Tatsuki Tahara Radio Research Institute, National Institute of Information and Communications Technology (NICT), Koganei, Tokyo, Japan

Naoki Takada Kochi University, Kochi, Japan

Fan Wang Graduate School of Engineering, Chiba University, Chiba, Japan

Yota Yamamoto Faculty of Engineering, Tokyo University of Science, Katsushika-ku, Tokyo, Japan

Part I
Introduction to Holography

Part I consists of three chapters. In computer holography, diffraction calculations play an important role. First, diffraction calculations are explained, followed by the principles of holography, computer holograms, and digital holography.

Chapter 1
Light Wave, Diffraction, and Holography

Takashi Kakue

Abstract Optics can be mainly classified into three fields: geometric, wave, and quantum optics. Holography is based on wave optics and treats interference and diffraction, which are basic phenomena of light in wave optics. This chapter first describes how to express light waves mathematically based on wave optics. Then, the properties of light waves such as interference and diffraction and the principle of holography are described. For detailed descriptions in this chapter, please refer to the following references [1–3].

1.1 Expression of Light

Light is an electromagnetic wave generated by synchronized oscillations of electric and magnetic fields. Their oscillation is orthogonal and perpendicular to the propagation direction of light. When the **scalar approximation** holds, light can be described by only one of the electric or magnetic field; the electric field is generally used for the description.

1.1.1 Scalar Waves

The wave equation with scalar representation is given by

$$\frac{\partial^2 E}{\partial z^2} - \frac{1}{c^2}\frac{\partial^2 E}{\partial t^2} = 0. \tag{1.1}$$

Then, based on Eq. (1.1), linearly polarized light, which propagates along the z-direction, can be described as follows:

T. Kakue (✉)
Graduate School of Engineering, Chiba University, Chiba, Japan
e-mail: t-kakue@chiba-u.jp

T. Shimobaba and T. Ito (eds.), *Hardware Acceleration of Computational Holography*,
https://doi.org/10.1007/978-981-99-1938-3_1

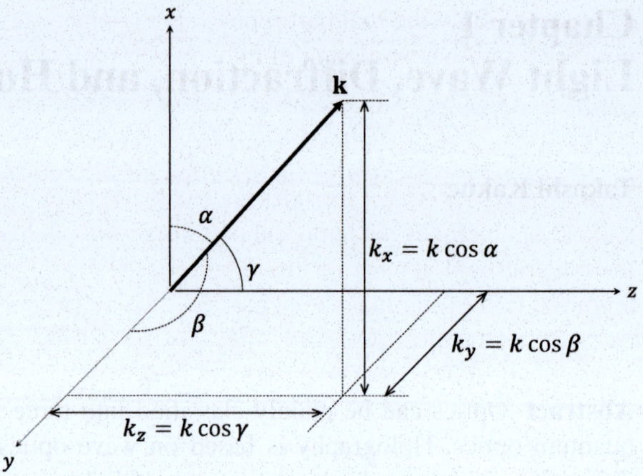

Fig. 1.1 Relationship of the angles between wave vector and xyz-axes

$$E(x, y, z, t) = E(\mathbf{r}, t) = A\cos(\mathbf{k}\cdot\mathbf{r} - \omega t - \theta_0). \tag{1.2}$$

Here, $E(x, y, z, t)$ indicates the electric field at the point $\mathbf{r} = (x, y, z)$ at t. c is the speed of light in a vacuum. A is the amplitude of the electric field. $\mathbf{k} = (k_x, k_y, k_z)$ indicates the **wave vector**. ω denotes the angular frequency of light and can be expressed as follows using the frequency f or the wavelength λ of light:

$$\omega = 2\pi f = \frac{2\pi c}{\lambda}. \tag{1.3}$$

θ_0 denotes the initial phase.

The wave vector can be described using the unit vector $\mathbf{p} = (\cos\alpha, \cos\beta, \cos\gamma)$ along the propagation direction as follows:

$$\mathbf{k} = k\mathbf{p} = (k\cos\alpha, k\cos\beta, k\cos\gamma), \tag{1.4}$$

where k is defined as the wave number and can be described by

$$k = |\mathbf{k}| = \frac{2\pi}{\lambda}. \tag{1.5}$$

Because α, β, and γ denote the angles between the wave vector and x-, y-, and z-axes, respectively, as shown in Fig. 1.1, and $\cos\alpha$, $\cos\beta$, and $\cos\gamma$ are defined as the **direction cosines**.

The expression of light of Eq. (1.2) can be given in complex-number form:

$$E(x, y, z, t) = A\exp[i(\mathbf{k}\cdot\mathbf{r} - \omega t - \theta_0)]. \tag{1.6}$$

In this case, only the real part represents the physical wave. By using the complex-number form, the expression of light can be separated into two exponential parts:

$$E(x, y, z, t) = A\exp[i(\mathbf{k}\cdot\mathbf{r} - \theta_0)]\exp(-i\omega t). \tag{1.7}$$

Here, $\exp[i(\mathbf{k}\cdot\mathbf{r} - \theta_0)]$ includes the spatial part of the electric field only. In contrast, $\exp(-i\omega t)$ includes the temporal part only. In holography, the spatial distribution of the electric field is used for light calculations for simplicity. Then, henceforth, the temporal part of light can be neglected, and light can be described using the spatial part only of the electric field:

$$E(x, y, z) = A\exp[i(\mathbf{k}\cdot\mathbf{r} - \theta_0)] = A\exp(i\theta). \tag{1.8}$$

Here, $A\exp(i\theta)$ is the **complex amplitude** of light. θ is defined as the phase of light.

1.1.2 Plane Waves and Spherical Waves

Assuming that $\mathbf{k}\cdot\mathbf{r}$ is constant in Eq. (1.8), we get

$$\mathbf{k}\cdot\mathbf{r} = c_{kr}, \tag{1.9}$$

where c_{kr} is constant. Equation (1.9) indicates that point $\mathbf{r}$ is perpendicular to the unit vector $\mathbf{p}$. Then, the wavefront of light, which satisfies Eq. (1.9), becomes a plane. This wave is called the **plane wave**. Because the wave phases are equal at the wavefront, the wavefront is called the equiphase surface.

Then, we represent light with the following equation:

$$E(x, y, z) = \frac{A}{|\mathbf{r} - \mathbf{r}_\mathrm{s}|}\exp[i(\mathbf{k}\cdot(\mathbf{r} - \mathbf{r}_\mathrm{s}) - \theta_0)], \tag{1.10}$$

where $\mathbf{r}_\mathrm{s} = (x_s, y_s, z_s)$. The wave expressed by Eq. (1.10) is called the **spherical wave**. It diverges from or converges to the source point $\mathbf{r}_\mathrm{s}$. The amplitude of the spherical wave attenuates according to the distance from the source point $\mathbf{r}_\mathrm{s}$. Equation (1.10) also shows the plane wave can be expressed by Eq. (1.10) when the source point $\mathbf{r}_\mathrm{s}$ is at infinity; we can regard $(\mathbf{r} - \mathbf{r}_\mathrm{s})$ as a constant when the source point $\mathbf{r}_\mathrm{s}$ is at infinity.

1.2 Coherence of Light

Coherence is defined as the interference capacity of light. The **coherence** can be classified into temporal and spatial coherence. The **temporal coherence** indicates the relationship between waves generated at different times. The **spatial coherence** indicates the relationship between the waves at the different parts of a light source or a wavefront of light. The detail of coherence is described in Refs. [1, 2].

1.3 Interference of Light

Interference is produced by the correlation between the individual waves. However, no interference pattern can be observed even if the waves emitted from the two light bulbs or light-emitted diodes (LEDs) are used as the light source. We can observe only the uniform brightness pattern according to the sum of brightness of the light source. This is because the waves emitted from the light bulb and LED are based on spontaneous emission to generate light and have no or little correlation. Light interference requires correlation (or coherence) between individual waves. The most common light sources with high coherence are laser sources, which are based on stimulated emission to generate light.

If the following coherent light waves are superposed at a point (x, y, z)

$$\begin{aligned} E_1(x, y, z) &= A_1 \exp(i\theta_1), \\ E_2(x, y, z) &= A_2 \exp(i\theta_2). \end{aligned} \tag{1.11}$$

Because the superposed complex amplitude is $E_1(x, y, z) + E_2(x, y, z)$, its intensity is given by

$$\begin{aligned} I(x, y, z) &= |E_1(x, y, z) + E_2(x, y, z)|^2 \\ &= A_1^2 + A_2^2 + 2A_1A_2\cos(\theta_2 - \theta_1) \\ &= I_1 + I_2 + 2\sqrt{I_1 I_2}\cos(\Delta\theta). \end{aligned} \tag{1.12}$$

Here, $I_1 = A_1^2$, $I_2 = A_2^2$, and $\Delta\theta = \theta_2 - \theta_1$. Equation (1.12) implies that the intensity of the superposed waves increases or decreases depending on $\Delta\theta$, which is the phase difference between the two waves. This phenomenon indicates light interference. When $\Delta\theta = 2n\pi$, where n is an integer, the intensity is maximum, which is constructive interference. In contrast, when $\Delta\theta = (2n + 1)\pi$, where n is an integer, the intensity is minimum, which is destructive interference.

1.4 Diffraction of Light

The situation, as shown in Fig. 1.2, is considered; a screen is set behind an obstacle, and a light wave illuminates the screen through the obstacle. Because light is blocked by the obstacle, the obstacle forms a shadow on the screen. Here, based on geometrical optics, the edge of the shadow should be sharp. However, in practice, the edge becomes blurred, indicating that the light wave goes around the obstacle

Fig. 1.2 Formation of edge-blurred shadow on screen by obstacle

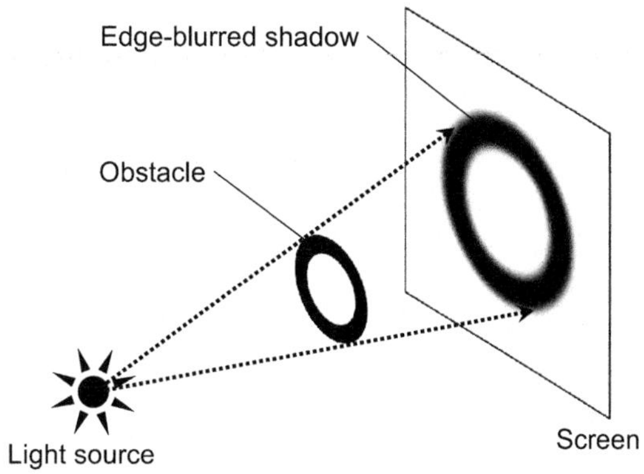

against the behavior defined by geometrical optics. This is called **diffraction** and can be explained using wave optics. Diffraction can be mathematically expressed by diffraction integrals [1–3].

1.4.1 Sommerfeld Diffraction Integral

Diffraction integrals express light propagation from the source plane to the destination plane. Let us assume that the function of the aperture pattern at $\mathrm{A}_1(x_1, y_1)$ on the source plane is $u_1(x_1, y_1)$. When a light wave is introduced into the source plane from $-z$-direction in Fig. 1.3, diffraction occurs by the aperture pattern $u_1(x_1, y_1)$. The diffraction pattern at $A_2(x_2, y_2)$ on the destination plane, $u_2(x_2, y_2)$, is observed

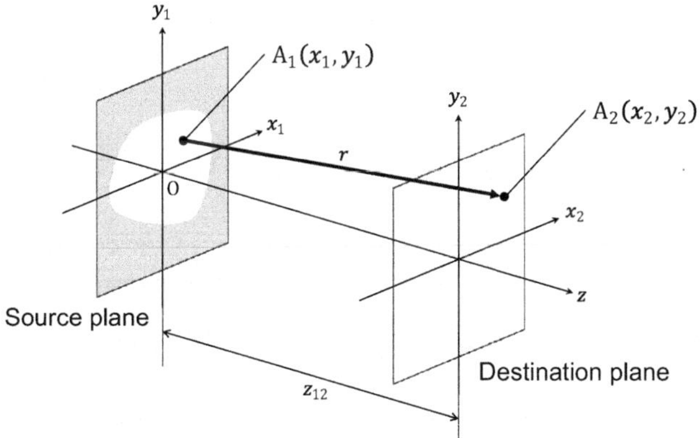

Fig. 1.3 Diffraction integral between source and destination planes

at the destination plane. In this section, several algorithms for calculating diffraction integrals are derived based on the **Sommerfeld diffraction** integral.

The Sommerfeld diffraction integral can be described as follows:

$$u_2(x_2, y_2) = \frac{1}{i\lambda} \int\int u_1(x_1, y_1) \frac{\exp(ikr)}{r} \cos\phi \, dx_1 dy_1, \tag{1.13}$$

where r denotes the distance between A_1 and A_2 and can be expressed as

$$r = \sqrt{(x_2 - x_1)^2 + (y_2 - y_1)^2 + z_{12}^2}. \tag{1.14}$$

Here, z_{12} represents the distance between the source and destination planes, and ϕ is the angle between the normal of the source plane and the line segment $\overline{A_1 A_2}$, as shown in Fig. 1.3. $\cos\phi$ is called the **inclination factor** or obliquity factor and can be expressed by $\cos\phi = z_{12}/r$. Then, Eq. (1.13) can be also expressed as

$$u_2(x_2, y_2) = \frac{1}{i\lambda} \int\int u_1(x_1, y_1) \frac{\exp(ikr)}{r} \frac{z_{12}}{r} dx_1 dy_1. \tag{1.15}$$

1.4.2 Angular Spectrum Method

The **angular spectrum method** (or **plane wave expansion method**) is mainly used to calculate diffraction integrals on a computer and can be derived based on Eq. (1.15). First, in this book, two-dimensional (2D) **Fourier transform** and inverse 2D Fourier transform are, respectively, defined as follows:

$$\begin{aligned} U(f_x, f_y) &= \int_{-\infty}^{\infty}\int_{-\infty}^{\infty} u(x, y) \exp[-i2\pi(f_x x + f_y y)] dx dy \\ &= \mathcal{F}[u(x, y)], \end{aligned} \tag{1.16}$$

$$\begin{aligned} u(x, y) &= \int_{-\infty}^{\infty}\int_{-\infty}^{\infty} U(f_x, f_y) \exp[i2\pi(f_x x + f_y y)] df_x df_y \\ &= \mathcal{F}^{-1}[U(f_x, f_y)]. \end{aligned} \tag{1.17}$$

Here, $\mathcal{F}[\,]$ and $\mathcal{F}^{-1}[\,]$ indicates the operators of 2D Fourier and inverse 2D Fourier transforms, respectively. (f_x, f_y) denote x- and y-coordinates in the frequency domain. Although the coefficient $1/(2\pi)$ or $1/\sqrt{2\pi}$ is generally set before the integrals of Eqs. (1.16) and (1.17), we omit it for simplicity.

The convolution integral can be described as

$$\begin{aligned} u_{conv}(x_2, y_2) &= \int_{-\infty}^{\infty}\int_{-\infty}^{\infty} u_1(x_1, y_1) u_2(x_2 - x_1, y_2 - y_1) dx_1 dy_1 \\ &= u_1(x_1, y_1) \otimes u_2(x_2, y_2), \end{aligned} \tag{1.18}$$

where $\otimes$ represents the **convolution** operator. Using the **convolution theorem** [2], Eq. (1.18) can also be described using Fourier transforms as follows:

$$u_{conv}(x_2, y_2) = \mathcal{F}^{-1}[\mathcal{F}[u_1(x_1, y_1)]\mathcal{F}[u_2(x_1, y_1)]]. \tag{1.19}$$

Applying Eqs. (1.19) to (1.15), the following relationship can be obtained:

$$\begin{aligned} u_2(x_2, y_2) &= \int\int u_1(x_1, y_1)\left[\frac{z_{12}}{i\lambda}\frac{\exp(ikr)}{r^2}\right]dx_1dy_1 \\ &= \mathcal{F}^{-1}\left[\mathcal{F}[u_1(x_1, y_1)]\mathcal{F}\left[\frac{z_{12}}{i\lambda}\frac{\exp(ikr)}{r^2}\right]\right] \\ &= \mathcal{F}^{-1}[\mathcal{F}[u_1(x_1, y_1)]\mathcal{F}[h(x_1, y_1)]] \\ &= \mathcal{F}^{-1}[\mathcal{F}[u_1(x_1, y_1)]H(f_x, f_y)]]. \end{aligned} \tag{1.20}$$

Here,

$$h(x_1, y_1) = \frac{z_{12}}{i\lambda}\frac{\exp(ikr)}{r^2}, \tag{1.21}$$

$$H(f_x, f_y) = \mathcal{F}[h(x_1, y_1)] = \mathcal{F}\left[\frac{z_{12}}{i\lambda}\frac{\exp(ikr)}{r^2}\right]. \tag{1.22}$$

$h(x_1, y_1)$ is called the **impulse response**. $H(f_x, f_y)$ is called the **transfer function** and can be calculated analytically as follows [2]:

$$H(f_x, f_y) = \exp\left(i2\pi z_{12}\sqrt{\frac{1}{\lambda^2} - f_x^2 - f_y^2}\right). \tag{1.23}$$

Therefore, the angular spectrum method can be described by

$$\begin{aligned} u_2(x_2, y_2) &= \mathcal{F}^{-1}\left[\mathcal{F}[u_1(x_1, y_1)]\exp\left(i2\pi z_{12}\sqrt{\frac{1}{\lambda^2} - f_x^2 - f_y^2}\right)\right] \\ &= \mathcal{F}^{-1}\left[U(f_x, f_y)\exp\left(i2\pi z_{12}\sqrt{\frac{1}{\lambda^2} - f_x^2 - f_y^2}\right)\right]. \end{aligned} \tag{1.24}$$

Here,

$$\begin{aligned} U(f_x, f_y) &= \mathcal{F}[u_1(x_1, y_1)] \\ &= \int\int u_1(x_1, y_1)\exp[-i2\pi(f_x x_1 + f_y y_1)]dx_1dy_1. \end{aligned} \tag{1.25}$$

$U(f_x, f_y)$ is called the **angular spectrum**. The inverse Fourier transform of the angular spectrum can be given by

$$u_1(x_1, y_1) = \mathcal{F}^{-1}[U(f_x, f_y)] \\ = \int\int U(f_x, f_y)\exp[i2\pi(f_x x_1 + f_y y_1)]df_x df_y. \tag{1.26}$$

Meanwhile, a plane wave $u_{plane}(x, y, z)$, which propagates along the wave vector $\boldsymbol{k} = (k_x, k_y, k_z)$, is considered. Assuming that the amplitude of the plane wave is a, it can be expressed by

$$u_{plane}(x, y, z) = a\exp[i\boldsymbol{k} \cdot \boldsymbol{r}] \\ = a\exp[i(k_x x + k_y y + k_z z)]. \tag{1.27}$$

Here, a plane wave at $(x_1, y_1, 0)$ is defined as $u_{plane}(x_1, y_1, 0) = u_{plane}(x_1, y_1)$. Comparing Eq. (1.26) with (1.27) using $u_{plane}(x_1, y_1, 0) = u_{plane}(x_1, y_1)$, $u_1(x_1, y_1)$ is expressed as the sum of plane waves with various spatial frequencies and whose amplitudes are $U(f_x, f_y)$. The relationship between the wave vector and spatial frequency can be expressed as

$$U(f_x, f_y)\exp[i2\pi(f_x x_1 + f_y y_1)] = a\exp[i(k_x x_1 + k_y y_1)]. \tag{1.28}$$

Then, using the **direction cosines** and $\sqrt{\cos^2\alpha + \cos^2\beta + \cos^2\gamma} = 1$, the following equations can be obtained:

$$\cos\alpha = \lambda f_x, \\ \cos\beta = \lambda f_y, \\ \cos\gamma = \sqrt{1 - (\lambda f_x)^2 - (\lambda f_y)^2}. \tag{1.29}$$

The angular spectrum method can calculate the Sommerfeld diffraction integral using Fourier transforms with no approximation. Moreover, the angular spectrum method has a short computational time because fast Fourier transform can be used to perform the Fourier transform on a computer.

1.4.3 *Fresnel Diffraction*

The **Fresnel diffraction** can be derived from Eq. (1.15) by approximation. Based on **Taylor expansion**, Eq. (1.14) can be expressed by

$$
\begin{aligned}
r &= \sqrt{(x_2 - x_1)^2 + (y_2 - y_1)^2 + z_{12}^2} \\
&= z_{12}\sqrt{1 + \frac{(x_2 - x_1)^2 + (y_2 - y_1)^2}{z_{12}^2}} \\
&\approx z_{12} + \frac{(x_2 - x_1)^2 + (y_2 - y_1)^2}{2z_{12}} - \frac{\left[(x_2 - x_1)^2 + (y_2 - y_1)^2\right]^2}{8z_{12}^3} + \cdots .
\end{aligned} \tag{1.30}
$$

Equation (1.30) is approximated using only its first and second terms and omitting terms after the third in wave optics:

$$
r \approx z_{12} + \frac{(x_2 - x_1)^2 + (y_2 - y_1)^2}{2z_{12}}. \tag{1.31}
$$

This approximation is called the Fresnel or **paraxial approximation**. Using Eqs. (1.31) and (1.15), it can be rewritten as

$$
\begin{aligned}
u_2(x_2, y_2) &= \frac{1}{i\lambda} \int\int u_1(x_1, y_1) \frac{\exp(ikr)}{r} \frac{z_{12}}{r} dx_1 dy_1 \\
&\approx \frac{1}{i\lambda} \int\int u_1(x_1, y_1) \frac{\exp\left\{ik\left[z_{12} + \frac{(x_2-x_1)^2+(y_2-y_1)^2}{2z_{12}}\right]\right\}}{z_{12}} \frac{z_{12}}{z_{12}} dx_1 dy_1 .
\end{aligned} \tag{1.32}
$$

Here, approximation of $r \approx z_{12}$ is applied, except for the inside of its exponential term. r inside the exponential term must be calculated accurately according to Eq. (1.31) because it affects the phase of a light wave. In contrast, the others affect not the phase, but the amplitude of a light wave. Finally, the Fresnel diffraction can be described as

$$
\begin{aligned}
u_2(x_2, y_2) \approx & \frac{\exp(i\frac{2\pi}{\lambda}z_{12})}{i\lambda z_{12}} \\
& \times \int\int u_1(x_1, y_1) \exp\left\{i\frac{\pi}{\lambda z_{12}}\left[(x_2 - x_1)^2 + (y_2 - y_1)^2\right]\right\} dx_1 dy_1 .
\end{aligned} \tag{1.33}
$$

The Fresnel approximation is satisfied when the terms of Eq. (1.30) after the second term are sufficiently smaller than λ:

$$
\frac{\left[(x_2 - x_1)^2 + (y_2 - y_1)^2\right]^2}{8z_{12}^3} \ll \lambda. \tag{1.34}
$$

For example, if the maximum value of $|x_2 - x_1|$ and $|y_2 - y_1|$ is 2 cm and $\lambda =$ 550[nm],

$$z_{12}^3 \gg \frac{[(2\times10^{-2})^2+(2\times10^{-2})^2]^2}{8\times550\times10^{-9}} = \frac{6.4\times10^{-7}}{4.4\times10^{-6}} \approx 1.5\times10^{-1}[\mathrm{m}^3]. \quad (1.35)$$

Then,

$$z_{12} \gg (1.5\times10^{-1})^{\frac{1}{3}} \approx 0.5[\mathrm{m}]. \quad (1.36)$$

For computational time, the Fresnel diffraction is calculated on a computer using convolution and Fourier transform expressions.

1.4.4 *Fresnel Diffraction Based on Convolution Expression*

The Fresnel diffraction can be described using a convolution integral:

$$\begin{aligned} u_2(x_2,y_2) &= \frac{\exp(i\frac{2\pi}{\lambda}z_{12})}{i\lambda z_{12}} \\ &\times \int_{-\infty}^{\infty}\int_{-\infty}^{\infty} u_1(x_1,y_1)\exp\left\{i\frac{\pi}{\lambda z_{12}}\left[(x_2-x_1)^2+(y_2-y_1)^2\right]\right\}dx_1dy_1 \\ &= \frac{\exp(i\frac{2\pi}{\lambda}z_{12})}{i\lambda z_{12}}\left\{u_1(x_2,y_2)\otimes\exp[i\frac{\pi}{\lambda z_{12}}(x_2^2+y_2^2)]\right\} \\ &= \frac{\exp(i\frac{2\pi}{\lambda}z_{12})}{i\lambda z_{12}}[u_1(x_2,y_2)\otimes h_f(x_2,y_2)]. \end{aligned} \quad (1.37)$$

Here, the impulse response $h_f(x_2, y_2)$ is defined as

$$h_f(x_2,y_2) = \exp\left[i\frac{\pi}{\lambda z_{12}}(x_2^2+y_2^2)\right]. \quad (1.38)$$

Using the convolution theorem, the Fresnel diffraction based on the convolution expression can be derived as

$$\begin{aligned} u_2(x_2,y_2) &= \frac{\exp(i\frac{2\pi}{\lambda}z_{12})}{i\lambda z_{12}}[u_1(x_2,y_2)\otimes h_f(x_2,y_2)] \\ &= \frac{\exp(i\frac{2\pi}{\lambda}z_{12})}{i\lambda z_{12}}\mathcal{F}^{-1}[\mathcal{F}[u_1(x_2,y_2)]\mathcal{F}[h_f(x_2,y_2)]] \\ &= \mathcal{F}^{-1}\left[\mathcal{F}[u_1(x_2,y_2)]\mathcal{F}\left[\frac{\exp(i\frac{2\pi}{\lambda}z_{12})}{i\lambda z_{12}}h_f(x_2,y_2)\right]\right] \\ &= \mathcal{F}^{-1}[\mathcal{F}[u_1(x_2,y_2)]H(f_x,f_y)]. \end{aligned} \quad (1.39)$$

Here, $H_f(f_x, f_y)$ is defined as follows and can be calculated analytically:

$$\begin{aligned} H_f(f_x, f_y) &= \mathcal{F}\left[\frac{\exp\left(i\frac{2\pi}{\lambda}z_{12}\right)}{i\lambda z_{12}}h_f(x_2, y_2)\right] \\ &= \exp\left(i\frac{2\pi}{\lambda}z_{12}\right)\exp\left[i\pi\lambda z_{12}(f_x^2 + f_y^2)\right], \end{aligned} \tag{1.40}$$

where f_x and f_y denote the x- and y-coordinates in the frequency domain.

1.4.5 Fresnel Diffraction Based on Fourier transform Expression

The Fresnel diffraction based on the Fourier transform expression can be derived as follows. First, based on Eq. (1.33), the following equation can be obtained:

$$\begin{aligned} u_2(x_2, y_2) &= \frac{\exp(i\frac{2\pi}{\lambda}z_{12})}{i\lambda z_{12}}\int_{-\infty}^{\infty}\int_{-\infty}^{\infty}u_1(x_1, y_1) \\ &\quad\times \exp\left\{i\frac{\pi}{\lambda z_{12}}\left[(x_2 - x_1)^2 + (y_2 - y_1)^2\right]\right\}dx_1dy_1 \\ &= \frac{\exp(i\frac{2\pi}{\lambda}z_{12})}{i\lambda z_{12}}\int_{-\infty}^{\infty}\int_{-\infty}^{\infty}u_1(x_1, y_1) \\ &\quad\times \exp\left\{i\frac{\pi}{\lambda z_{12}}\left(x_2^2 - 2x_2x_1 + x_1^2 + y_2^2 - 2y_2y_1 + y_1^2\right)\right\}dx_1dy_1 \\ &= \frac{\exp(i\frac{2\pi}{\lambda}z_{12})}{i\lambda z_{12}}\exp\left[i\frac{\pi}{\lambda z_{12}}(x_2^2 + y_2^2)\right] \\ &\quad\times \int_{-\infty}^{\infty}\int_{-\infty}^{\infty}u_1(x_1, y_1)\exp\left[i\frac{\pi}{\lambda z_{12}}(x_1^2 + y_1^2)\right] \\ &\quad\times \exp\left[-i2\pi\left(\frac{x_1x_2}{\lambda z_{12}} + \frac{y_1y_2}{\lambda z_{12}}\right)\right]dx_1dy_1. \end{aligned} \tag{1.41}$$

Then, the following definitions are introduced:

$$u_1'(x_1, y_1) = u_1(x_1, y_1)\exp\left[i\frac{\pi}{\lambda z_{12}}(x_1^2 + y_1^2)\right], \tag{1.42}$$

$$x_2' = \frac{x_2}{\lambda z_{12}}, \qquad y_2' = \frac{y_2}{\lambda z_{12}}. \tag{1.43}$$

Then, from Eq. (1.41),

$$u_2(x_2, y_2) = \frac{\exp(i\frac{2\pi}{\lambda}z_{12})}{i\lambda z_{12}} \exp\left[i\frac{\pi}{\lambda z_{12}}(x_2^2 + y_2^2)\right] \times \int_{-\infty}^{\infty}\int_{-\infty}^{\infty} u_1'(x_1, y_1)\exp[-i2\pi(x_1x_2' + y_1y_2')]dx_1dy_1. \tag{1.44}$$

Here, comparing Eq. (1.16) with the integrals of Eq. (1.44), the following equation can be derived:

$$\int_{-\infty}^{\infty}\int_{-\infty}^{\infty} u_1'(x_1, y_1)\exp[-i2\pi(x_1x_2' + y_1y_2')]dx_1dy_1 = \mathcal{F}[u_1'(x_1, y_1)]. \tag{1.45}$$

Finally, applying Eqs. (1.45) to (1.44),

$$u_2(x_2, y_2) = \frac{\exp(i\frac{2\pi}{\lambda}z_{12})}{i\lambda z_{12}} \exp\left[i\frac{\pi}{\lambda z_{12}}(x_2^2 + y_2^2)\right]\mathcal{F}[u_1'(x_1, y_1)]. \tag{1.46}$$

The above indicates that the Fresnel diffraction can be calculated using single Fourier transform.

1.4.6 Fraunhofer Diffraction

The **Fraunhofer diffraction** is used when calculating the diffraction pattern far from the source plane. Let us assume that the phase of the exponential term of Eq. (1.42) is sufficiently smaller than 2π as follows:

$$\frac{\pi}{\lambda z_{12}}(x_1^2 + y_1^2) \ll 2\pi. \tag{1.47}$$

Then, the value of the exponential term is approximated as 1.

$$\exp\left[i\frac{\pi}{\lambda z_{12}}(x_1^2 + y_1^2)\right] \approx 1. \tag{1.48}$$

The Fraunhofer diffraction can be described by

$$u_2(x_2, y_2) = \frac{\exp(i\frac{2\pi}{\lambda}z_{12})}{i\lambda z_{12}} \exp\left[i\frac{\pi}{\lambda z_{12}}(x_2^2 + y_2^2)\right]\mathcal{F}[u_1(x_1, y_1)]. \tag{1.49}$$

The Fraunhofer diffraction can be calculated by Eq. (1.47) when the following inequality satisfies:

$$z_{12} \gg \frac{x_1^2 + y_1^2}{2\lambda}. \tag{1.50}$$

For example, if the maximum value of $|x_1|$ and $|y_1|$ is 2 cm and $\lambda = 550$[nm],

$$z_{12} \gg \frac{(2 \times 10^{-2})^2 + (2 \times 10^{-2})^2}{2 \times 550 \times 10^{-9}} = \frac{8 \times 10^{-4}}{1.1 \times 10^{-6}} \approx 730[\mathrm{m}]. \tag{1.51}$$

Optical experiments of the Fraunhofer diffraction based on the condition of Eq. (1.51) are difficult to perform. In fact, lenses are used for obtaining the Fraunhofer diffraction because the diffraction pattern at the focal plane of a lens corresponds to the Fraunhofer diffraction pattern [2].

1.4.7 Special Diffraction Calculations

Although the angular spectrum method and Fresnel diffraction are mainly used to calculate diffraction patterns, they have the following limitations:

1. The source and destination planes are parallel.
2. The optical axes of the source and destination planes are identical.
3. The sampling intervals of the source and destination planes are not determined freely.

Recently, several algorithms have been proposed for overcoming these limitations [4–20]. In this book, the **shifted and scaled diffractions** are described. The former can overcome the second limitation, and the latter can overcome the third limitation. Let us introduce the parameters s and (o_x, o_y), which determine the scale and shift rates between the source and destination planes, to express the shifted and scaled diffraction. Then, as shown in Fig. 1.4, the sampling intervals of the destination and

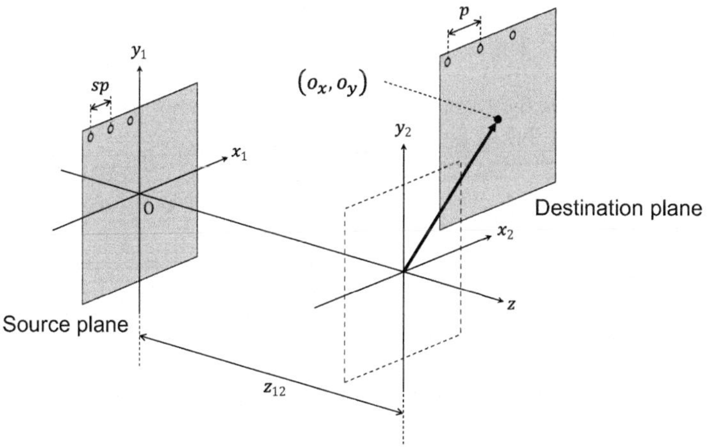

Fig. 1.4 Scaled and shifted diffractions

source planes are defined as p and sp, respectively. When $s = 1$, this is the same situation as normal Fresnel diffraction. When $s > 1$, the area of the source plane is larger than that of the destination plane. In contrast, when $s < 1$, the area of the source plane is smaller than that of the destination plane. The Fresnel diffraction is described by Eq. (1.33). Here, the coordinates (x_1, y_1) are s times greater, and the origin of the destination plane, which indicates the position of the optical axis, is shifted to (o_x, o_y). Then, $(x_2 - x_1)^2$ and $(y_2 - y_1)^2$ of Eq. (1.33) can be considered as

$$\begin{aligned}&(x_2 - sx_1 + o_x)^2\\&\quad = s(x_2 - x_1)^2 + (s^2 - s)x_1^2 + (1 - s)x_2^2 + 2o_x x_2 - 2so_x x_1 + o_x^2,\end{aligned} \tag{1.52}$$

$$\begin{aligned}&(y_2 - sy_1 + o_y)^2\\&\quad = s(y_2 - y_1)^2 + (s^2 - s)y_1^2 + (1 - s)y_2^2 + 2o_y y_2 - 2so_y y_1 + o_y^2.\end{aligned} \tag{1.53}$$

Applying Eqs. (1.52) and (1.53) to (1.33), the following equation can be obtained:

$$\begin{aligned}u_2(x_2, y_2) = C_z \iint & u_1(x_1, y_1)\\&\times \exp\left\{i\frac{\pi}{\lambda z_{12}}\left[(s^2 - s)x_1^2 - 2so_x x_1 + (s^2 - s)y_1^2 - 2so_y y_1\right]\right\}\\&\times \exp\left\{i\frac{\pi}{\lambda z_{12}}\left[s(x_2 - x_1)^2 + s(y_2 - y_1)^2\right]\right\}dx_1 dy_1.\end{aligned} \tag{1.54}$$

Here,

$$\begin{aligned}C_z = \frac{\exp\left(i\frac{2\pi}{\lambda}z_{12}\right)}{i\lambda z_{12}} &\exp\left\{i\frac{\pi}{\lambda z_{12}}\left[(1 - s)x_2^2 + 2o_x x_2 + o_x^2\right]\right\}\\\times &\exp\left\{i\frac{\pi}{\lambda z_{12}}\left[(1 - s)y_2^2 + 2o_y y_2 + o_y^2\right]\right\}.\end{aligned} \tag{1.55}$$

Because Eq. (1.54) has the form of the convolution integral, it can be described as follows using the convolution theorem:

$$u_2(x_2, y_2) = C_z \mathcal{F}^{-1}\left[\mathcal{F}[u_1(x_1, y_1)\exp(i\phi_u)]\mathcal{F}[\exp(i\phi_h)]\right], \tag{1.56}$$

where

$$\begin{aligned}\exp(i\phi_u) = \exp&\left\{i\frac{\pi}{\lambda z_{12}}\left[(s^2 - s)x_1 - 2so_x x_1\right]\right\}\\\times \exp&\left\{i\frac{\pi}{\lambda z_{12}}\left[(s^2 - s)y_1^2 - 2so_y y_1\right]\right\},\end{aligned} \tag{1.57}$$

$$\exp(i\phi_h) = \exp\left\{i\frac{\pi}{\lambda z_{12}}\left[sx_1^2 + sy_1^2\right]\right\}. \tag{1.58}$$

1.5 Holography

Holography was proposed in 1948 by **Dennis Gabor** as a technique to record a wavefront of light [21]. Although he invented this technique to improve the spatial resolution of electron microscopy, he was unable to realize his aim. This failure was owing to the insufficient coherence of light sources. After the invention of holography in 1960, lasers were developed, which have high coherence. In 1962, Leith and Upatnieks proposed a novel holographic recording method using lasers [22]. In this book, we first consider the method proposed by Leith and Upatnieks.

1.5.1 Recording of the Hologram

Figure 1.5 shows the holography recording process. A light wave, which is emitted from an optical source with laser-like coherence, is split into two optical paths by a beam splitter. One is introduced into a three-dimensional (3D) object after expanding its beam diameter using lenses. Light waves reflected and/or diffused by the 3D object arrive at a recording material and are called object waves. Another light wave from the beam splitter is directly introduced into the recording material after expanding its beam diameter using lenses and is called a reference wave. Because a high coherence light source is used, the object waves and reference wave interfere. Here, the distribution of the object waves can be described as

$$O(x, y) = A_O(x, y) \exp[i\theta_O(x, y)]. \tag{1.59}$$

Here, $A_O(x, y)$ and $\theta_O(x, y)$ represent the amplitude and phase of the object wave at the coordinates of (x, y). Similarly, the distribution of the reference wave can be expressed by

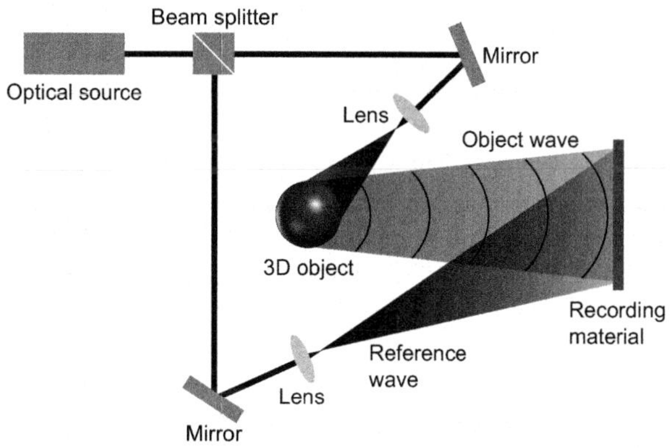

Fig. 1.5 Recording process in holography

$$R(x, y) = A_R(x, y) \exp[i\theta_R(x, y)], \tag{1.60}$$

where $A_R(x, y)$ and $\theta_R(x, y)$ represent the amplitude and phase of the reference wave at the coordinates of (x, y). Then, the intensity distribution generated by interference between the object and reference waves can be described by

$$\begin{aligned} I(x, y) &= |O(x, y) + R(x, y)|^2 \\ &= [O(x, y) + R(x, y)][O(x, y) + R(x, y)]^* \\ &= |O(x, y)^2| + |R(x, y)|^2 + O(x, y)R^*(x, y) + O^*(x, y)R(x, y). \end{aligned} \tag{1.61}$$

Here, * indicates the complex conjugate of a complex number. $I(x, y)$ is called a **hologram**. Holograms have information of the 3D object as interference patterns.

When photosensitive materials such as silver-halide emulsion, a photopolymer, and a photoresist are used as hologram recording materials, development (and/or bleaching) processes are necessary. When image sensors, such as CCDs and CMOSs, are used, a development process is unnecessary, and the image the sensors record corresponds to a hologram.

1.5.2 Reconstruction of Hologram

When reconstructing a hologram, a light wave identical to the reference wave, called the hologram-illumination wave, is introduced into the hologram (Fig. 1.6). Mathematically, this phenomenon corresponds to multiplying the amplitude transmission of the hologram by the hologram-illumination wave. Then, the following equation can describe hologram reconstruction:

$$\begin{aligned} &I(x, y) \times R(x, y) \\ &\quad = [|O(x, y)^2| + |R(x, y)|^2 + O(x, y)R^*(x, y) + O^*(x, y)R(x, y)] \times R(x, y) \\ &\quad = [|O(x, y)^2| + |R(x, y)|^2]R(x, y) + A_R^2(x, y)O(x, y) + O^*(x, y)R^2(x, y). \end{aligned} \tag{1.62}$$

The first term of the right-hand side of Eq. (1.62) includes the hologram-illumination wave multiplied by $|O(x, y)|^2 + |R(x, y)|^2$. This is called the **zeroth-order diffraction** (or non-diffraction) wave. The second term includes the **object wave**, which forms the virtual image behind the hologram. The coefficient $A_R^2(x, y)$ represents the intensity of the light introduced into the hologram, which affects the intensity (or brightness) of the reconstructed object wave. The third term includes the complex conjugate of the object wave multiplied by $R^2(x, y)$. This is called the **conjugate wave** and forms the real image in front of the hologram.

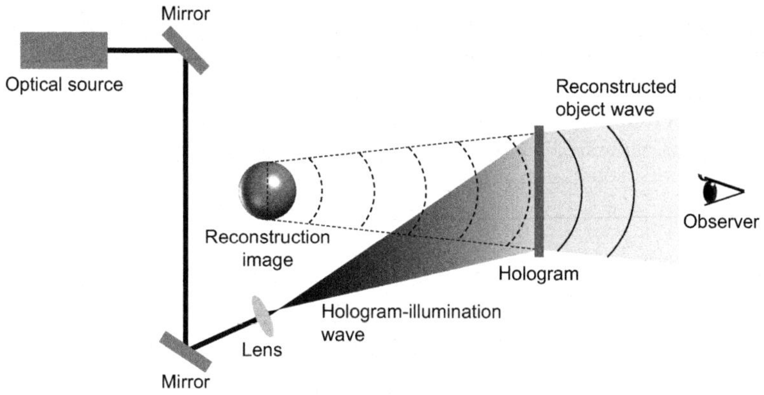

Fig. 1.6 Reconstruction process in holography

1.5.3 In-line Holography and Off-Axis Holography

Holography can be categorized into two types: in-line and off-axis holography. As shown in Fig. 1.7a, in-line holography uses a reference wave introduced perpendicular to the recording material. In contrast, off-axis holography uses a reference wave that is obliquely introduced into a recording material (Fig. 1.7b). The angle between the normal of the recording material and propagation direction of the reference wave

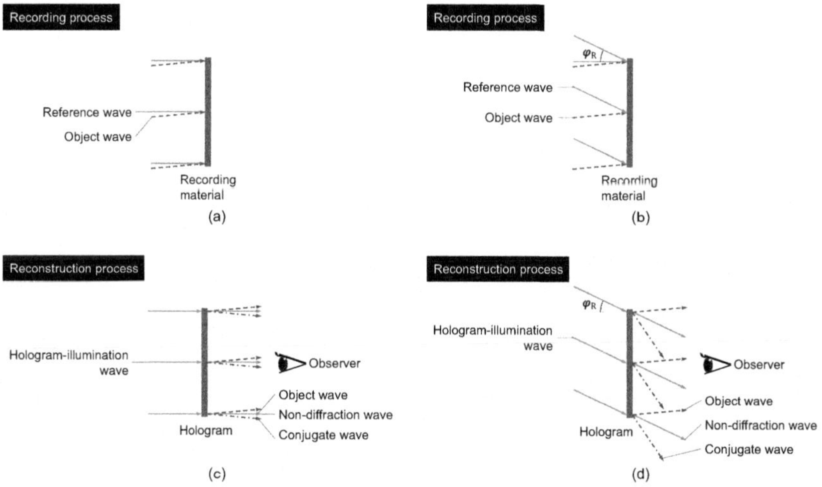

Fig. 1.7 Two types of holography. **a** and **b** represent the recording process in in-line and off-axis holography, respectively. **c** and **d** represent the reconstruction processes in in-line and off-axis holography, respectively

is defined by φ_R. Then, $\varphi_R = 0°$ indicates in-line holography, and $\varphi_R \neq 0°$ indicates off-axis holography.

First, **in-line holography** is considered. Because $\varphi_R = 0°$ corresponds to $\alpha = 90°$, $\beta = 90°$, and $\gamma = 0°$ in Fig. 1.1, the reference wave at the recording material ($z = 0$) can be expressed based on Eqs. (1.8) and (1.60) as

$$R(x, y) = A_R(x, y) \exp[i(k \cdot 0 - \theta_0)] = A_R(x, y) \exp(-i\theta_0). \tag{1.63}$$

Then, assuming that a plane wave whose amplitude and initial phase at the recording material are 1 and 0, respectively, is used as a reference wave for simplicity, $R(x, y)$ can be described by

$$R(x, y) = 1. \tag{1.64}$$

The intensity distribution of interference patterns can be expressed by

$$\begin{aligned} I(x, y) &= |O(x, y) + R(x, y)|^2 \\ &= |O(x, y)^2| + 1 + O(x, y) + O^*(x, y). \end{aligned} \tag{1.65}$$

For hologram reconstruction, $R(x, y) = 1$ is used as the reconstruction light:

$$\begin{aligned} I(x, y) \times R(x, y) &= I(x, y) \\ &= |O(x, y)^2| + 1 + O(x, y) + O^*(x, y). \end{aligned} \tag{1.66}$$

The first and second terms indicate the zeroth-order diffraction wave. The third and fourth terms describe the object and conjugate waves, respectively. Although only the third term contributes to the reconstruction of the object image, the other terms are also reconstructed simultaneously. In in-line holography, as shown in Fig. 1.7c, not only the object wave but also the zeroth-order diffraction and conjugate waves arrive at the observer's eyes. This situation suggests that the zeroth-order diffraction and conjugate waves prevent an observer from observing the object wave only. Hence, the quality of the reconstructed object image is degraded. This degradation is called the twin-image problem and a major problem of in-line holography.

Off-axis holography can overcome this problem. Here, $\alpha = 90° - \varphi_R$, $\beta = 90°$, and $\gamma = 0°$ are considered $\varphi_R \neq 0°$ for simplicity. The reference wave at the recording material can be expressed as

$$\begin{aligned} R(x, y) &= A_R(x, y) \exp\left\{i[k(x \cos\alpha + 0) - \theta_0]\right\} \\ &= A_R(x, y) \exp\left[i(kx \sin\varphi_R - \theta_0)\right]. \end{aligned} \tag{1.67}$$

Then, assuming that a plane wave whose amplitude and initial phase at the recording material are 1 and 0, respectively, is used as a reference wave for simplicity, $R(x, y)$ can be described by

$$R(x, y) = \exp(ikx \sin\varphi_R). \tag{1.68}$$

The intensity distribution of interference patterns can be expressed by

$$\begin{aligned} I(x, y) &= |O(x, y) + R(x, y)|^2 \\ &= |O(x, y)^2| + 1 \\ &\quad + O(x, y)\exp(-ikx\sin\varphi_R) + O^*(x, y)\exp(ikx\sin\varphi_R). \end{aligned} \tag{1.69}$$

For hologram reconstruction, $R(x, y) = \exp(ikx\sin\varphi_R)$ is used as the reconstruction light:

$$\begin{aligned} I(x, y) \times R(x, y) = &(|O(x, y)|^2 + 1)\exp(ikx\sin\varphi_R) \\ &+ O(x, y) + O^*(x, y)\exp(i2kx\sin\varphi_R). \end{aligned} \tag{1.70}$$

The second term describes the object wave. The first term indicates the zeroth-order diffraction wave and propagates along φ_R. Meanwhile, the third term indicates the conjugate wave, and it propagates along $2\varphi_R$ under the rough approximation of $2\sin\varphi_R \approx 2\varphi_R$. Therefore, in off-axis holography, the object wave is not superposed on the zeroth-order diffraction and conjugate waves (Fig. 1.7d).

1.5.4 Types of Holograms

Holograms can be categorized according to how they are recorded. In-line and off-axis holograms are one of the categories in terms of incident angles of the reference wave. In terms of distances between the object and recording material, holograms can be categorized into the following three types: Fresnel, image, and Fraunhofer holograms. **Fresnel holograms** can be obtained when the object wave can be described by the Fresnel diffraction. This situation implies that they can be recorded using the standard optical setup shown in Fig. 1.5. **Fraunhofer holograms** can be obtained when the object wave can be described by the Fraunhofer diffraction, indicating that they can be recorded when the object is extremely far from the recording material. However, as described in Sect. 1.4.6, it is difficult to realize the Fraunhofer diffraction. Then, a convex lens is used to record Fraunhofer holograms. A spherical wave from a point-light source positioned at the focal point of a convex lens is converted into a plane wave or a collimated wave by the lens. This function makes it possible to realize the Fraunhofer diffraction approximately. As shown in Fig. 1.8, an object is positioned at the front focal point of a convex lens with a focal length of f, and a recording material is set at the back focal point of the lens. Because light waves from the object become collimated waves after passing through the lens, they can be regarded as Fraunhofer-diffraction-based light waves. The lens function shown in Fig. 1.8 can be described and calculated using a Fourier transform. Hence, holograms recorded by the optical setup of Fig. 1.8 are called Fourier holograms.

Image holograms can be obtained when an object is positioned near the recording material. However, positioning an object near the recording material is difficult

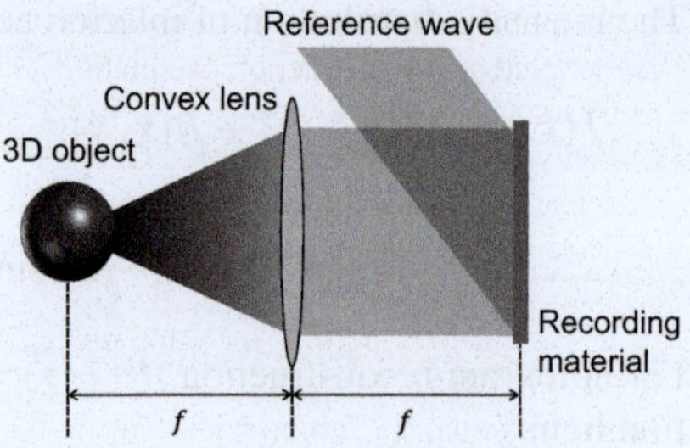

Fig. 1.8 Recording of Fourier hologram

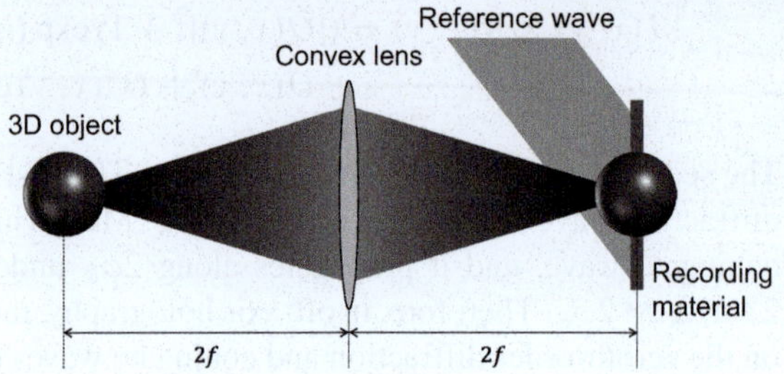

Fig. 1.9 Recording of image hologram

because an actual object has size and/or volume. Then, as shown in Fig. 1.9, a convex lens with a focal length of f is used to record image holograms. An object and recording material are positioned at $2f$ behind and in front of the lens, respectively. A real image of the object is formed near the recording material by the lens function. Then, image holograms can be obtained using the real image as the object wave.

References

1. Born, M. Wolf, M.: Principles of Optics. Cambridge University Press, Cambridge (1999).
2. Goodman, J.W.: Introduction to Fourier Optics. Roberts and Company Publishers, Greenwood Village (2005).
3. Shimobaba, T., Ito, T.: Computer Holography, CRC Press (2019).
4. Muffoletto, R.P. Tyler, J.M., Tohline, J.E.: Shifted Fresnel diffraction for computational holography. Opt. Express **15**, 5631–5640 (2007).
5. Shimobaba, T., Kakue, T., Okada, N., Oikawa, M., Yamaguchi, M., Ito, T.: Aliasing-reduced Fresnel diffraction with scale and shift operations. J. Opt. **15**, 075405 (2007).
6. Matsushima, K.: Shifted angular spectrum method for off-axis numerical propagation. Opt. Express **18**, 18453–18463 (2010).
7. Shimobaba, T., Matsushima, K. Kakue, T., Masuda, M., Ito, T.: Scaled angular spectrum method. Opt. Lett. **37**, 4128–4130 (2012).
8. Yamamoto, K., Ichihashi, Y., Senoh, T., Oi, R., Kurita, T.: Calculating the Fresnel diffraction of light from a shifted and tilted plane. Opt. Express **20**, 12949–12958 (2012).
9. Yu, X., Xiahui, T., Yingxiong, Q., Hao, P., Wei W.: Band-limited angular spectrum numerical propagation method with selective scaling of observation window size and sample number. J. Opt. Soc. Am. A **29**, 2415–2420 (2012).

10. Shimobaba, T., Kakue, T., Oikawa, M., Okada, N., Endo, Y., Hirayama, R., Ito, T.: Nonuniform sampled scalar diffraction calculation using nonuniform fast Fourier transform. Opt. Lett. **38**, 5130–5133 (2013).
11. Nguyen, G.-N., Heggarty, K., Gérard, P.., Serio, B. Meyrueis, P.: Computationally efficient scalar nonparaxial modeling of optical wave propagation in the far-field. Appl. Opt. **53**, 2196–2205 (2014).
12. Kim, Y.-H., Byun, C.-W., Oh, H., Pi, J.-E., Choi, J.-H., Kim, G.H., Lee, M.-L., Ryu, H., Hwang, C.-S.: Off-axis angular spectrum method with variable sampling interval. Opt. Commun. **348**, 31–37 (2015).
13. Matsushima, K., Schimmel, H., Wyrowski, F.: Fast calculation method for optical diffraction on titled planes by use of the angular spectrum of plane waves. J. Opt. Soc. Am. A **20**, 1755–1762 (2003).
14. De Nicola S., Finizio, A., Pierattini, G., Ferraro, P. Alfieri, D.: Angular spectrum method with correction of anamorphism for numerical reconstruction of digital holograms on tilted planes. Opt. Express **13**, 9935–9940 (2005).
15. Jeong, S.J., Hong, C.K.: Pixel-size-maintained image reconstruction of digital holograms on arbitrarily tilted planes by the angular spectrum method. Appl. Opt. **47**, 3064–3071 (2008).
16. Chang, C., Xia, J., Wu, J., Lei, W., Xie, Y., Kang, M., Zhang, Q.: Scaled diffraction calculation between tilted planes using nonuniform fast Fourier transform. Opt. Express **22**, 17331–17340 (2014).
17. Stock, J., Worku, N.G., Gross, H.: Coherent field propagation between tilted planes. J. Opt. Soc. Am. A **34**, 1849–1855 (2017).
18. Tommasi, T., Bianco, B.: Frequency analysis of light diffraction between rotated planes. Opt. Lett. **17**, 556–558 (1992).
19. Matsushima, K.: Formulation of the rotational transformation of wave fields and their application to digital holography. Appl. Opt. **47**, D110–D116 (2008).
20. Kim, Y.-H., Kim, H.G., Ryu, H., Chu, H.-Y., Hwang, C.-S.: Exact light propagation between rotated planes using non-uniform sampling and angular spectrum method. Opt. Commun. **344**, 1–6 (2015).
21. Gabor D.: A New Microscopic Principle. Nature **166**, 777–778 (1948).
22. Leith, E.N., Upatnieks, J.: Reconstructed Wavefronts and Communication Theory. J. Opt. Soc. Am. **52**, 1123–1130 (1962).

Chapter 2
Computer-Generated Hologram

Yasuyuki Ichihashi

Abstract This chapter explains the principles of computer-generated holograms based on the point-cloud method, which is required for implementation into central processing units (CPUs) in Chap. 9, graphics processing units (GPUs) in Chap. 10, and field programmable gate arrays (FPGAs) in Chap. 19.

2.1 Computer-Generated Amplitude Hologram

As described in the principle of holography in Chap. 1, the light intensity distribution I on a hologram [1] is expressed by the following equation:

$$I = |O + R|^2 = |O|^2 + |R|^2 + OR^* + O^*R, \tag{2.1}$$

where O is the object light, R is the reference light, and $*$ represents the complex conjugate. When the reference light is parallel light incident with amplitude R_0 at incident angle θ, the light distribution of the reference light on the hologram is expressed by

$$R(x_\alpha, y_\alpha) = R_0 e^{jkx_\alpha \sin\theta}, \tag{2.2}$$

where k is the wave number and $j = \sqrt{-1}$.

Considering the object light as a collection (**point cloud**) of point light sources emitted as spherical waves, $O(x_\alpha, y_\alpha)$ can be expressed as

$$O(x_\alpha, y_\alpha) = \sum_{i=1}^{N} \frac{A_i}{r_{\alpha i}} e^{jkr_{\alpha i}}, \tag{2.3}$$

Y. Ichihashi (✉)
Radio Research Institute, National Institute of Information and Communications Technology (NICT), 4-2-1 Nukuikitamachi, Koganei, Tokyo 184–8795, Japan
e-mail: y-ichihashi@nict.go.jp

T. Shimobaba and T. Ito (eds.), *Hardware Acceleration of Computational Holography*,
https://doi.org/10.1007/978-981-99-1938-3_2

where A_i is the amplitude of the object light and the distance between a hologram pixel and an object point expressed as

$$r_{\alpha i} = \sqrt{(x_\alpha - x_i)^2 + (y_\alpha - y_i)^2 + z_i^2}. \tag{2.4}$$

In Eq. (2.1), the first and second terms do not contribute to holographic reconstruction, the third term is the reconstructed object light we require, and the fourth term is the conjugate light. Ignoring the first and second terms in Eq. (2.1) and substituting Eqs. (2.2) and (2.3) into Eq. (2.1), we obtain the **amplitude hologram** as

$$\begin{aligned} I(x_\alpha, y_\alpha) &= \sum_{i=1}^{N} \frac{A_i R_0}{r_{\alpha i}} e^{jk(r_{\alpha i} - x_\alpha \sin\theta)} + \frac{A_i R_0}{r_{\alpha i}} e^{-jk(r_{\alpha i} - x_\alpha \sin\theta)} \\ &= \sum_{i=1}^{N} \frac{2A_i R_0}{r_{\alpha i}} \cos\left(k(r_{\alpha i} - x_\alpha \sin\theta)\right). \end{aligned} \tag{2.5}$$

As can be seen from Eq. (2.5), the calculation of a **computer-generated hologram** (**CGH**) includes trigonometric functions and square roots [2]. The calculation cost is proportional to the number of object points and the resolution of the CGH, M:

$$\text{Calculation cost} = M \times N.$$

When the number of object points is 10,000 and the resolution of the hologram is $1,920 \times 1,080$ pixels (approximately 2 million pixels), about 20 billion calculations of trigonometric functions and square roots must be performed. Furthermore, when the number of object points reaches 100,000, the number of calculations will be 200 billion. Therefore, methods for reducing this enormous number of calculations are described next.

2.2 Fresnel CGH

First, to simplify the calculation, the hologram is irradiated vertically with the reference light ($\theta = 0$), and the object points are gathered in a specific range. When the distance between the point cloud and the hologram plane is sufficiently long, the change in $R_{\alpha i}$ is small. The coefficient $2R_0/R_{\alpha i}$ of the cosine function can be ignored. Therefore, Eq. (2.5) can be simplified to

$$I(x_\alpha, y_\alpha) = \sum_{i=1}^{N} A_i \cos(kr_{\alpha i}), \tag{2.6}$$

$r_{\alpha i}$ can be approximated as follows by using the **Taylor expansion** with the **binomial theorem**:

$$
\begin{aligned}
r_{\alpha i} &= z_i \sqrt{1 + \frac{(x_\alpha - x_i)^2 + (y_\alpha - y_i)^2}{{z_i}^2}} \\
&= z_i + \frac{(x_\alpha - x_i)^2 + (y_\alpha - y_i)^2}{2z_i} - \frac{\left\{(x_\alpha - x_i)^2 + (y_\alpha - y_i)^2\right\}^2}{8{z_i}^3} \cdots \\
&\approx z_i + \frac{(x_\alpha - x_i)^2 + (y_\alpha - y_i)^2}{2z_i},
\end{aligned}
\tag{2.7}
$$

where we assume that ${z_i}^2$ is much larger than $(x_\alpha - x_i)^2 + (y_\alpha - y_i)^2$. This is referred to as the **Fresnel approximation**.

2.3 Recurrence Algorithm

To further reduce the calculation cost, rather than calculating the distance between each object point and each hologram pixel one by one, a method that uses a **recurrence formula** [3, 4] with the phase difference between adjacent hologram pixels is described.

By writing $r_{\alpha i} = \Theta(x_\alpha, y_\alpha)$ in Eq. (2.6) and considering a point $(x_\alpha + n, y_\alpha)$, where $n = p, 2p, 3p \cdots$ and p is the sampling interval of the hologram, away from an arbitrary point (x_α, y_α) on the hologram plane in the x-axis direction, the distance Θ_n between the point $(x_\alpha + n, y_\alpha)$ and the object point (x_i, y_i, z_i) can be expressed by

$$
\begin{aligned}
\Theta_n &= \Theta\left(x_\alpha + n, y_\alpha\right) = z_i + \frac{(x_\alpha + n - x_i)^2 + (y_\alpha - y_i)^2}{2z_i} \\
&= z_i + \frac{(x_\alpha - x_i)^2 + (y_\alpha - y_i)^2}{2z_i} + \frac{2n\left(x_\alpha - x_i\right) + n^2}{2z_i}.
\end{aligned}
\tag{2.8}
$$

Similarly, Θ_{n-1} is expressed by

$$
\begin{aligned}
\Theta_{n-1} &= \Theta\left(x_\alpha + n - 1, y_\alpha\right) = z_i + \frac{(x_\alpha + n - 1 - x_i)^2 + (y_\alpha - y_i)^2}{2z_i} \\
&= z_i + \frac{(x_\alpha - x_i)^2 + (y_\alpha - y_i)^2}{2z_i} + \frac{2\left(n - 1\right)\left(x_\alpha - x_i\right) + (n - 1)^2}{2z_i}.
\end{aligned}
\tag{2.9}
$$

Therefore, the difference between Eqs. (2.8) and (2.9) is expressed by

$$\Theta_n - \Theta_{n-1} = \frac{2n\,(x_\alpha - x_i) + n^2}{2z_i} - \frac{2\,(n-1)\,(x_\alpha - x_i) + (n-1)^2}{2z_i}$$
$$= \frac{2\,(x_\alpha - x_i) + 2n - 1}{2z_i} = \frac{2\,(x_\alpha - x_i) + 1}{2z_i} + \frac{n-1}{z_i}. \tag{2.10}$$

Since Eq. (2.10) is a recurrence formula, Θ_0 is expressed by

$$\Theta_0 = z_i + \frac{(x_\alpha - x_i)^2 + (y_\alpha - y_i)^2}{2z_i}. \tag{2.11}$$

Equation (2.11) matches Eq. (2.7). In addition, the following two formulas are introduced:

$$\Delta_0 = \frac{2(x_\alpha - xi)}{2zi} \tag{2.12}$$

$$\Gamma = \frac{1}{z_i}. \tag{2.13}$$

By substituting Eqs. (2.12) and (2.13) into Eq. (2.10), the following equation is obtained:

$$\Theta_n - \Theta_{n-1} = \Delta_0 + (n-1)\Gamma. \tag{2.14}$$

Here, Δ_{n-1} is defined as

$$\Delta_{n-1} = \Delta_0 + (n-1)\Gamma. \tag{2.15}$$

Thus, Eq. (2.14) can be expressed as

$$\Theta_n = \Theta_{n-1} + \Delta_{n-1}. \tag{2.16}$$

Equation (2.16) means that Θ_n can be obtained from Θ_{n-1} for the previous point and Δ_{n-1}. Furthermore, $\Delta_n - \Delta_{n-1}$ is calculated from Eq. (2.15) as follows:

$$\Delta_n - \Delta_{n-1} = \Delta_0 + n\Gamma - \Delta_0 - (n-1)\Gamma = \Gamma \leftrightarrow \Delta_n = \Delta_{n-1} + \Gamma. \tag{2.17}$$

Equation (2.17) is a recurrence formula for Δ. In summary, the hologram calculation procedure is as follows.

First, the distance $r_{\alpha i} = \Theta(x_\alpha, y_\alpha)$ between an arbitrary point (x_i, y_i, z_i) in the three-dimensional image and a point (x_α, y_α) on an arbitrary hologram plane is obtained using Eq. (2.11). Next, $\Theta_0 = \Theta(x_\alpha, y_\alpha)$ and $\Theta_1 = \Theta(x_\alpha + 1, y_\alpha)$ are obtained from Eqs. (2.16) and (2.17). The subsequent hologram pixels from Θ_2 to Θ_n are then obtained by applying Eqs. (2.16) and (2.17). Then, the light intensity at all points on the hologram plane is obtained using Eq. (2.6).

The square root calculation is eliminated by approximation using the binomial theorem as shown in Eq. (2.7). As a result, not only the CPU but also hardware accelerators such as graphics processing units (GPUs) [5, 6] and field programmable gate arrays (FPGAs) [7, 8] can be used to speed up the calculation effectively. The method using the recurrence formulas shown in Eqs. (2.16) and (2.17) can achieve even higher speeds when implemented in hardware accelerators such as FPGAs using fixed-point numbers. These implementation methods are described in Chaps. 9 and 19.

2.4 Kinoform (Phase-Only Hologram)

Equation (2.1) represents an **amplitude hologram**. On the other hand, there is a **phase-only hologram** called a **kinoform**. A kinoform can be obtained from the phase distribution of object light on the hologram plane and is represented by the following equation:

$$\Phi\left(x_\alpha, y_\alpha\right) = \arg\left\{O\left(x_\alpha, y_\alpha\right)\right\}, \tag{2.18}$$

where $arg\{\cdot\}$ represents the operator calculating the argument of the complex amplitude $O\left(x_\alpha, y_\alpha\right)$. A kinoform has the disadvantage that the amplitude cannot be controlled but the advantage that it has high light efficiency (theoretically 100%) and does not contain conjugate light, in contrast to the amplitude hologram expressed by Eq. (2.1). A kinoform can be displayed as a hologram on a phase-modulated spatial light modulator.

References

1. T. C. Poon (ed.): Digital holography and three-dimensional display: Principles and Applications, Springer, Heidelberg (2006).
2. M. Lucente: Interactive computation of holograms using a look-up table, Journal of Electronic Imaging, **2**, 28–34 (1993).
3. T. Shimobaba, and T. Ito: An efficient computational method suitable for hardware of computer-generated hologram with phase computation by addition, Comput. Phys. Commun., **138**, 44–52 (2001).
4. T. Shimobaba, S. Hishinuma, and T. Ito:Special-purpose computer for holography HORN-4 with recurrence algorithm, Comput. Phys. Commun., **148**, 160–170 (2002).
5. N. Masuda, T. Ito, T. Tanaka, A. Shiraki, and T. Sugie: Computer generated holography using a graphics processing unit, Opt. Express, **14**, 603–608 (2006).
6. H. Kang, F. Yaraş, and L. Onural: Graphics processing unit accelerated computation of digital holograms, Apl. Opt., **48**, H137–H143 (2009).
7. T. Ito, N. Masuda, K. Yoshimura, A. Shiraki, T. Shimobaba, T. Sugie: Special-purpose computer HORN-5 for a real-time electroholography, Opt. Express, **13**, 1923–1932 (2005).
8. T. Sugie, T. Akamatsu, T. Nishitsuji, R. Hirayama, N. Masuda, H. Nakayama, Y. Ichihashi, A. Shiraki, M. Oikawa, N. Takada, Y. Endo, T. Kakue, T. Shimobaba, and T. Ito: High-performance parallel computing for next-generation holographic imaging, Nature Electronics, **1**, 254–259 (2018).

Chapter 3
Basics of Digital Holography

Takashi Kakue

Abstract Holograms can be recorded by image sensors, such as CCDs and CMOSs as digital 2D image data. In this book, a holography for digital holograms is called digital holography. Digital holography requires no mechanical movement of optical elements when acquiring 3D information of an object; the amplitude (or intensity) and phase information of the object wave can be dynamically obtained simultaneously. For detailed descriptions, please refer to [1].

3.1 Digital Holography

Figure 3.1 shows an optical setup of digital holography [2–4]. A laser beam is expanded by the beam expander and introduced into a beam splitter and split into two paths. One illuminates the object at z_O from the image sensor. Here, transparent objects, such as microorganisms and biological cells, are assumed because microscopy based on **digital holography** is an example. Then, the transmitted and diffracted waves correspond to the object wave. The object wave includes amplitude (or intensity) and phase information. The amplitude information corresponds to the transparency of the object. The phase information corresponds to the thickness of the object. The object wave is introduced into the image sensor via the beam combiner such as a half mirror. Another beam from the beam splitter is used as the reference wave and introduced into the image sensor via the beam combiner. The object and reference waves interfere at the image sensor plane and intensity distribution of interference fringes, which corresponds to a hologram, is recorded as a digital image by the image sensor. Assuming that the image sensor plane is defined as $z = 0$, the hologram image $I(x_I, y_I, 0)$ can be expressed by

$$\begin{aligned} I(x_I, y_I, 0) = {} & |O(x_I, y_I, 0)|^2 + |R(x_I, y_I, 0)|^2 \\ & + O(x_I, y_I, 0)R^*(x_I, y_I, 0) + O^*(x_I, y_I, 0)R(x_I, y_I, 0), \end{aligned} \tag{3.1}$$

T. Kakue (✉)
Graduate School of Engineering, Chiba University, Chiba, Japan
e-mail: t-kakue@chiba-u.jp

T. Shimobaba and T. Ito (eds.), *Hardware Acceleration of Computational Holography*,
https://doi.org/10.1007/978-981-99-1938-3_3

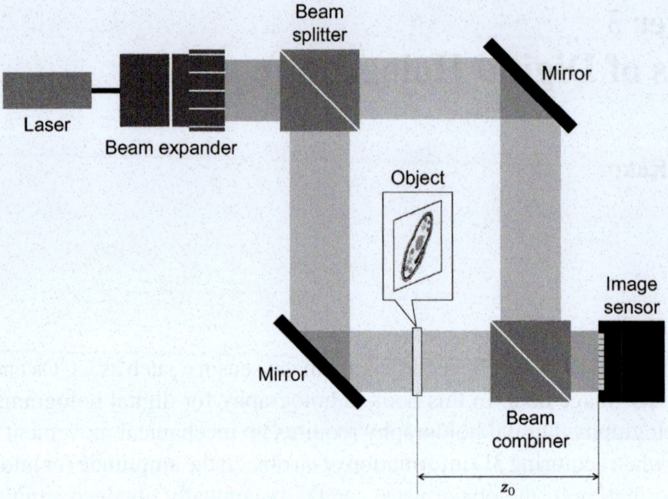

Fig. 3.1 Recording process in digital holography

where $O(x_I, y_I, 0)$ and $R(x_I, y_I, 0)$ are considered the object and reference waves at the image sensor plane, respectively.

The object wave at $z = z_O$, $O(x_O, y_O, z_O)$ can be described by

$$O(x_O, y_O, z_O) = A_O(x_O, y_O, z_O) \exp [i\theta_O(x_O, y_O, z_O)]. \tag{3.2}$$

Here, $A_O(x_O, y_O, z_O)$ and $\theta_O(x_O, y_O, z_O)$ correspond to the amplitude and phase information of the object wave, respectively. Because $O(x_O, y_O, z_O)$ propagates to the image sensor plane, $O(x_I, y_I, 0)$ can be expressed by

$$O(x_I, y_I, 0) = \mathrm{Prop}[O(x_O, y_O, z_O); -z_O], \tag{3.3}$$

where Prop[ ] denotes the operator of the diffraction calculation and can be described using the Fresnel diffraction, for example, by

$$\begin{aligned} \mathrm{Prop}[u_1(x_1, y_1, 0); z_{12}] &= u_2(x_2, y_2, z_{12}) \\ &= \frac{\exp\left(i\frac{2\pi}{\lambda}z_{12}\right)}{i\lambda z_{12}} \times \iint u_1(x_1, y_1, 0) \\ &\quad \times \exp\left\{i\frac{\pi}{\lambda z_{12}}\left[(x_2 - x_1)^2 + (y_2 - y_1)^2\right]\right\} dx_1 dy_1. \end{aligned} \tag{3.4}$$

The operator has the following properties:

$$\text{Prop}[u_1(x, y, z) + u_2(x, y, z); d] = \text{Prop}[u_1(x, y, z); d] + \text{Prop}[u_2(x, y, z); d], \tag{3.5}$$

$$\text{Prop}[C \times u(x, y, z)] = C \times \text{Prop}[u(x, y, z); d], \tag{3.6}$$

$$\text{Prop}[u(x, y, z); d_1 + d_2] = \text{Prop}\left[\text{Prop}[u(x, y, z); d_1]; d_2\right]. \tag{3.7}$$

The reconstruction process of holograms in digital holography is performed computationally. The reconstruction of the constructed hologram can be mathematically described by

$$\begin{aligned} &I(x_I, y_I, 0) \times R(x_I, y_I, 0) \\ &= \left(|O(x_I, y_I, 0)|^2 + |R(x_I, y_I, 0)|^2\right) R(x_I, y_I, 0) \\ &\quad + O(x_I, y_I, 0) + O^*(x_I, y_I, 0) R^2(x_I, y_I, 0) \\ &= D(x_I, y_I, 0) R(x_I, y_I, 0) + O(x_I, y_I, 0) + O^*(x_I, y_I, 0) R^2(x_I, y_I, 0), \end{aligned} \tag{3.8}$$

where

$$D(x_I, y_I, 0) = |O(x_I, y_I, 0)|^2 + |R(x_I, y_I, 0)|^2, \tag{3.9}$$

and the intensity of the reference wave is assumed as 1. Although Eq. (3.8) includes the object wave in the second term, it is not $O(x_O, y_O, z_O)$ but $O(x_I, y_I, 0)$. Then, as shown in Fig. 3.2, backward-diffraction calculation is required for obtaining $O(x_O, y_O, z_O)$. The reconstruction process of holograms in digital holography can be expressed by

$$\begin{aligned} U_O(x_O, y_O, z_O) &= \text{Prop}[I(x_I, y_I, 0) \times R(x_I, y_I, 0); z_O] \\ &= \text{Prop}[D(x_I, y_I, 0) R(x_I, y_I, 0); z_O] + O(x_O, y_O, z_O) \\ &\quad + \text{Prop}[O^*(x_I, y_I, 0) R^2(x_I, y_I, 0); z_O]. \end{aligned} \tag{3.10}$$

Because the second term indicates the object wave at $z = z_O$, the object wave can be reconstructed by backward-diffraction calculation. As backward-diffraction calculation, not only the Fresnel diffraction but also the angular spectrum method can be applied.

Because $A_O(x_O, y_O, z_O)$ and $\theta_O(x_O, y_O, z_O)$ of the reconstructed object wave have the relationship shown in Fig. 3.3, they can be calculated using

$$A_O(x_O, y_O, z_O) = \sqrt{\{\text{Re}\,[O(x_O, y_O, z_O)]\}^2 + \{\text{Im}\,[O(x_O, y_O, z_O)]\}^2}, \tag{3.11}$$

$$\theta_O(x_O, y_O, z_O) = \arg\left[\frac{\text{Im}[O(x_O, y_O, z_O)]}{\text{Re}[O(x_O, y_O, z_O)]}\right]. \tag{3.12}$$

Here, Re[ ] and Im[ ] denote the operators that describe the real and imaginary parts of a complex number, respectively. arg[ ] indicates the operator that describes the argu-

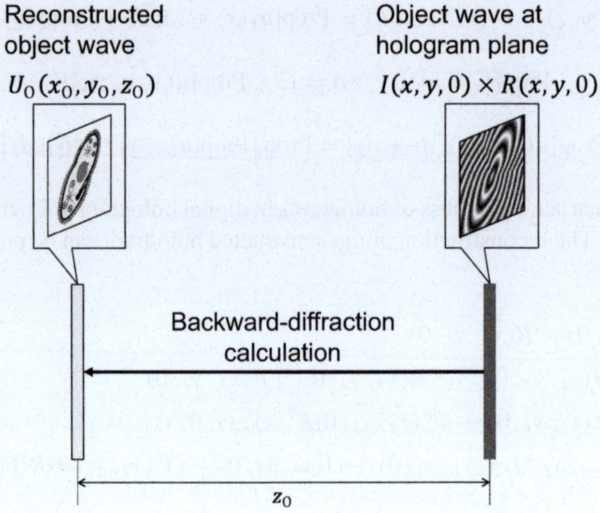

Fig. 3.2 Reconstruction process in digital holography

Fig. 3.3 Amplitude and phase of light wave

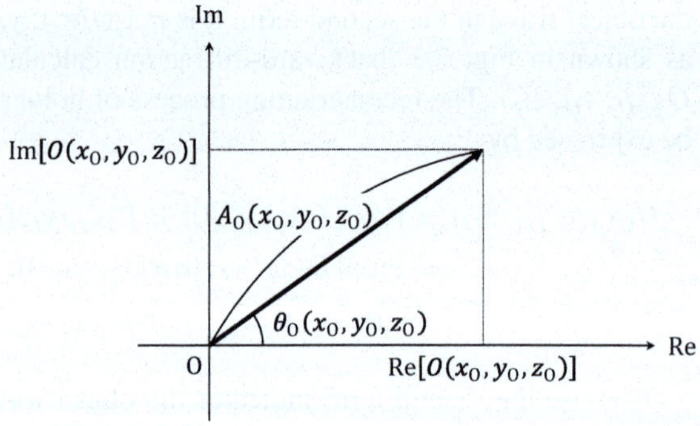

ment of a complex number. To calculate the argument on a computer, arctangent2, which has a range of $(-\pi, \pi]$, is generally used.

3.2 Off-Axis Digital Holography

As described in Eq. (1.5.3), **off-axis holography** can overcome the **twin-image problem**. In digital holography, off-axis holography is also preferred to in-line holography in terms of image quality [5–12]. However, for off-axis digital holography, the inci-

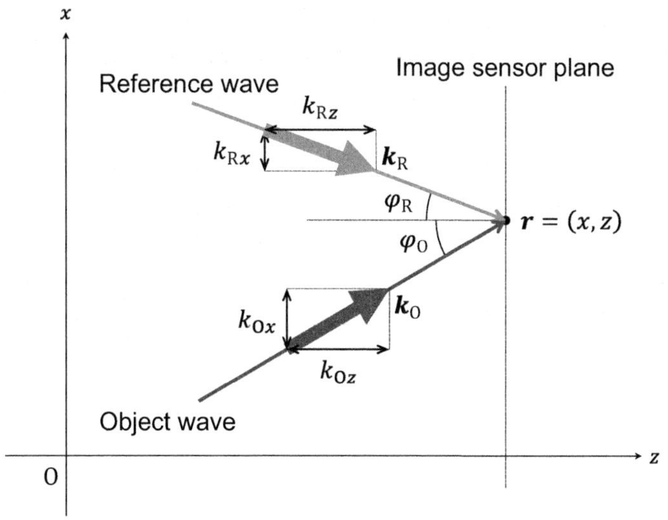

Fig. 3.4 Wave vectors in off-axis digital holography

dent angle of the reference wave cannot be large due to the insufficient performance of image sensors.

Let us consider the period of interference fringes formed by the reference and object waves. For simplicity, the x-z-plane is shown in Fig. 3.4. The wave number k is defined by Eq. (1.5.3). The object wave $O_{\boldsymbol{r}}$, emitted from a point of the object, at $\boldsymbol{r} = (x, z)$ on the image sensor plane is described by

$$O_{\boldsymbol{r}} = A_O \exp(i\boldsymbol{k}_O \cdot \boldsymbol{r}). \tag{3.13}$$

Here, $\boldsymbol{k}_O$ denotes the wave vector of the object wave and can be expressed by

$$\boldsymbol{k}_O = (k_{Ox}, k_{Oz}) = (k \sin\varphi_O, k \cos\varphi_O). \tag{3.14}$$

The reference wave $R_{\boldsymbol{r}}$ at $\boldsymbol{r}$ can be described by

$$R_{\boldsymbol{r}} = A_R \exp(i\boldsymbol{k}_R \cdot \boldsymbol{r}). \tag{3.15}$$

Here, $\boldsymbol{k}_R$ denotes the **wave vector** of the object wave and can be expressed by

$$\boldsymbol{k}_R = (k_{Rx}, k_{Rz}) = (k \sin\varphi_R, k \cos\varphi_R). \tag{3.16}$$

Then, the interference fringe pattern at $\boldsymbol{r}$ can be expressed by

$$I(\boldsymbol{r}) = |O_r + R_r|^2 = A_O^2 + A_R^2 + 2A_O A_R \cos[(\boldsymbol{k}_O - \boldsymbol{k}_R)\cdot \boldsymbol{r}] \\ = A_O^2 + A_R^2 + 2A_O A_R \cos\varphi_{OR}, \tag{3.17}$$

$$\varphi_{OR} = (\boldsymbol{k}_O - \boldsymbol{k}_R)\cdot \boldsymbol{r} \\ = k(\sin\varphi_O + \sin\varphi_R)x + k(\cos\varphi_O + \cos\varphi_R)z. \tag{3.18}$$

Here, obtaining the **spatial frequency** of a signal is considered. The spatial frequency of a one-dimensional signal $\exp(i2\pi fx)$ is f. By defining $\theta(x)$ as $\theta(x) = 2\pi fx$, the spatial frequency can also be calculated as follows:

$$\frac{1}{2\pi}\frac{d\theta(x)}{dx} = f. \tag{3.19}$$

Applying Eqs. (3.19)–(3.18), the spatial frequency f_x of the interference fringe pattern $I(\boldsymbol{r})$ can be calculated by

$$f_x = \frac{1}{2\pi}\frac{d\varphi_{OR}}{dx} = \frac{k(\sin\varphi_O + \sin\varphi_R)}{2\pi} = \frac{\sin\varphi_O + \sin\varphi_R}{\lambda}. \tag{3.20}$$

The period of the interference fringe pattern, d_x, can be described by the reciprocal of the spatial frequency:

$$d_x = \frac{1}{f_x} = \frac{\lambda}{\sin\varphi_O + \sin\varphi_R}. \tag{3.21}$$

For example, $d_x \approx 1\,\mu\text{m}$ assuming $\lambda = 532\,\text{nm}$, $\phi_O = 0°$, and $\phi_R = 30°$. Based on the **sampling theorem**, the interference fringes of d_x must be detected by a sampling interval of less than $d_x/2$. In the condition described above, a sampling interval of approximately 500 nm is required for recording hologram materials. This can be easily satisfied by photosensitive materials, such as silver-halide emulsion, a photopolymer, and a photoresist. In contrast, the pixel pitch of even leading-edge image sensors is no more than 1 μm. This limited leading edge makes setting a large incident angle of the reference wave difficult in digital holography. Thus, the viewing zone and spatial resolution of the reconstructed images of the object decrease.

Then, spectra or spatial frequencies of the object, reference, and zeroth-order diffraction waves are considered. By applying Fourier transform to Eq. (3.1), the following equation can be obtained:

$$\mathcal{F}[I(x,y)] = \mathcal{F}\left[|O(x,y)|^2 + |R(x,y)|^2\right] + \mathcal{F}[O(x,y)R^*(x,y)] \\ + \mathcal{F}[O^*(x,y)R(x,y)]. \tag{3.22}$$

Here, the z-coordinate is omitted for simplicity. The first term of the right-hand side of Eq. (3.22) denotes the spectrum of the zeroth-order diffraction term, and it consists of low-frequency components. The second term includes the spectrum of the object wave. Assuming that the reference wave is a plane wave and can be described by

$$R(x, y) = \exp\left[i(kx \sin\varphi_{Rx} + ky \sin\varphi_{Ry})\right], \tag{3.23}$$

the second term of Eq. (3.22) can be expressed by

$$\mathcal{F}[O(x, y)R^*(x, y)] = \mathcal{F}\left[O(x, y)\exp\left[-i(kx \sin\varphi_{Rx} + ky \sin\varphi_{Ry})\right]\right]. \tag{3.24}$$

Here, the following frequency-shift property of Fourier transforms is considered by assuming $U(f_x, f_y) = \mathcal{F}[u(x, y)]$:

$$\mathcal{F}[u(x - s, y - t)] = U(f_x, f_y)\exp\left[-i2\pi(f_x s + f_y t)\right], \tag{3.25}$$

$$\mathcal{F}^{-1}[u(x, y)\exp\left[i2\pi(sx + ty)\right]] = U(f_x - s, f_y - t)]. \tag{3.26}$$

Then, applying Eqs. (3.25) and (3.24),

$$\mathcal{F}[O(x, y)R^*(x, y)] = \tilde{O}\left(f_x + \frac{\sin\varphi_{Rx}}{\lambda}, f_y + \frac{\sin\varphi_{Ry}}{\lambda}\right), \tag{3.27}$$

where $\tilde{O}$ indicates the spectrum of the object wave and $\tilde{O}(f_x, f_y) = \mathcal{F}[O(x, y)]$. Equation (3.27) implies that the spectrum of the object wave shifts by

$$(\lambda \sin\varphi_{Rx}, \lambda \sin\varphi_{Ry}), \tag{3.28}$$

from $\tilde{O}(f_x, f_y)$ in the frequency domain according to the incident angle of the reference wave. Similarly, the third term of Eq. (3.22) can be described by

$$\begin{aligned}\mathcal{F}[O^*(x, y)R(x, y)] &= \mathcal{F}\left[O^*(x, y)\exp\left[i(kx \sin\varphi_{Rx} + ky \sin\varphi_{Ry})\right]\right]\\ &= \tilde{O}^*\left(f_x - \frac{\sin\varphi_{Rx}}{\lambda}, f_y - \frac{\sin\varphi_{Ry}}{\lambda}\right),\end{aligned} \tag{3.29}$$

where $\tilde{O}^*$ indicates the spectrum of the conjugate wave and $\tilde{O}^*(f_x, f_y) = \mathcal{F}[O^*(x, y)]$. Equation (3.29) indicates the spectrum of the conjugate wave shifts by

$$(\lambda \sin\varphi_{Rx}, \lambda \sin\varphi_{Ry}), \tag{3.30}$$

from $\tilde{O}^*(f_x, f_y)$ in the frequency domain.

As described previously, the spectra of the object, reference, and zeroth-order diffraction waves can be separated in the frequency domain. Then, it is easy to extract the spectrum of the object wave only. After the extraction of the spectrum of the object wave, it is shifted to the origin of the frequency domain. Finally, applying inverse

Fourier transform to the shifted spectrum, the object wave only can be obtained in the spatial domain. Certainly, a diffraction calculation is required for the obtained object wave.

3.3 Phase-Shifting Digital Holography

Phase-shifting digital holography [13–21] can also overcome the twin-image problem. This technique is a type of in-line holography. Figure 3.5 shows an optical setup of phase-shifting digital holography. This optical setup has a phase-shifting device to shift the phase of the reference wave in nanometer order. A mirror mounted on a piezoelectric element or a phase retarder, such as a wave plate and a phase-modulation-type spatial-light modulator, can be used as a phase-shifting device. Here, assuming the phase of the reference wave as θ_R, the complex amplitude of a planar reference wave at the image sensor plane can be described as a function of θ_R by

$$R(\theta_R) = A_R \exp(i\theta_R). \tag{3.31}$$

Then, a hologram, $I(\theta_R)$, formed by $R(\theta_R)$ and the object wave O can be described by

$$\begin{aligned} I(\theta_R) &= |O + R(\theta_R)|^2 \\ &= |O|^2 + |R(\theta_R)|^2 + OR^*(\theta_R) + O^*R(\theta_R) \\ &= |O|^2 + |A_R|^2 + OA_R \exp(-i\theta_R) + O^*A_R \exp(i\theta_R). \end{aligned} \tag{3.32}$$

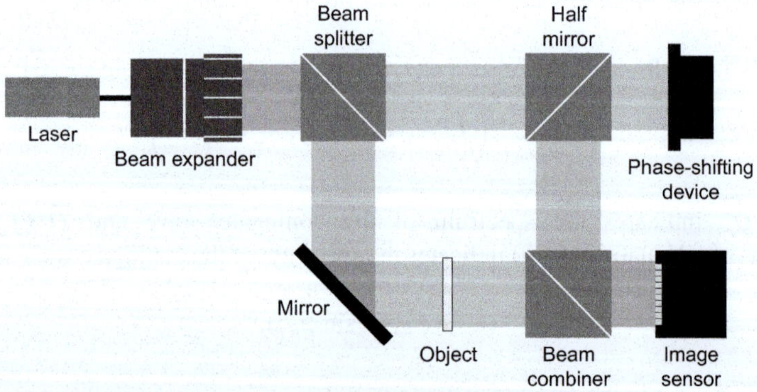

Fig. 3.5 Optical setup of phase-shifting digital holography

3.3.1 Four-Step Phase-Shifting Digital Holography

Let us consider reference waves with four different phase shifts of 0, $\pi 2$, π, and $3\pi/2$ [13, 15]. Four holograms recorded by the four reference waves can be described by

$$I(0) = |O|^2 + A_R^2 + OA_R + O^* A_R, \tag{3.33}$$

$$I\left(\frac{\pi}{2}\right) = |O|^2 + A_R^2 - iOA_R + iO^* A_R, \tag{3.34}$$

$$I(\pi) = |O|^2 + A_R^2 - OA_R - O^* A_R, \tag{3.35}$$

$$I\left(\frac{3\pi}{2}\right) = |O|^2 + A_R^2 + iOA_R - iO^* A_R. \tag{3.36}$$

The real part of the object wave can be obtained using Eqs. (3.33) and (3.35):

$$\mathrm{Re}[O] = \frac{1}{4A_R}[I(0) - I(\pi)]. \tag{3.37}$$

Meanwhile, the imaginary part of the object wave can be obtained using Eqs. (3.34) and (3.36):

$$I(\frac{\pi}{2}) - I(\frac{3\pi}{2}) = -i2OA_R + i2O^* A_R = -i2A_R(O - O^*) = 4A_R\mathrm{Im}[O]. \tag{3.38}$$

Then,

$$\mathrm{Im}[O] = \frac{1}{4A_R}\left[I(\frac{\pi}{2}) - I(\frac{3\pi}{2})\right]. \tag{3.39}$$

Finally, O can be calculated by

$$O = \frac{1}{4A_R}\left\{[I(0) - I(\pi)] + i\left[I\left(\frac{\pi}{2}\right) - I\left(\frac{3\pi}{2}\right)\right]\right\}. \tag{3.40}$$

The number of phase shifts affects the robustness of results in practice; it can be enhanced according to the number of phase shifts. In contrast, more time is required to record holograms necessary for calculating the phase-shifting method in proportion to the number of phase shifts.

3.3.2 Three-Step Phase-Shifting Digital Holography

Let us consider reference waves with three different phase shifts [15]. The values of 0, $\pi 2$, and π are assumed as three phase shifts. Then, as with the four-step case, three holograms recorded by the three reference waves can be described by

$$I(0) = |O|^2 + A_R^2 + OA_R + O^* A_R, \tag{3.41}$$

$$I\left(\frac{\pi}{2}\right) = |O|^2 + A_R^2 - iOA_R + iO^* A_R, \tag{3.42}$$

$$I(\pi) = |O|^2 + A_R^2 - OA_R - O^* A_R. \tag{3.43}$$

Here, the first and second terms, which correspond to the zeroth-order diffraction waves, on the right-hand side are common. This equality implies that they can be cancelled by subtraction. First, focusing on Eqs. (3.41) and (3.43), the following equation can be obtained:

$$\begin{aligned} I(0) - I(\pi) &= 2OA_R + 2O^* A_R \\ &= 2A_R(O + O^*) \\ &= 4A_R \mathrm{Re}[O]. \end{aligned} \tag{3.44}$$

Then,

$$\mathrm{Re}[O] = \frac{1}{4A_R}[I(0) - I(\pi)]. \tag{3.45}$$

Because A_R can be considered constant when the reference wave is assumed as a plane wave, the coefficient $1(4A_R)$ can be omitted for simplicity when it is calculated on a computer. Therefore, Eq. (3.45) indicates that the real part of the object wave can be calculated using $I(0)$ and $I(\pi)$. Similarly, using the three holograms, the following equation can be obtained:

$$\begin{aligned} I(0) - 2I\left(\frac{\pi}{2}\right) + I(\pi) &= i2OA_R - i2O^* A_R \\ &= i2A_R(O - O^*) \\ &= -4A_R \mathrm{Im}[O]. \end{aligned} \tag{3.46}$$

Then,

$$\mathrm{Im}[O] = -\frac{1}{4A_R}\left[I(0) - 2I\left(\frac{\pi}{2}\right) + I(\pi)\right]. \tag{3.47}$$

Equation (3.47) indicates that the imaginary part of the object wave can be calculated using $I(0)$, $I(\pi 2)$, and $I(\pi)$. Finally, O can be calculated by

$$O = \frac{1}{4A_R}\left\{[I(0) - I(\pi)] - i\left[I(0) - 2I\left(\frac{\pi}{2}\right) + I(\pi)\right]\right\}. \tag{3.48}$$

3.4 Single-Shot Phase-Shifting Digital Holography

Although phase-shifting digital holography can reconstruct the object wave without the zeroth-order and conjugate waves, it requires the sequential recording of multiple holograms for calculating the phase-shifting method. Therefore, it is difficult for phase-shifting digital holography to record a moving object or dynamic phenomenon. To overcome this problem, **single-shot phase-shifting digital holography** was proposed [22–32]. There are some techniques to realize single-shot phase-shifting digital holography. Here, a method using space-division multiplexing of holograms is considered [22, 23]. Figure 3.6 shows an optical setup for single-shot phase-shifting digital holography using space-division multiplexing. A light wave from the optical source is split into two paths by the polarization-beam splitter. One is introduced into the object, and another is used as the reference wave. The object wave and reference waves are combined by the polarization-beam combiner. After passing through the quarter-wave plate in front of the image sensor, the two waves are introduced into the image sensor plane, and interference fringes are recorded by the image sensor. Here, the image sensor has a pixel-by-pixel micro-polarizer array (Fig. 3.6). This micro-polarizer array can select four polarization axes of 0°, 45°, 90°, and 135° for a set of 2×2 pixels. Owing to the micro-polarizer array, an interference fringe pattern that includes four pixel-by-pixel phase-shifted holograms can be recorded with a single-shot exposure.

Let us consider the relationship between the polarization directions and phase-shift values. Figure 3.7a shows the polarization states of the object and reference waves in Fig. 3.6. After passing through the polarization-beam splitter, their polarization states are linear and orthogonal. This orthogonality is kept after passing through the polarization-beam combiner. The two waves with orthogonal linear polarization are

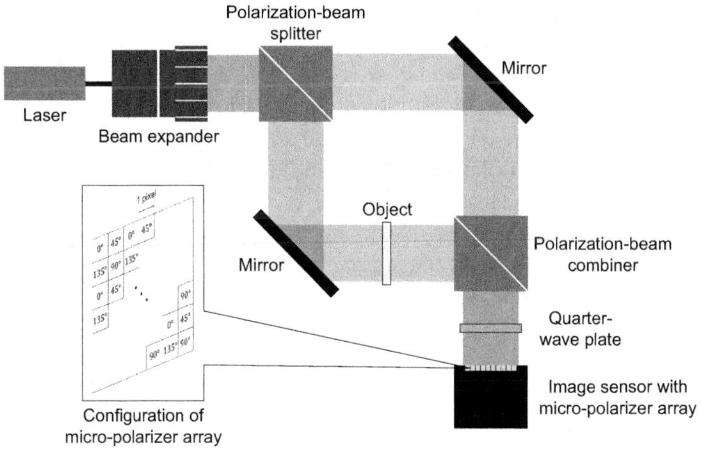

Fig. 3.6 Optical setup of single-shot phase-shifting digital holography

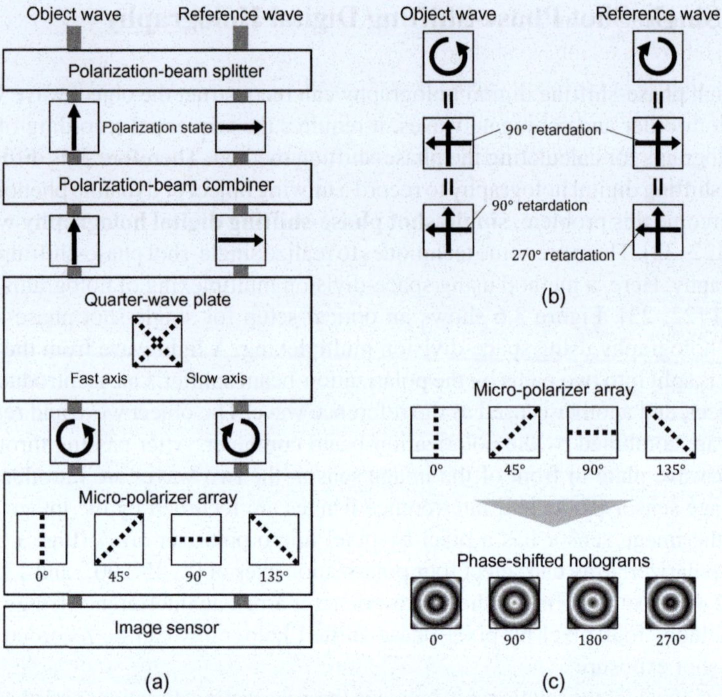

Fig. 3.7 Phase shift using polarization. **a** Transition of polarization states of object and reference waves. **b** Expression of circular polarization using two orthogonal linear polarizations. **c** Phase-shifted holograms detected by the micro-polarizer array

introduced into the quarter-wave plate. Because the fast axis of the quarter-wave plate is inclined to 45° relative to the polarization directions of the reference and object waves, the polarization states of the two waves are converted into circular polarization by the function of the quarter-wave plate. Their rotating directions are opposite. This is because the polarization directions of the object and reference waves before passing through the quarter-wave plate are orthogonal. The object and reference waves with circular polarization are introduced into the image sensor with the micro-polarizer array. Because the micro-polarizer array can independently detect four linear polarization directions pixel by pixel, we consider circular polarizations as combinations of two orthogonal linear polarizations, as shown in Fig. 3.7b. The horizontal component of the polarization of the object wave has 90° retardation relative to its vertical component. Conversely, the horizontal component of the polarization of the reference wave has 270° retardation relative to its vertical component. Then, if micro-polarizers with vertical (0°) and horizontal (90°) orientations are used to detect the object and reference waves, holograms with 0° and 180° phase shifts, respectively, can be recorded (Fig. 3.7c). Similarly, micro-polarizers with 45° and 135° orientations can record holograms with 90° and 270° phase shifts, respectively.

Therefore, using the micro-polarizer array, an interference fringe pattern including four pixel-by-pixel phase-shifted holograms can be recorded. Single-shot phase-shifting digital holography can be realized by other optical setups: using the Talbot effect (or self-imaging phenomenon), a reference wave with random phases, and a reference wave with an inclined angle (corresponding to the off-axis setup).

References

1. Shimobaba, T., Ito, T.: Computer Holography, CRC Press (2019).
2. Goodman, J. W. and Lawrence, R. W.: Digital image formation from electronically detected holograms. Appl. Phys. Lett. **11**, 77–79 (1967).
3. Kronrod, M. A. and Merzlyakov, N. S. and Yaroslavskii, L. P.: Reconstruction of a hologram with a computer. Sov. Phys. Tech. Phys. **17**, 333–334 (1972).
4. Schnars, U. and Jüptner, W.: Direct recording of holograms by a CCD target and numerical reconstruction. Appl. Opt. **33**, 179–181 (1994).
5. Cuche, E. and Marquet, P. and Depeursinge, C.: Spatial filtering for zero-order and twin-image elimination in digital off-axis holography. Appl. Opt. **39**, 4070–4075 (2000).
6. Pavillon, N. and Seelamantula, C. S. and Kühn, J. and Unser, M. and Depeursinge, C.: Suppression of the zero-order term in off-axis digital holography through nonlinear filtering. Appl. Opt. **48**, H186–H195 (2009).
7. Verrier, N. and Atlan, M.: Off-axis digital hologram reconstruction: some practical considerations. Appl. Opt. **50**, H136–H146 (2011).
8. Sánchez-Ortiga, E. and Doblas, A. and Saavedra, G. and Martínez-Corral, M. and Garcia-Sucerquia, J.: Off-axis digital holographic microscopy: practical design parameters for operating at diffraction limit. Appl. Opt. **53**, 2058–2066 (2014).
9. Rubin, M. and Dardikman, G. and Mirsky, S. K. and Turko, N. A. and Shaked, N. T.: Six-pack off-axis holography. Opt. Lett. **42**, 4611–4614 (2017).
10. Mirsky, S. K. and Shaked, N. T.: First experimental realization of six-pack holography and its application to dynamic synthetic aperture superresolution. Opt. Express **27**, 26708–26720 (2019).
11. Yao, L. and Yu, L. and Lin, X. and Wu, Y. and Chen, J. and Zheng, C. and Wu, X. and Gao, X.: High-speed digital off-axis holography to study atomization and evaporation of burning droplets. Combust. Flame **230**, 111443 (2021).
12. Anzai, W. and Kakue, T. and Shimobaba, T. and Ito, T.: Temporal super-resolution high-speed holographic video recording based on switching reference lights and angular multiplexing in off-axis digital holography. Opt. Lett. **47**, 3151–3154 (2022).
13. Yamaguchi, I. and Zhang, T.: Phase-shifting digital holography. Opt. Lett. **22**, 1268–1270 (1997).
14. Cai, L. Z. and Liu, Q. and Yang, X. L.: Generalized phase-shifting interferometry with arbitrary unknown phase steps for diffraction objects. Opt. Lett. **29**, 183–185 (2004).
15. Schnars, U. and Jueptner, W.: Digital Holography. Springer Berlin, Heidelberg (2005).
16. Meng, X. F. and Cai, L. Z. and Xu, X. F. and Yang, X. L. and Shen, X. X. and Dong, G. Y. and Wang, Y. R.: Two-step phase-shifting interferometry and its application in image encryption. Opt. Lett. **31**, 1414–1416 (2006).
17. Liu, J.-P. and Poon, T.-C.: Two-step-only quadrature phase-shifting digital holography. Opt. Lett. **34**, 250–252 (2009).
18. Kikuchi, Y. and Barada, D. and Kiire, T. and Yatagai, T.: Doppler phase-shifting digital holography and its application to surface shape measurement. Opt. Lett. **35**, 1548–1550 (2010).
19. Yoshikawa, N: Phase determination method in statistical generalized phase-shifting digital holography. Appl. Opt. **52**, 1947–1953 (2013).

20. Xia, P. and Wang, Q. and Ri, S. and Tsuda, H.: Calibrated phase-shifting digital holography based on a dual-camera system. Opt. Lett. **42**, 4954–4957 (2017).
21. Chen, L. and Singh, R. K. and Chen, Z. and Pu, J.: Phase shifting digital holography with the Hanbury Brown–Twiss approach. Opt. Lett. **45**, 212–215 (2020).
22. Millerd, J. E. and Brock, N. J. and Hayes, J. B. and North-Morris, M. B. and Novak, M. and Wyant, J. C.: Pixelated phase-mask dynamic interferometer. Proc. SPIE **5531**, 304 (2004).
23. Awatsuji, Y. and Sasada, M. and Kubota, T.: Parallel quasi-phase-shifting digital holography. Appl. Phys. Lett. **85**, 1069 (2004).
24. Awatsuji, Y. and Fujii, A. and Kubota, T. and Matoba, O.: Parallel three-step phase-shifting digital holography. Appl. Opt. **45**, 2995–3002 (2006).
25. Awatsuji, Y. and Tahara, T. and Kaneko, A. and Koyama, T. and Nishio, K. and Ura, S. and Kubota, T. and Matoba, O.: Parallel two-step phase-shifting digital holography. Appl. Opt. **47**, D183–D189 (2008).
26. Martínez-León, L. and Araiza-E, M. and Javidi, B. and Andrés, P. and Climent, V. and Lancis, J. and Tajahuerce, E.: Single-shot digital holography by use of the fractional Talbot effect. Opt. Express **17**, 12900–12909 (2009).
27. Murata S. and Harada D. and Tanaka, Y.: Spatial Phase-Shifting Digital Holography for Three-Dimensional Particle Tracking Velocimetry. Jpn. J. Appl. Phys. **35**, 09LB01 (2009).
28. Nomura, T. and Imbe, M.: Single-exposure phase-shifting digital holography using a random-phase reference wave. Opt. Lett. **35**, 2281–2283 (2010).
29. Tahara, T. and Ito, K. and Fujii, M. and Kakue, T. and Shimozato, Y. and Awatsuji, Y. and Nishio, K. and Ura, S. and Kubota, T. and Matoba, O.: Experimental demonstration of parallel two-step phase-shifting digital holography. Opt. Express **18**, 18975–18980 (2010).
30. Kakue, T. and Yonesaka, R. and Tahara, T. and Awatsuji, Y. and Nishio, K. and Ura, S. and Kubota, T. and Matoba, O.: High-speed phase imaging by parallel phase-shifting digital holography. Opt. Lett. **36**, 4131–4135 (2011).
31. Xia, P. and Awatsuji, Y. and Nishio, K. and Matoba, O.: One million fps digital holography. Electron. Lett. **50**, 1693–1694 (2014).
32. Kakue, T. and Endo, Y. and Nishitsuji, T. and Shimobaba, T. and Masuda, N. and Ito, T.: Digital holographic high-speed 3D imaging for the vibrometry of fast-occurring phenomena. Sci. Rep. **7**, 10413 (2017).

Part II
Introduction to Hardware

Part II consists of four chapters. Computer holography requires many computationally time-consuming calculations of light waves. In computer holography, knowledge of hardware accelerators that speed up these calculations is important. In Part II, an overview of each hardware accelerator is given.

Chapter 4
Basic Knowledge of CPU Architecture for High Performance

Takashige Sugie

Abstract In this chapter, we show the advantages and disadvantages of the pipeline processing architecture. We also discuss the effects of instruction and program algorithm on processing speed of the central processing unit (CPU). In order to increase the computation speed of a CPU, it is important to know what functions the CPU is equipped with for computation. On the basis of this, the mechanism of functions related mainly to parallel processing is explained. We show that as the degree of parallel processing increases, the amount of data required per unit time also increases. This implies that it is not enough to only process arithmetic instructions efficiently, but it is also important to transfer data at high speed. Finally, we discuss different memory types in which data are stored and the data structure suitable for the CPU.

4.1 Introduction

Broadly speaking, there exist two methods for speeding up numerical computations on **central processing units** (**CPU**s). The first method involves a soft approach that can help enhance CPU performance by improving computational algorithms. The second method is developing a program that can maximize CPU's peak performance. When a CPU is working at its peak performance, all of its computing units are performing valid computations. Therefore, for speeding up a CPU, we must understand what functions are implemented in the CPU, and then ensure that all those functions perform valid computations at all times. Moreover, the programs and data that make up these functions can be rearranged into appropriate calculation procedures and data formats, respectively, in such a way that the functions become easy to process. Such a method can be called a hard approach to speeding up a CPU's numerical computations because it involves modifications suitable to the computer's architecture. This chapter introduces the basic knowledge of the CPU architecture needed to implement the hard approach. Pipeline processing, parallel computer architecture,

T. Sugie (✉)
Chiba University, 1-33 Yayoi-cho, Inege-ku, Chiba 263-8522, Japan
e-mail: myrhrk@gmail.com

T. Shimobaba and T. Ito (eds.), *Hardware Acceleration of Computational Holography*,
https://doi.org/10.1007/978-981-99-1938-3_4

and cache memory architecture all affect the speedup of numerical calculations. This chapter explains the CPU architecture to focus on these three points.

Pipeline processing is an essential technology for achieving high-speed processing in CPUs. The scheduling of pipeline processing is very complicated, and in certain situations, the desired speedup effect cannot be obtained. Fortunately, the compiler optimization and the instruction scheduler inside CPUs have become highly intelligent in recent years, and therefore we hardly need to design a program while considering pipeline schedules. In the first place, this problem is so complex not to be solved at the level of software programming. However, we prefer to know the pipeline architecture because it is a technology that is adopted always in current CPUs. Here, we elaborate on developing efficient programs that can realize high-speed calculations in CPUs.

The CPU comprises multiple circuits that can execute program instructions. Generally only one of these circuits works unless we create a program that uses multiple circuits. As the number of the circuits used in computers is increasing year by year, it is important that we program on the premise of parallel processing.

The parallelization of circuits allows us to perform multiple calculations at once. For performing calculations smoothly, all operands must have been prepared to the arithmetic circuits. One downside of parallelization of circuits, however, is that the supply of data cannot keep up with high calculation speeds. Therefore, considering efficient data processing is critical for achieving peak CPU performance. As a countermeasure against this problem, the CPU has a built-in high-speed memory device called cache memory. Creating a data processing procedure that can use the cache memory effectively is key for achieving speedup.

Certain concepts in this chapter are explained using the C programming language. Furthermore, notation methods such as hexadecimal numbers, which follow the C programming language format, are also used.

4.2 Pipeline Processing

4.2.1 Fundamentals of CPU Architecture

Before explaining pipeline processing, we describe the basic operation of the CPU. The CPU is designed performing general-purpose calculations. Not all calculations are implemented through the hardware. In fact, complex calculations are realized by repeating simple operations. The basic functions that the CPU performs as hardware are relatively simple. These are: four arithmetic operations, bit operations, and data load/store [1]. For complex but frequently used mathematical formulas such as trigonometric functions and square roots, specialized arithmetic circuits are implemented [2].

When performing a calculation, operands are loaded into the memory called a **register**. A register can only store about 64 bits in general. The register is the only memory that is closely connected to the arithmetic circuit, and it operates at the highest speed. The number of mounted multi-purpose registers in recent CPUs is about 16 [3], which is not a very high number. Current computers use **dynamic random access memories** (**DRAM**s), which have a larger storage capacity than registers, but their communication speed is relatively slow.

A program is a list of instructions in a pre-determined order. The CPU uses various advanced technologies to execute programs. Therefore the instruction processing is highly complicated [4]. Instruction processing in CPUs can be roughly divided into five steps:

(1) IF: Instruction Fetch First, the CPU fetches instructions from the system memory (DRAMs).
(2) ID: Instruction Decode The CPU decodes the instruction to interpret what it means.
(3) EX: Execution The CPU executes the instruction.
(4) MEM: Memory Read/Write If the instruction requires access to the system memory, the CPU communicates with an external memory device (e.g., DRAMs or Hard disk drives).
(5) WB: Write Back Finally, the result of the processing is written to the specified register according to each instruction.

Types of instruction processing include numerical operations, data transfer, and program flow control. Depending on the content of the instruction, not all of the aforementioned steps are necessary.

4.2.2 Pipeline Architecture

When a control circuit of the CPU processes instructions sequentially, the execution efficiency is poor because the next instruction cannot be processed until all steps of the first instructions are completed. In general, we divide the processing into different steps in such a way that each step can be executed independently. For example, for this study, we prepare a controller specialized for fetching instructions in the IF step and a controller that has only the function that can decode instructions in the ID step. These controllers can operate independently in other steps as well. In this way, if the control circuit of the IF step fetches the instruction from the system memory and passes it to the ID step, the control circuit of the IF step can immediately start fetching the next instruction. In the ID step, when the processing is taken over to the appropriate step by the decoded instruction, the decoding of the next instruction can start. The same applies to other steps. If instructions can be processed continuously, each step has an intermediate state of the instruction processing, as shown in Fig. 4.1. If the processing time of each step is the same and instructions can be processed continuously, we can obtain instruction-processing results at equal time intervals. The biggest advantage is that the instruction-processing time becomes from 5 steps

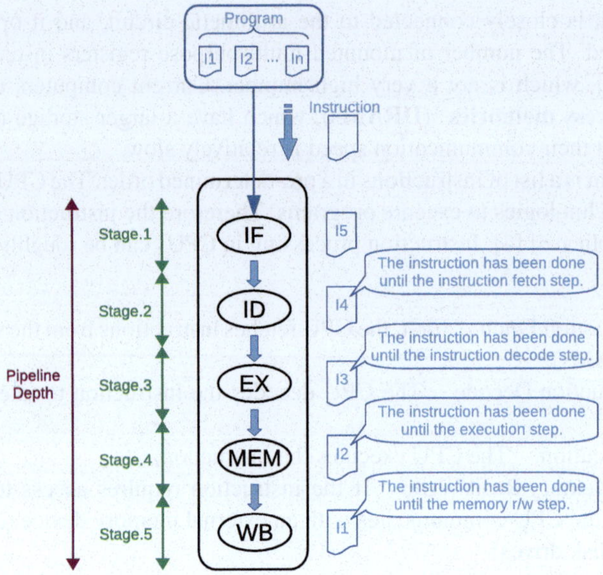

Fig. 4.1 Processing of instructions using the pipeline processing technology. It looks like that processing of the instruction has the write back step only

to 1 step. In the last step (the WB step), the instruction has been processed until the fourth step. Therefore, it looks like that the processing of the instruction has been completed by 1 step. As each step is always ready for execution, the operation rate of the circuit increases and execution efficiency improves. In other words, pipeline processing is the method that involves dividing the problem into small pieces and stacking their processing results. An individual piece is called a **pipeline stage**. The number of divisions is called the **pipeline depth**. Moreover, the time taken to obtain the final result in the pipeline architecture is called the **pipeline delay**. In Fig. 4.1, the pipeline depth is 5. Due to the use of advanced technologies in current CPUs, the number of divisions increases, and consequently the pipeline becomes deeper. The pipeline depth of recent Intel CPUs is about 14 [5].

Pipeline processing is an indispensable technology for current computers. However, in practice, it is difficult to perform pipeline processing efficiently. First, there is no guarantee that instructions can be continuously input into the pipeline. In addition, the circuit area is enlarged because the processing circuit cannot be simply divided. Furthermore, the very structure of the pipeline gives rise to other problems as described in the next paragraph.

Listing 4.1 is part of a C program that finds the minimum from the two-dimensional array psi. The element of the two-dimensional array psi has an 8-bit integer type with HEIGHT × WIDTH as the number of elements. Figure 4.2 shows how this program is processed by the pipeline processing. The horizontal axis denotes time and the vertical axis denotes the program flow. The values of the variables i, j, and

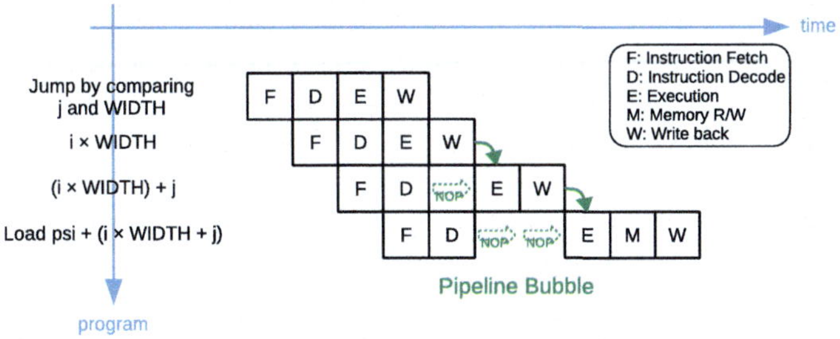

Fig. 4.2 Image diagram of the pipeline bubble

WIDTH and the pointer to the two-dimensional array psi are assumed to be loaded into registers. The second line of the program compares j and WIDTH. If it is true, the for statement continues, and if false, it ends. If true, the CPU loads psi[i][j] to check the condition of the if statement on the third line. The variable psi is a pointer indicating the start address of a two-dimensional array. We need to calculate the address of psi[i][j] by psi + (i × WIDTH + j). The calculation of i × WIDTH must be completed in order to calculate (i × WIDTH) + j. In other words, the EX stage of the (i × WIDTH) + j instruction can be processed only after the WB stage of the i × WIDTH instruction is completed. In such a case, a process called **NOP** (**No Operation**), which basically does nothing, is automatically inserted. Although the NOP results in the intended processing, the original pipeline delay increases from 4 to 5. In the next load instruction, two NOP stages are inserted. The pipeline delay becomes 7. Such a hazard is called a **pipeline bubble**. Even if there is no dependency relationship between instructions, a pipeline bubble occurs. The same stage overlaps at the same time by depending on the timing because different instructions require different numbers of pipeline stages. In a conditional branch instruction or the like, any one program is speculatively executed according to the conditional result. If the result is not expected, all speculatively executed instructions are invalidated. Such a hazard is called a **pipeline stall**. In a pipeline with multiple stages, a stall in a deep stage can be a serious hazard.

Listing 4.1 Sample program that may cause pipeline stall

```
1 for (i = 0; i < HEIGHT; i++) {
2   for (j = 0; j < WIDTH; j++) {
3     if (psi[i][j] < psi_min) psi_min = psi[i][j];
4   }
5 }
```

Although the pipeline architecture is of high-speed processing performance, if the pipeline processing flow is obstructed, the performance drops. One way to speed up computations on the CPU is to design programs that can keep the pipeline processing flow as smooth as possible. In the past, improvements could be made by making adjustments such as rearranging instructions to minimize pipeline hazards. At present, however, CPUs come with excellent compiler optimization, because of which there is no need to consider the pipeline flow while designing programs. The CPU is highly parallelized and can execute the many programs in parallel. Compared with the programs we design, compilers can design better, that is, more efficient and safer programs. The scheduler in the current CPUs is also highly functional and can change the order of instructions to some extent [6–8]. Therefore, we appropriate to take the stance that we help the compiler to achieve its full performance. For example, we can give optimization hints to the compiler.

Listing 4.2 is a program that counts the number of zero values from the two-dimensional array psi. If the two-dimensional array psi is a dense matrix, the evaluation of the condition of the if statement is likely to be false. If we know that the probability of branching is biased, we can give a hint to the compiler. We can use the built-in function __builtin_expect [9] provided by the C language compiler of the GCC (Gnu Compiler Collection) [10]. If we know the two-dimensional array psi is sparse, we can hint to the compiler using __builtin as in Listing 4.3. However, in Listing 4.3 we cannot obtain the effect because the program is too simple. We can also use __builtin_expect_with_probability [9] if we know the probability of branching is biased. In the first place, we may be able to substitute branch instructions with arithmetic instructions. It is also important that we carefully consider whether a branch instruction is really needed. Reducing branch instructions is helpful for the compiler to work efficiently.

Listing 4.2 Sample program of counting the number of zero values

```
for (i = 0; i < HEIGHT; i++) {
  for (j = 0; j < WIDTH; j++) {
    if (psi[i][j] == 0) n++;
  }
}
```

Listing 4.3 Sample program of counting the number of zero values using __builtin_expect

```
for (i = 0; i < HEIGHT; i++) {
  for (j = 0; j < WIDTH; j++) {
    if (__builtin_expect(psi[i][j] == 0, 0)) n++;
  }
}
```

At the software programming stage, we basically do not have to consider the pipeline processing of the instructions. The most appropriate way to deal with pipeline hazards is to use the CPU's compiler optimization and scheduler. However, we should know that pipeline technology is used to process instructions. Because it helps to understand other technologies such as atomic processing in parallel programming. Being aware of the advantages and disadvantages of pipeline processing can help us deepen our understanding of computers, and using it in a precise manner can help us speed up CPU computation.

4.2.3 Instruction Latency and Throughput

We consider the following two programs: Listing 4.4 and Listing 4.5. Because the two instructions used in Listing 4.4 have no dependency on each other, both can be executed simultaneously. Listing 4.5 cannot execute the next instruction without completing the first instruction. The first line in the both lists is the same instruction. However, the instruction on the first line affects the instruction on the second line differently. Regarding this difference, Intel defines two values for instructions: latency and throughput [11]. **Latency** is the number of the clock cycles until the contents of a instruction are completed, such as until the result of a trigonometric function is obtained, or until the data is completely written to the system memory. **Throughput** is the number of clock cycles required to wait before the pipeline is free to accept the same instruction again. For example, if the operand is given to the trigonometric function arithmetic circuit, other trigonometric function arithmetic instructions can be started without waiting for the result. In Listing 4.4, the processing of the instruction on the second line can start after the throughput time of the instruction on the first line. In Listing 4.5, the processing of the instruction on the second line must wait for the calculation result of the instruction on the first line. That is, the processing of the instruction on the second line can start after the latency of the instruction on the first line.

Listing 4.4 Two instructions are independent mutually

```
xaj = xa - xj;
yaj = ya - yj;
```

Listing 4.5 Second instruction can calculate to get the result of the first instruction

```
xaj = xa - xj;
xaj2 = xaj * xaj;
```

To make calculations efficient, we need to build a program that ensures that as many instructions as possible are processed with the throughput time. Depending on the order in which the instructions are processed, latency can be hidden. Rather than continuously executing instructions with high latency, concealment can be realized by putting some instructions with low latency between instructions of high latency. The thing we need to keep in mind is that there should be no dependency between these instructions. For example, the CPU first processes the load instructions of data used later. Next, the CPU executes other low-latency instructions unrelated to the previous load instructions. In this way, data arrives in the registers while the CPU processes the low-latency instructions. The processing speed increases because the latency seems to approach the throughput. That said, as with pipeline processing, compiler optimization in current CPUs can determine a better instruction order than what we can. In other words, the performance improvement achieved by manually changing the order of the instructions is rarely significant. Another issue with changing the order of the instructions manually is that it can make reading of the source code difficult. Therefore, the order of the instructions should be left to the compiler.

Considering whether we can replace high latency instructions with some low-latency instructions is also one way.

4.3 Parallel Architecture

4.3.1 Single Instruction Multiple Data (SIMD) Architecture

It is a well-known fact that the processing performance of the CPU increases as the clock frequency increases. Almost all computer circuits operate in synchronization with the clock. Therefore, the computer performance is directly proportional to the clock frequency. In fact, in the 1990s, to achieve higher speedups, clock frequency of the computers was significantly increased. To maintain a high clock frequency, the pipeline stage is further subdivided, which in turn creates a deeper pipeline. For example, the CPU called Prescott made by Intel has 31 stages. In addition, the increase in power consumption and heat generation becomes significant as the frequency increases. With times, these became fatal problems and the increase in the clock frequency in computers stagnated.

To address this problem, a method was developed to effectively utilize circuits that perform meaningless operations. A so-called 32-bit CPU is equipped with 32-bit registers. These registers could also be used for operations of short word lengths (such as 8-bit operations). Figure 4.3 shows a simple addition operation of 8-bit data. At this time, only the lower 8 bits are performing valid operations. The upper 24 bits only perform 0 + 0 operations. Here, we assume that 32-bit data are composed of four 8-bit data. Subsequently, we perform four operations with one instruction. Such an operation can be realized using a special register that ignores the carry at the main bit position that is the power of 2 such as 8-bit and 16-bit, as shown in Fig. 4.4. This method is called saturation calculation. It is also called the **single instruction multiple data** (**SIMD**) method because the CPU calculates multiple data using only one instruction.

Intel's SIMD technology continues to evolve to MMX, streaming SIMD extensions (SSE), and advanced vector extensions (AVX). Currently, we can perform saturation calculations using 512-bit SIMD instructions [12]. This means the ability

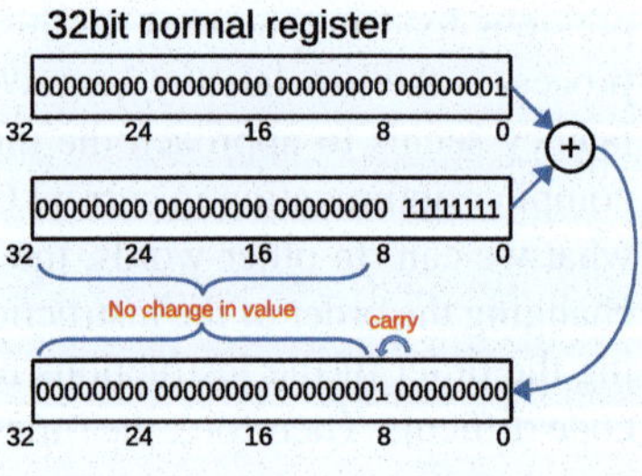

Fig. 4.3 Image diagram of addition calculation of 8-bit data using the 32-bit normal register

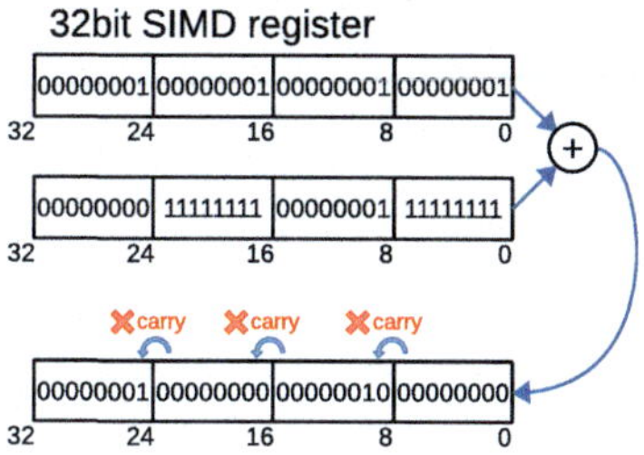

Fig. 4.4 Image diagram of four addition calculations of 8-bit data using the 32-bit SIMD register

to process 16 float-type calculations simultaneously. Therefore, we must design the part of the program that deals with numerical calculations on the assumption that SIMD is used.

4.3.2 Superscalar and Vector Processor

In Sect. 4.2.1, we explained that the CPU processes instructions in five steps. EX in the third step processes arithmetic and logical operations in the execution unit according to the instructions. For general-purpose calculations, the execution unit has various types of calculator circuits. Normally, it processes the instructions one by one in order. A processor that has one instruction pipeline and executes one basic operation according to the instruction is called a **scalar computer**.

Independent operations, that is, operations with no dependencies between them, can be processed in parallel, for example, calculating the distance between two points. The formula for calculating the distance between two points (x_α, y_α) and (x_j, y_j) is $\sqrt{(x_\alpha - x_j)^2 + (y_\alpha - y_j)^2}$. We can see that the terms x and y are not related. Therefore, if there exist multiple execution units, they can be processed in parallel. A processor with multiple execution units is called a **superscalar computer**. As shown in red and green in Fig. 4.5, $(x_\alpha - x_j)^2$ and $(y_\alpha - y_j)^2$ can be assigned to different execution units. Hence, the instruction related to y can start processing without waiting for the execution of the instruction related to x. The current CPU has 8, 10, or 12 execution units [13–15].

In addition to a scalar processor, there also exist a **vector processor**. In the execution unit of the vector processor, arithmetic units are arranged according to a specific mathematical expression. For example, the vector processor used for calculating the distance between two points is shown in Fig. 4.6. The rounded rectangle of "sqrt" in the figure is a square root operation. Since it is a circuit that can calculate only a specific mathematical formula, no instruction is required. In a vector processor, pipeline processing is used in the arithmetic circuit, and the operands input directly. Therefore, in vector processors, if the operands can be input continuously, the expected calculation time is 1 clock. In such case, vector processors can calculate faster than scalar processors.

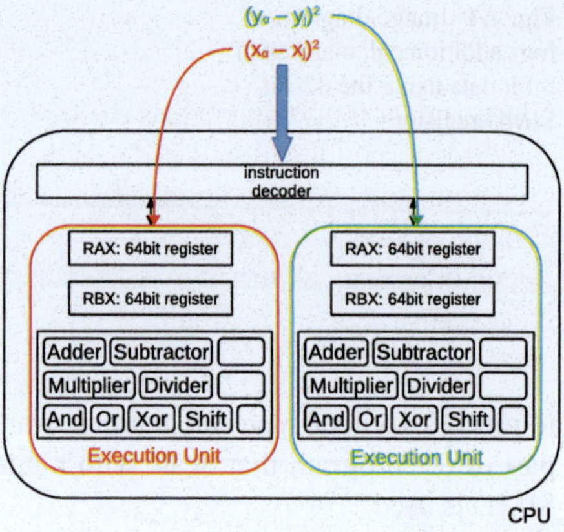

Fig. 4.5 Superscalar processor with two execution units

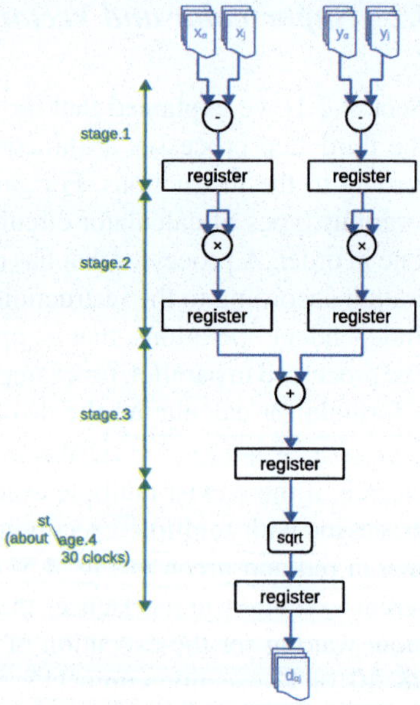

Fig. 4.6 Calculation pipeline for finding the distance between two points in a vector processor

Takashige Sugie

The circuit of "sqrt" is actually a vector-type circuit in also scalar processors. Intel CPUs have a latency of about 30 [16]. This means that these CPUs consist of approximately 30 stages of pipeline. Even though the current general CPUs are superscalar processor, it is highly parallelized, it's not too wrong to say a vector processor. The current CPUs achieve general-purpose computations by inheriting scalar processors and use vector processors to improve computational performance for algorithms with high computational costs. The current CPUs have a flexible design that can calculate various mathematical formulas at high speed.

4.3.3 Logical CPU

A superscalar CPU can process multiple instructions as long as the number of available execution units allows it. In a general program, it is difficult to always use all execution units. Some programs may be able to assign valid instructions to all execution units, but it is a very specific algorithm. For example, a program that requires the calculation result of the previous instruction, such as a recurrence formula, cannot process instructions in parallel. In a memory-dependent algorithm, most execution units may process NOP instructions described in Sect. 4.2.2 for a long time due to multiple memory access instructions. In such a program, many execution units do not perform valid operations. If a program different from the currently executing program can be processed in the instruction pipeline, there is a possibility to assign another instruction to an empty execution unit. For this purpose, the input port of the instruction decoder is expanded so that two programs can be accepted. We can virtually make it look like there are two CPUs. This technology is called **Hyper-Threading Technology** [17] in Intel's CPUs. Using two physical CPUs, we can expect nearly double the performance. However, this technology cannot always give such performance. Fortunately, computer-generated hologram (CGH) calculations can slightly improve performance using logical CPUs.

4.3.4 Multi-Core CPU

Remarkable advancements have been made in the microfabrication technology, and nowadays it is possible to develop a CPU with an extremely small process size of 2-nm [18]. As the processor size decreases, more semiconductors and wiring, and thus instruction-processing circuits, can be mounted in the same area. This, consequently, increases CPU cache memory capacity and execution units.

A circuit part corresponding to a single CPU is called a "physical core" or simply a "**core**", and a CPU with four cores or more on a single chip is called **multi-core CPU** or **many-core CPU**. When we take virtual CPUs such as the hyper-threading technology into account, we use the term logical-core to prevent misunderstandings. In the multi-core CPU shown in Fig. 4.7, there are four physical cores or eight logical cores. Normally, an execution program created in C language does not include parallel processing. Therefore, even if we use a multi-core CPU, only a single-core is used.

All CPU cores, including logical cores, must be used to achieve high-speed numerical calculations. The simplest way to achieve this is to run the program for the number of CPU cores without using parallel processing. The operating system (OS) normally used in computers these days is a multitasking OS. Using the multitasking OS, we can run multiple programs. In UNIX, a task is traditionally called a process. The shell (e.g., bash, tcsh, or zsh) calls the child processes created by itself "job", and manages it. These were designed in the age of single-core CPUs and are not suitable for usage that distributes the load in the program. Therefore, "thread" that was suitable parallel

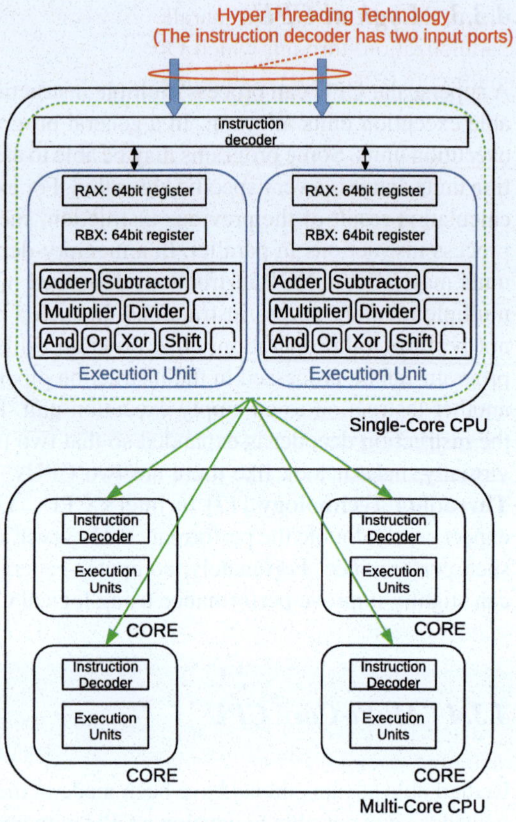

Fig. 4.7 Single-core CPU with the hyper-threading technology and a multi-core CPU that has four physical cores

distributed processing was developed. The thread is designed to reduce the overhead required for parallel distributed processing.

One of the most common application program interfaces (API) for multi-threading programming is pthreads (POSIX Threads). One of the well-known standard C library, the GNU C Library (glibc) [20], includes NPTL (Native POSIX Thread Library) [21] as pthreads. We can write a program that each thread directly can access all variables in source codes at the time of programming. It happens that multiple threads access the same variable at a given point in time. In such a case, a function to protect the consistency of variables is also provided by pthreads. We can also control the execution of the thread by some condition, and specify the CPU core that executes the thread. However, it is difficult to program because executing multiple programs simultaneously in parallel is a complicated task.

OpenMP [22] provides an excellent feature that helps parallel programming. Compilers such as GCC [10] support OpenMP. There is no need to prepare a special development environment for OpenMP. We only need to add one compiler option (-fopenmp) to enable the compiler's OpenMP processing. The OpenMP header file is included in the source code, and a parallelization method that uses the pragma directive in the part we want to parallel is used. Furthermore, OpenMP supports SIMD parallelization, through which OpenMP manages all controls of load balancing in multi-core CPUs and parallel processing.

4.4 Memory Architecture

4.4.1 Continuity of Data

The CPU always performs calculations while communicating with the system memory to compensate for the small memory capacity of the registers. If communication with the system memory is interrupted, the processing speed of the CPU is significantly affected. For example, Intel Core X-Series processor has four channels to the system memory [23]. This processor can request different communications for the four memory modules. In other words, we can expect up to four times the communication speed of a single memory module using the Intel Core X-Series processor. In this way, the CPU and the system memory are connected by a dedicated high-speed communication path. However, achieving peak performance is impossible with inefficient processing. Inefficient processing is a discontinuous access that accesses distant addresses. Therefore, it is important to pay attention to whether discontinuous access is occurring, and try to prevent it from happening as much as possible.

Even if we write a program that intends to access data continuously, there is actually a rudimentary misunderstanding that it is discontinuous access. Listings 4.6 and 4.7 involve finding the maximum and minimum values from the two-dimensional array psi with WIDTH × HEIGHT elements. Both programs use the variables h and w as loop counters. Since these variables increase one by one, the two-dimensional array psi is continuously referenced. The difference is whether they access sequentially to the column direction or to the row direction. Figures 4.8 and 4.9 are image diagrams of column- and row-major order access. The square block shows the elements of the two-dimensional array psi. The red arrow indicates the access order. Access starts from psi[0][0]. Both appear to be correct, but the problem arises when we consider how multi-dimensional arrays are allocated in memory. The CPU manages data using addresses. Since the address is basically comprised of one integer value, the data are

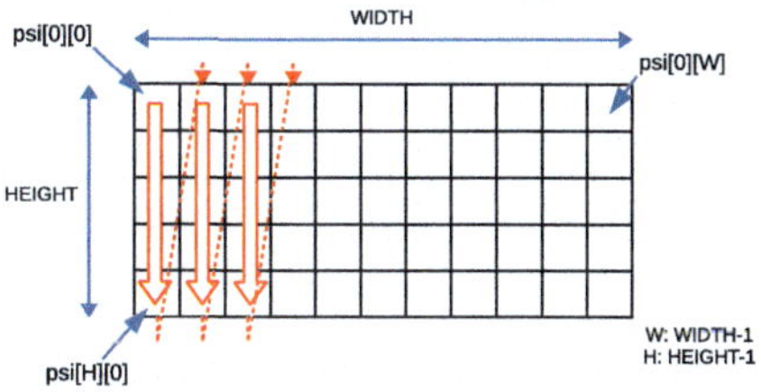

Fig. 4.8 Image diagram of Listing 4.6 (column-major order access)

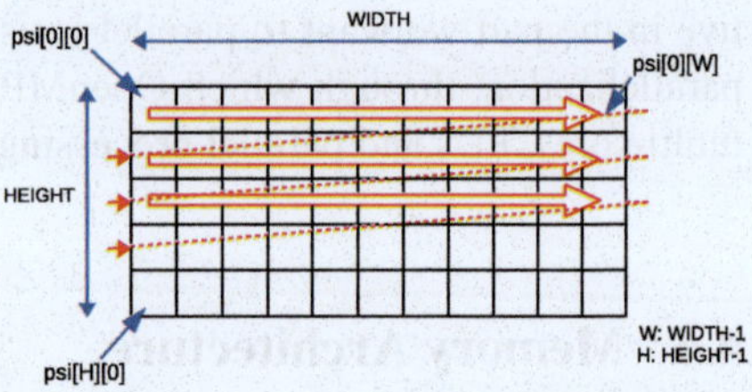

Fig. 4.9 Image diagram of Listing 4.7 (row-major order access)

managed in a one-dimensional array. A multi-dimensional array in a program is in reality a very long one-dimensional array.

Listing 4.6 Access program example to the column direction.

```
void loop_column(int (*psi)[WIDTH])
{
  int w, h, max = INT32_MIN, min = INT32_MAX;
  struct timespec start_time, diff_time;

  stopwatch_get_time(&start_time);
  for (w = 0; w < WIDTH; w++) { // difference from Listing 4.7
    for (h = 0; h < HEIGHT; h++) { // difference from Listing 4.7
      if (psi[h][w] < min) min = psi[h][w];
      if (max < psi[h][w]) max = psi[h][w];
    }
  }
  stopwatch_diff_from(&start_time, &diff_time);
  printf("[%-16s] %'ld.%09ld sec. (min: %d) (max: %d)\n", "loop
      column", diff_time.tv_sec, diff_time.tv_nsec, min, max);
}
```

Listing 4.7 Access program example to the row direction.

```
void loop_row(int (*psi)[WIDTH])
{
  int w, h, max = INT32_MIN, min = INT32_MAX;
  struct timespec start_time, diff_time;

  stopwatch_get_time(&start_time);
  for (h = 0; h < HEIGHT; h++) { // difference from Listing 4.6
    for (w = 0; w < WIDTH; w++) { // difference from Listing 4.6
      if (psi[h][w] < min) min = psi[h][w];
      if (max < psi[h][w]) max = psi[h][w];
    }
  }
  stopwatch_diff_from(&start_time, &diff_time);
  printf("[%-16s] %'ld.%09ld sec. (min: %d) (max: %d)\n", "loop row
      ", diff_time.tv_sec, diff_time.tv_nsec, min, max);
}
```

Figure 4.10 is a representation of Fig. 4.8 as a one-dimensional array. The variable psi is a pointer indicating the starting address of its own array. There are data in the

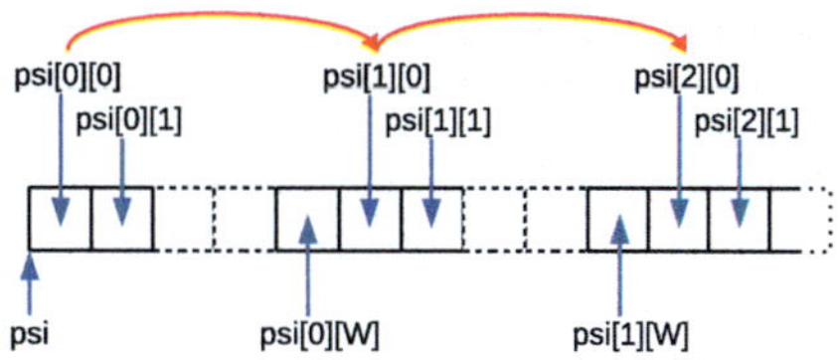

Fig. 4.10 Image diagram of the column-major order access on the memory space

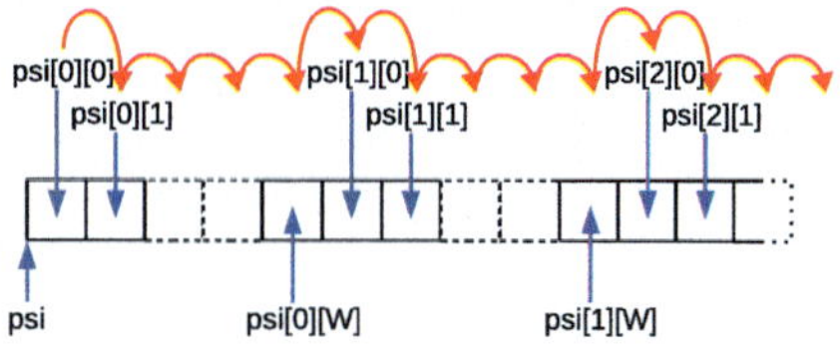

Fig. 4.11 Image diagram of the row-major order access on the memory space

order of psi[0][0], psi[0][1], psi[0][2], ..., psi[0][w] from the position indicated by the variable psi. Next data is followed by psi[1][0], psi[1][1], psi[1][2], ..., psi[1][w] corresponding to the second row of the two-dimensional array. Hence, the program in Listing 4.6 accesses data of the two-dimensional array psi in the order shown by the red line in Fig. 4.10. Access to the column direction is discontinuous access in the memory even if the index is continuous. The CPU regenerates the address and issues a memory access instruction each time variable h changes. The overhead for communication increases, and the access speed to the two-dimensional array psi decreases.

Figure 4.11 shows the access in the row direction, that is, access in the order of the one-dimensional array. If the data to be accessed are continuous, the CPU uses a communication method called **burst transfer** to communicate with the system memory. The burst transfer can read and write several consecutive data from a specified address. Some CPUs support 8-cycle burst transfer [24]. The CPU can transfer at higher speeds than when communicating while always distant addresses such as Listing 4.6.

Data continuity also affects the performance of SIMD processing. Operands used in SIMD instructions are packed multiple data. Therefore, the instructions called **gather-scatter**, which create packed data by gathering data from discontinuous addresses and scatter the packed data to discontinuous addresses, are implemented. Although gather-scatter instructions can access discontinuous data efficiently, we can perform more high-speed processing by arranging data sequentially in the order in which they are used and by accessing them in that order. We should design the data structure suitable for the memory architecture.

4.4.2 Cache Hierarchy

Due to advances in microfabrication technology, general CPUs are equipped with multiple circuits called cores that correspond to the functions of a single CPU. Some AMD CPUs have up to 64 cores [25]. Even though such a large-scale circuit is implemented, the circuits of the CPU chip are not filled with the core alone. A faster memory than the system memory called the **cache memory** is implemented. Installing a cache memory between the low-speed system memory and the register that operates at the highest speed reduces the performance difference between the two. When the CPU saves data from the register, the CPU only needs to write the data to the cache memory. The writing is completed in a much shorter time than writing to the system memory. The cache memory controller sends data to the system memory at an appropriate time. When data are loaded into a register, it can be retrieved immediately if the data exist in the cache memory. In this way, the CPU does not necessarily need to access the system memory.

The cache memory has a three-level hierarchical structure. As an example, Fig. 4.12 shows a block diagram of Skylake architecture. A separate cache memory for instructions and data is provided in the core so that it can quickly send instruction to the pipeline and data to the registers. The capacity of these cache memories is 32 KiB and is called the first cache (L1: Level.1). Next, a mid-level cache (L2: Level.2) is connected, which can store 1 MiB mixture of instructions and data used in the CPU core. Furthermore, the low-level cache (L3: Level.3) shared by multiple CPU

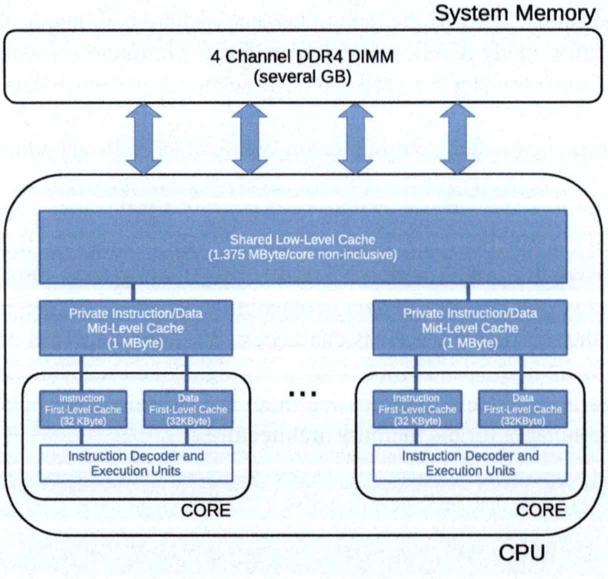

Fig. 4.12 Cache hierarchy of sixth-generation Intel core X-series processor families

cores follows, followed by the system memory. In Linux [19], we can confirm the cache memory size using the command "lscpu".

While the cache memory can transfer data faster as it gets closer to the execution unit, it has the trade-off, namely, that the storage capacity decreases. The latency of the L1 cache is 4–6 clocks [26]. Access to the L2 cache takes 2–3 times longer compared with the L1 cache. The L3 cache takes 4–5 times longer compared with the L2 cache, and it has 8–15 times higher latency than the L1 cache. One of the important points of speeding up is whether or not we can write a program that the operands necessary for the processing of the instruction exist in a low-level cache memory.

Algorithms that access addresses that change discontinuously one after another in a wide space exceeding the cache memory size are not recommended. Since the probability of existing the data in the cache memory is extremely low, communication with the system memory occurs, and the cache memory does not function effectively. The matrix computation algorithm, which is highly memory dependent and cannot access addresses continuously, is an example of such an algorithm. Therefore, a technique called **cache blocking** is often used to increase the cache hit rate. The cache blocking technique does not deal with a big program as it is, and performs processing in small programs wherein the amount of data required for calculation is less than the cache memory size. In the case of CGH calculation, instead of calculation using all object points simultaneously, object points are divided into smaller sets within the cache memory size. When using this method, it is important to pay attention to the cache level. As mentioned earlier, the cache memory has different levels that have different properties. For example, the L2 cache is dedicated to the CPU core, and the L3 cache is common to multiple CPU cores. Furthermore, it is important to understand the characteristics of data, that is, whether the data are used only by a specific CPU core or referenced by multiple CPU cores. For algorithms where there is no common data, we divide the data using the L2 + L3 cache size as the cache size per CPU core [27].

A multi-core CPU requires large volumes of operand data to execute instructions. The number of instructions and operands is multiplied by the number of cores, and using SIMD computations increases the operand size per instruction. Therefore, it is very important to design cache efficiency well for CPUs that are heavily dependent on system memory.

4.4.3 Cache Line

Here we consider a case where an instruction that uses one float-type variable is executed. Since the float type is 32 bits, 4 bytes are loaded from the system memory into the register through the cache memory. In reality, only 4 bytes are not loaded from the system memory at this time. The cache memory manages data in a unit called **cache line**, and the size of the minimum data that is loaded is equal to the size of the cache line. In recent CPUs, it is 64 bytes. In Linux, the cache line size is described in /sys/devices/system/cpu/cpu0/cache/index0/coherency_line_size. In other words, a 64-byte communication always occurs for data access of 64 bytes or less. At first glance, it may seem like a disadvantage, but there is greater merit by arranging the data in the order used by the program.

The problem occurs when the program has a calculation that requires more than contiguous 64 bytes of data. For example, if all the float-type variables used in the calculation are arranged at positions separated by 64 bytes or more, the communication amount is 16 times in the worst case. If there are many types of variables necessary for the calculation, there is a possibility that the data loaded at the beginning is deleted from the cache memory. If there is another variable to be used in the deleted cache line, the same cache line data must be loaded again, increasing unnecessary load.

Therefore, it is critical to design the data format by considering the boundary of the cache line and the order in which the data are used, rather than arranging the data randomly. In general, data optimization for the cache line is sometimes discussed. Fortunately, in CGH calculations, we can easily determine by using the calculation that continuously uses the coordinate data of the object. Section 9.2.3 describes the detail.

References

1. Intel. Intel 64 and IA-32 Architectures Software Developer's Manual. **1**, 5:5 (April 2022). https://www.intel.com/content/www/us/en/developer/articles/technical/intel-sdm.html (Retrieved September 5, 2022).
2. Intel. Intel 64 and IA-32 Architectures Software Developer's Manual. **1**, 5:12–13 (April 2022). https://www.intel.com/content/www/us/en/developer/articles/technical/intel-sdm.html (Retrieved September 5, 2022).
3. Intel. Intel 64 and IA-32 Architectures Software Developer's Manual. **1**, 3:11–12 (April 2022). https://www.intel.com/content/www/us/en/developer/articles/technical/intel-sdm.html (Retrieved September 5, 2022).
4. Intel. Intel 64 and IA-32 Architectures Optimization Reference Manual. E:40–41 (February 2022) https://www.intel.com/content/www/us/en/developer/articles/technical/intel-sdm.html (Retrieved September 5, 2022).
5. Intel. Intel 64 and IA-32 Architectures Software Developer's Manual. **1**, 2:10 (April 2022). https://www.intel.com/content/www/us/en/developer/articles/technical/intel-sdm.html (Retrieved September 5, 2022).
6. Intel. Intel 64 and IA-32 Architectures Optimization Reference Manual. 2:7 (February 2022) https://www.intel.com/content/www/us/en/developer/articles/technical/intel-sdm.html (Retrieved September 5, 2022).
7. Intel. Intel 64 and IA-32 Architectures Optimization Reference Manual. 2:12 (February 2022) https://www.intel.com/content/www/us/en/developer/articles/technical/intel-sdm.html (Retrieved September 5, 2022).
8. Intel. Intel 64 and IA-32 Architectures Optimization Reference Manual. 2:26 (February 2022) https://www.intel.com/content/www/us/en/developer/articles/technical/intel-sdm.html (Retrieved September 5, 2022).
9. Richard M. Stallman and the GCC Developer Community. Using the GNU Compiler Collection (GCC) For gcc version 13.0.0 (pre-release). 742–743 (2022). https://gcc.gnu.org/onlinedocs/gcc.pdf (Retrieved September 5, 2022).

10. The GNU Compiler Collection. https://gcc.gnu.org/ (Retrieved September 5, 2022).
11. Intel. Intel 64 and IA-32 Architectures Optimization Reference Manual. D:2 (February 2022). https://www.intel.com/content/www/us/en/developer/articles/technical/intel-sdm.html (Retrieved September 5, 2022).
12. Intel. Intel 64 and IA-32 Architectures Optimization Reference Manual. 2:34–35 (February 2022). https://www.intel.com/content/www/us/en/developer/articles/technical/intel-sdm.html (Retrieved September 5, 2022).
13. Intel. Intel 64 and IA-32 Architectures Optimization Reference Manual. 2:25 (February 2022). https://www.intel.com/content/www/us/en/developer/articles/technical/intel-sdm.html (Retrieved September 5, 2022).
14. Intel. Intel 64 and IA-32 Architectures Optimization Reference Manual. 2:11 (February 2022). https://www.intel.com/content/www/us/en/developer/articles/technical/intel-sdm.html (Retrieved September 5, 2022).
15. Intel. Intel 64 and IA-32 Architectures Optimization Reference Manual. 2:6 (February 2022). https://www.intel.com/content/www/us/en/developer/articles/technical/intel-sdm.html (Retrieved September 5, 2022).
16. Intel. Intel 64 and IA-32 Architectures Optimization Reference Manual. 18:74 (February 2022). https://www.intel.com/content/www/us/en/developer/articles/technical/intel-sdm.html (Retrieved September 5, 2022).
17. Intel. Intel 64 and IA-32 Architectures Optimization Reference Manual. 2:31–34 (February 2022). https://www.intel.com/content/www/us/en/developer/articles/technical/intel-sdm.html (Retrieved September 5, 2022).
18. IBM Introduces the World's 2-nm Node Chip. *IEEE Spectrum*, (2021). https://spectrum.ieee.org/ibm-introduces-the-worlds-first-2nm-node-chip (Retrieved September 5, 2022).
19. The Linux Kernel Archives. https://www.kernel.org/ (Retrieved September 5, 2022).
20. The GNU C Library. https://www.gnu.org/software/libc/ (Retrieved September 5, 2022).
21. Ulrich Drepper (Red Hat, Inc.) and Ingo Molnar (Red Hat, Inc.). The Native POSIX Thread Library. (February 21, 2005). https://akkadia.org/drepper/nptl-design.pdf (Retrieved September 5, 2022).
22. OpenMP. https://www.openmp.org/ (Retrieved September 5, 2022).
23. Intel. Intel Core X-Series Processor Datasheet Revision 005. **1**, 8 (2020). https://www.intel.com/content/www/us/en/products/docs/processors/core/core-technical-resources.html (Retrieved September 5, 2022).
24. Intel. Intel Core X-Series Processor Datasheet Revision 005. **1**, 9 (2020). https://www.intel.com/content/www/us/en/products/docs/processors/core/core-technical-resources.html (Retrieved September 5, 2022).
25. AMD. AMD EPYC 7003 SERIES PROCESSORS. (2022). https://www.amd.com/system/files/documents/amd-epyc-7003-series-datasheet.pdf (Retrieved September 5, 2022).
26. Intel. Intel 64 and IA-32 Architectures Optimization Reference Manual. 2:20 (February 2022). https://www.intel.com/content/www/us/en/developer/articles/technical/intel-sdm.html (Retrieved September 5, 2022).
27. Intel. Intel 64 and IA-32 Architectures Optimization Reference Manual. 2:21 (February 2022). https://www.intel.com/content/www/us/en/developer/articles/technical/intel-sdm.html (Retrieved September 5, 2022).

Chapter 5
Basics of CUDA

Minoru Oikawa

Abstract Graphics processing units (GPU) were originally developed to perform calculations related to graphics; however, GPUs are widely used to process scientific and technical calculations because they realize performance in highly parallel computing. NVIDIA's programming environment includes the compute unified device architecture (CUDA), which is a programming model that applies GPUs for general calculations. This chapter introduces the CUDA programming model and basic usage through sample programs.

5.1 Evolution of GPUs

In the late 2010s, in addition to general-purpose central processing units (CPU), GPUs were widely introduced in many consumer products, e.g., mobile phones, personal computers (PC), and video game consoles. A **GPU** is frequently mounted as an independent device; however, GPUs can be found on motherboards or may be integrated in the same package as a CPU. GPUs have been widely applied for image processing and general calculations due to their highly parallel processing performance, which is realized by the **compute unified device architecture** (**CUDA**) and various software libraries. Understanding how graphics processors came to be applied to scientific and engineering computing will help us understand CUDA.

The term GPU was first announced publicly in 1999 by NVIDIA. A GPU is a high-performance graphic accelerator chip with hardware 3D rendering functions but programmable features yet. The challenge of accelerating graphics drawing functions using special hardware goes back to the 1970s. Prior to the emergence of GPUs, such technology was referred to as "video display processors" or "graphics accelerators".

Supplementary Information The online version contains supplementary material available at https://doi.org/10.1007/978-981-99-1938-3_5.

M. Oikawa (✉)
Kochi University, 2-5-1 akebono-cho, kochi-shi, Kochi, Japan
e-mail: moikawa@kochi-u.ac.jp

T. Shimobaba and T. Ito (eds.), *Hardware Acceleration of Computational Holography*,
https://doi.org/10.1007/978-981-99-1938-3_5

Many modern computer display devices comprise a two-dimensional array of color point light sources called pixels. If each pixel can display 256 gradations of red, green, and blue (RGB), 3 bytes per pixel of data is required. A full high-definition (full HD) display has a resolution of 1,980 $\times$ 1,080 pixels; thus, the amount of data for a full color image displayed in full screen is approximately 6 MB (= 1,920 $\times$ 1,080 $\times$ 3). The display pattern of a graphical user interface (GUI) is updated instantly by operating of a keyboard, mouse, or other input devices. To make this function appear to run smoothly, it is necessary to update display pattern at a frequency of 30 times or more per second. As a result, the computer must update the display data up to 180 MB(= 6MB $\times$ 30) per second while calculating the pattern to display.

Updating display data continuously in real time is not so light task that can be neglected to burden the CPU running many other programs. Updating the display pattern on the screen is a limited (but a large amount) operation of functions, e.g., painting specific areas and drawing lines. Therefore, it is more efficient to offload these tasks from the CPU to a dedicated processing module. Thus, in the 1980s, 2D graphic accelerators were developed to speeds up planar graphics processing on PC.

In the same period, in terms of professional computing, Silicon Graphics, Inc. (SGI) announced many high-performance computers that focused on 3D graphics processing. To project 3D objects comprising many polygons on a 2D screen, a huge number of calculations are required for coordinate transformation, light scattering and reflection, texture mapping, etc. SGI made a significant contribution to the computer graphics field in terms of both hardware and software. For example, SGI developed a very large-scale integrated circuit chip dedicated to intensive graphics processing required for 3D graphics, and these chips were incorporated by them into their own computers, which were sold at high numbers. SGI also developed a high-performance graphics software library IRIS GL (integrated raster imaging system graphics library), which was later renamed OpenGL. OpenGL subsequently became a de facto standard graphics library.

In the 1990s, graphics-related functions and performance on PCs saw great progress. While GUIs began to increase in popularity, the resolution of display devices also increased higher; thus, more computational capacity was required for graphics processing. As a result of the standardization of OpenGL including the graphic interface API, graphics accelerator chips that were compatible with OpenGL could be developed by manufactures. Microsoft announced Direct3D API in 1996, which was a component of Microsoft's DirectX APIs, as its own 3D graphic interface for their Windows operating systems. Around this time, real-time 3D arcade games that made heavy use of polygons emerged, and the demand for such 3D games on PCs increased. Many manufacturers competed to provide faster graphic chips to accelerate 3D graphics functions. As a result, various manufacturers announced video cards with 3D accelerator functions for the PC market that were compliant with the standardized Direct3D or OpenGL. NVIDIA, which was founded in 1993, released a series of 3D graphics accelerator chips (at this time, these chips were not referred to as GPUs). The first graphics accelerator chip[1] from NVIDIA to be referred to as

[1] The product was called the GeForce256.

a GPU was released in 1999, and this chip had excellent features and demonstrated good performance at the time.

Early GPUs or graphics accelerators showed high performance; however, they were designed to execute predetermined graphics data processing, which is now referred to as a fixed-function graphics pipeline or fixed-function shader. While new graphics rendering algorithms or techniques were devised, it is inefficient for manufacturers to release new GPUs that support on hardware one by one. Eventually, a flexible feature was developed that allowed users to program a part of the graphics pipeline functions, which are referred to as programmable shaders. Programmable shaders are described using a special-purpose language called **shading language**, e.g., GLSL, HLSL, and Cg.[2] The programmable functions were used to describe graphics data processing; however, the computing power based on the massively parallel architecture of the GPU was attractive for researchers who required high computational capability at the time. They demonstrated that it is possible to execute scientific and technological calculations efficiently using GPUs and their graphics library interfaces. However, it was not easy to use for general-purpose programming.

5.2 Introduction to CUDA

In the 2000s, the fixed-function shader function was nearly replaced by the **programmable shader**. A GPU that realizes a programmable shader function should have a large number of small processor cores to process many polygons and pixels instantaneously and to support various types of shading functions. Gradually, many researchers used GPUs to process scientific and technical computing; however, since this was only possible by using the shader language created for graphics processing, it was not so easy to use environment. An environment in which the high computing power of the GPU could be used easily for general purposes was required. Thus, the CUDA development environment was announced by NVIDIA in 2006 in response to these demands.

CUDA is a general-purpose parallel computing platform, and its programming interface model on CUDA-capable GPUs was developed by the NVIDIA Corporation. The CUDA development environment is referred to as the CUDA toolkit, and it is freely available from NVIDIA's website [1]. The CUDA toolkit includes a compiler, math libraries, debuggers, profilers, and many sample programs. Essentially, the description syntax of CUDA is designed as an extension of C/C++. Since its first release in 2007, newer versions with new features are released each year; thus, CUDA supports many features.

[2] GLSL (OpenGL shading language) is part of OpenGL 2.0 (1992); HLSL (high-level shading language) is used with Direct3D 9.0 (2002); Cg(C for graphics) was released by NIVDIA in 2003.

5.3 Setting Up the CUDA Environment

Many versions of the CUDA toolkit [1] have been released by NVIDIA, and version 11.3.1 was released in 2021, as shown in Fig. 5.1. Generally, all versions of the CUDA toolkit support the Linux, Windows, Mac OSX (may not be supported by a particular version) operating systems. The basic installation procedure [3] is generally the same for any OS. First, the OS-specific installer must be downloaded from the NVIDIA site [1], and then the installer is executed. Detailed installation instructions [4–6] are provided by NVIDIA for each OS; thus, we omit this information here. However, we provide some general information in the following.

First, to install the CUDA toolkit, a CUDA-capable NVIDIA GPU must be installed. The CUDA toolkit can even be installed on mobile PCs as long as the GPU is made by NVIDIA. Unfortunately, the CUDA toolkit may not be compatible with low-cost or very old computers. Thus, it is necessary to check the hardware configuration of the target computer. CUDA-capable GPUs may be called "GeForce", "Quadro", "TITAN", or "Tesla" [2].

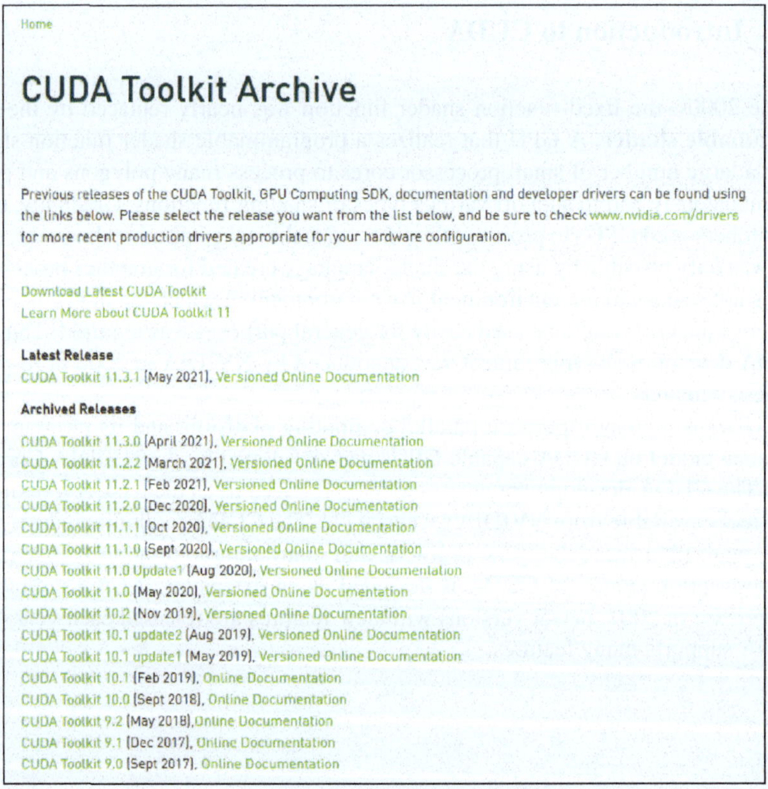

Fig. 5.1 CUDA toolkit download site [1]

Table 5.1 CUDA toolkit version, GPU product name, and other environments used in this chapter

CUDA toolkit	version 9.2 (released in 2018)
GPU product name	NVIDIA GeForce 1050Ti (Compute capability 6.1) CUDA cores: 768
Operation system	Linux Ubuntu 16.04 LTS (64-bit version)
C/C++ Compiler	GNU Compiler Collection (GCC) version 5.4.0

Second, the OS environment must be prepared appropriately. For example, for Windows or Mac OSX, the version of the OS should be determined. If you are running a Linux OS, the distribution name should be identified. The installation procedure varies depending on the OS and its version; thus, users should refer to the NVIDIA site for the procedure that matches the target environment.

Third, the C/C++ language development environment must be installed. Here, Microsoft Visual Studio, Xcode, and GNU Compiler Collection must be installed for Windows, Mac OSX, and Linux, respectively, prior to setting up the CUDA environment.

The CUDA installer can be executed once the above hardware and software requirements have been satisfied. Table 5.1 shows the computer environment we have used to run the sample programs discussed in the following.

After the installation is complete, we must confirm that the CUDA compiler can be used from the console by running the "nvcc" command as shown in Listing 5.1. If the installation was completed correctly, the version information of the CUDA toolkit should be displayed. If another message, e.g., "Command not found" is displayed, the installation was not completed as required; thus, the user should review the installation procedure.

Listing 5.1 nvcc command

```
$ nvcc --version
nvcc: NVIDIA (R) Cuda compiler driver
Copyright (c) 2005-2018 NVIDIA Corporation
Built on Tue_Jun_12_23:07:04_CDT_2018
Cuda compilation tools, release 9.2, V9.2.148
```

5.4 Hello World

Here, we explain the programming method of CUDA through some examples. Listing 5.2 shows the first CUDA sample code, which displays "Hello world" on the screen. While the original version in standard C language was intended to be executed by the CPU, this first program is executed by the GPU. This short program only involves a few lines; however, it includes important concepts of the CUDA programming model.

Listing 5.2 Sample program "hello.cu", which print messages with thread index numbers

```
1  #include <cuda_runtime.h>
2  #include <stdio.h>
3
4  __global__ void hello() //This funtion is executed by GPU.
5  {
6    printf("Hello world, block(%d,%d,%d),thread(%d,%d,%d).\n",
7          blockIdx.x, blockIdx.y, blockIdx.z, threadIdx.x, threadIdx.y, threadIdx.z);
8  }
9  int main()
10 {
11   dim3 grid(2,1,1); //define "grid" size for folloing hello().
12   dim3 block(3,1,1); //define "block" size for following hello().
13   hello <<<grid, block>>> ( ); // launch kernel function.
14   cudaDeviceSynchronize( ); // wait for completion of hello().
15   return 0;
16 }
```

The CUDA source code comprises two parts. The first common part is executed by the CPU, which is referred to as the **host code**, and the other part executed by the GPU, which is referred to as the **device code**. The host code is written in standard C/C++ (lines 9–16 in Listing 5.2). The device code is defined as functions that have a "__global__" declaration specifier at the front of its function name (line 4). Functions defined with "__global__" are referred to as **kernel functions**. A kernel function is device code that can only be invoked from the host code.

Figure 5.2 shows a schematic diagram of a typical hardware architecture, which illustrates the relationship between a CPU and GPU. Both have independent memory devices (DRAM) and are connected via a high-speed PCI Express interconnection.

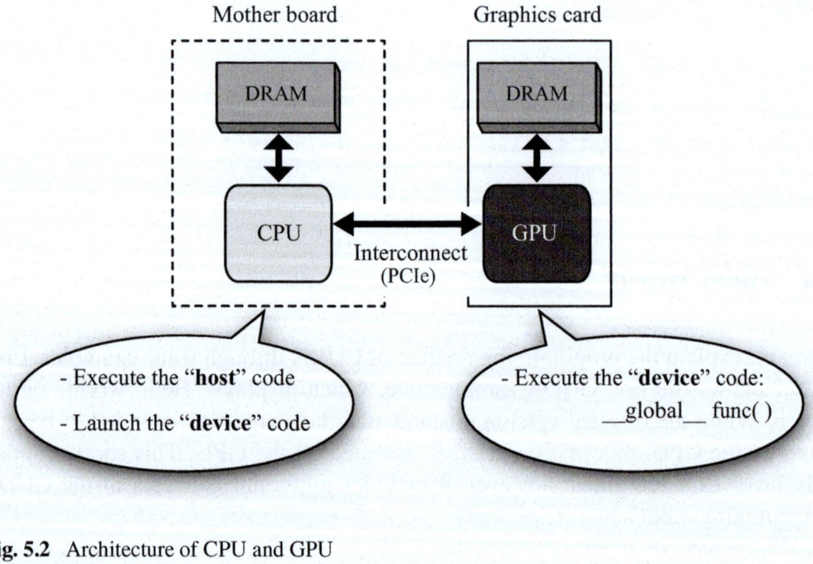

Fig. 5.2 Architecture of CPU and GPU

GPUs have simpler but many more processor cores than CPUs. By running many processor cores in parallel, the overall computational speed of a GPU is superior. Here, the host code is executed sequentially on the CPU, and the GPU code is executed in parallel on the GPU (Fig. 5.2). Therefore, it is necessary to specify how many threads are to be executed in parallel for the device code executed on the GPU (lines 11–13 in Listing 5.2). Note that two variables of "dim3" type can be seen in this region, i.e., "grid(2,1,1)" and "block(3,1,1)", and these variables are explained in the following.

5.4.1 CUDA Thread Construction

Figure 5.3 shows the **thread** configuration of the device code in the CUDA programming model. The CUDA threads model is defined hierarchically in two levels. Here, a "grid" is the highest level of the hierarchy, and a grid comprises multiple "thread blocks" (or simply referred to as a **block**). The thread blocks comprise multiple CUDA threads, which are represented as a single cube. The grid and thread blocks can be configured using one-dimensional (x), two-dimensional (x, y), or three-dimensional (x, y, z) indexes. For example, in the "hello world" program (Listing 5.2), the grid has a one-dimensional array of size two and the thread block has a one-dimensional array of size three. Therefore, a total of six CUDA threads are launched in parallel.

At line 13 in the program above, we launch the **kernel** functions which have six CUDA threads from the host code. Here, we observe the "<<<$grid, block$>>>" description between the device function name "hello" and its parameters list "( )". These constructs are not used in typically C/C++ language but are a part of the CUDA

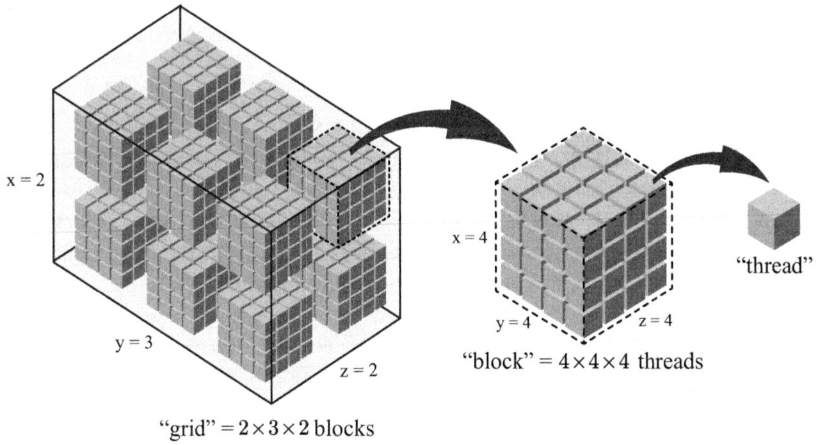

Fig. 5.3 Hierarchical CUDA thread construction

extension syntax. The first parameter inside the triple angle brackets "<<<. . .>>>" determines the dimension of the grid, and the second parameter determines the dimension of the thread block. On the right side of these, the parameters to be passed to the kernel function "hello()" are described in the same manner as ordinary C/C++.

5.4.2 Kernel Function

Here, we describe the kernel function that is the device code. In the hello() kernel function in Listing 5.2, we find the only "printf(...)", which outputs a message to the screen using the GPU. As described previously, the hello() kernel function is executed in six parallel CUDA threads comprising the grid and thread blocks. As a result, the hello world message is output six times because all parallel threads execute the same kernel function "hello()".

Each CUDA thread (Fig. 5.3) has its own unique thread ID, i.e., the "threadIdx" variables in following lines (Listing 5.3).

Listing 5.3 Inside the kernel function in "hello.cu"

```
printf("Hello world: block(%d,%d,%d),thread(%d,%d,%d).\n",
       blockIdx.x, blockIdx.y, blockIdx.z, threadIdx.x, threadIdx.y, threadIdx.z);
```

Here, the variable type of the threadIdx is "dim3"; thus, we can access threadIdx using three predefined variables, i.e., "threadIdx.x", "threadIdx.y", and "threadIdx.z". By using these variables with the unique values for each CUDA thread, the same device code and different calculations can be executed. As a result, the kernel function "hello()" runs on six parallel CUDA threads, each outputting a "Hello world" message and a three-dimensional number pair representing each thread ID.

5.4.3 Compilation and Execution

Now that we have introduced the basics of the CUDA programming model and syntax, we can actually execute it. Note that CUDA source code should have the ".cu" file extension to distinguish it from normal C/C++ source code. Please save the source code shown in Listing 5.2 with the filename "hello.cu". The CUDA source file can be compiled using the CUDA compiler "nvcc" from the command line as follows (Listing 5.4).

Listing 5.4 Compilation, execution and result of "hello.cu"

```
$ nvcc -o hello hello.cu
$ ./hello
Hello world: block(1,0,0), thread(0,0,0).
Hello world: block(1,0,0), thread(1,0,0).
Hello world: block(1,0,0), thread(2,0,0).
Hello world: block(0,0,0), thread(0,0,0).
```

```
Hello world: block(0,0,0), thread(1,0,0).
Hello world: block(0,0,0), thread(2,0,0).
```

Once executed, you should see the six "hello..." messages, i.e., the output of the six parallel CUDA threads. Here, we explain the timing chart to execute the host code and device code. The CUDA threads and host thread(s) are executed asynchronously with each other (Fig. 5.4); therefore, it is necessary to wait for the CUDA thread to finish before terminating the host thread. These two threads are synchronized by calling the "cudaDeviceSynchronize()" function (line 14, Listing 5.2).

5.5 Parallel Addition of Vectors

The second example program performs addition of two vectors on a GPU, as shown in Fig. 5.5. Listing 5.5 shows part of the source code in standard C language. Here, the result of adding two vectors **a** and **b** with four elements each is stored in vector **s**. The vectors are stored in a standard array. In the sequential programming method,

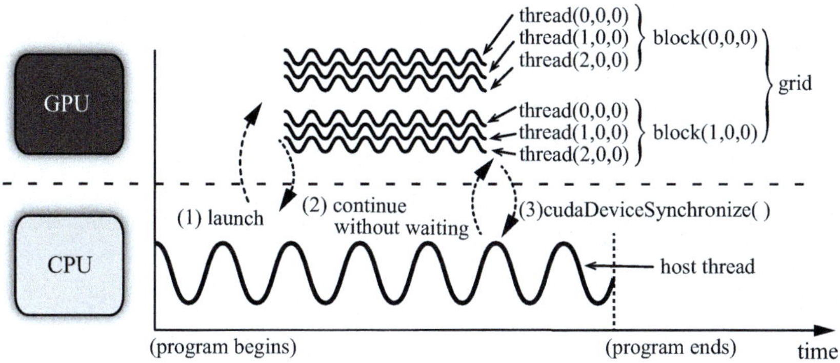

Fig. 5.4 Timing chart of each CPU/GPU threads execution

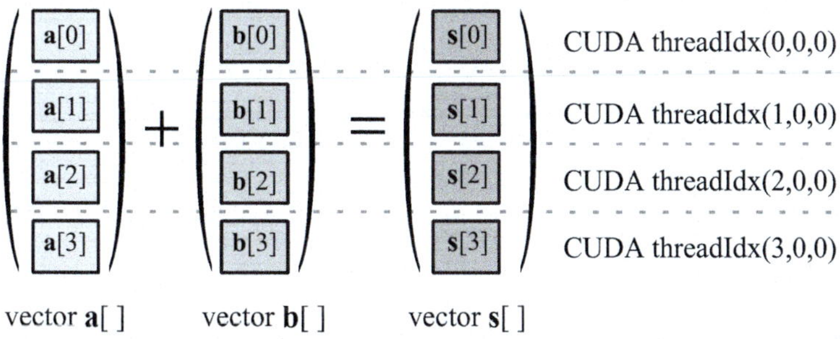

Fig. 5.5 Vector addition

its addition of the number of elements is generally performed sequentially using a loop syntax, e.g., “for” or “while”.

Listing 5.5 Sequential version for additon of vectors code on a CPU

```
int a[4]={1, 2, 3, 4};
int b[4]={10, 20, 30, 40};
int s[4];
for (int k=1; k<4; k++) s[k]=a[k]+b[k]; //Add one by one sequentially.
```

Listing 5.6 shows the CUDA version of Listing 5.5 for parallel addition on a GPU. Here, like a typical CUDA programming method, addition of each element is assigned to the CUDA thread to perform parallel computation (Fig. 5.5, right). This process is described in detail in the following.

Listing 5.6 Sample program “vector.cu”, which add each elements in parallel

```
#include <cuda_runtime.h>
#include <stdio.h>
const int L = 4; // Length of vector.

__global__ void AddVector(int *a, int *b, int *c)
{
    c[threadIdx.x] = a[threadIdx.x] + b[threadIdx.x];
}
int main()
{
    int host_a[L] = { 1, 2, 3, 4};
    int host_b[L] = {10, 20, 30, 40};
    int host_s[L] = { 0, 0, 0, 0};
    int *dev_a, *dev_b, *dev_s;
    // Allocate memory in GPU address space.
    cudaMalloc(&dev_a, sizeof(int)*L);
    cudaMalloc(&dev_b, sizeof(int)*L);
    cudaMalloc(&dev_s, sizeof(int)*L);
    // Transfer data to be calculated by GPU.
    cudaMemcpy(dev_a, host_a, sizeof(int)*L, cudaMemcpyHostToDevice);
    cudaMemcpy(dev_b, host_b, sizeof(int)*L, cudaMemcpyHostToDevice);
    // Invoke GPU threads.
    dim3 grid(1,1,1);
    dim3 block(L,1,1);
    AddVector<<<grid,block>>>(dev_a, dev_b, dev_s);
    // Transfer the results from GPU address space.
    cudaMemcpy(host_s, dev_s, sizeof(int)*L, cudaMemcpyDeviceToHost);
    // Print the results.
    printf("host_s[%d]={%2d, %2d, %2d, %2d}\n", L, host_s[0], host_s[1], host_s[2], host_s[3]);
    // Discard the allcated memory.
    cudaFree(dev_a);
    cudaFree(dev_b);
    cudaFree(dev_s);

    return 0;
}
```

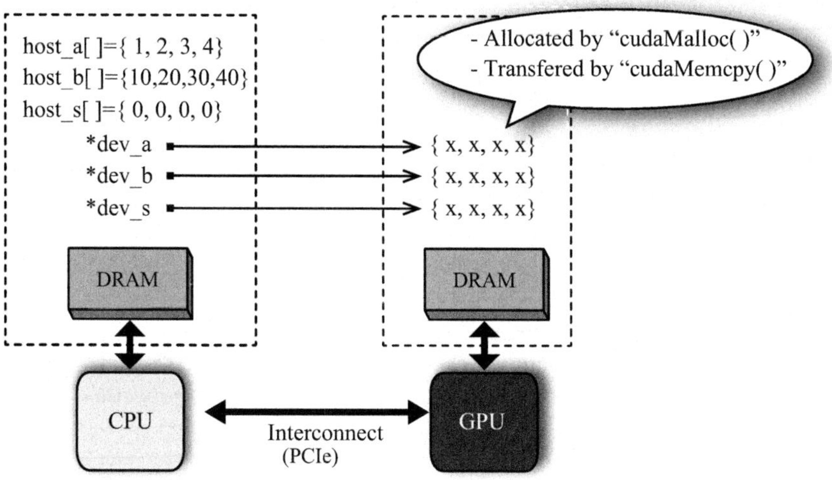

Fig. 5.6 Variables allocation at host and device

5.5.1 Data and Memory Management on GPU

As shown in Fig. 5.2, the memory spaces on the GPU and CPU sides are generally not shared, which mean that CUDA threads cannot directly access CPU address space (and vice versa). In other words, variables used by a CUDA thread must be allocated explicitly in GPU memory space.[3] For the data used on the GPU side, the programmer must clearly instruct the description that allocates the data area and then transfers the data from the host side, as shown in Fig. 5.6.

Typical functions that allocate and release memory on the GPU side include "cudaMalloc()" and "cudaFree()", which correspond to the malloc() and free() functions that perform equivalent processing on the host side, respectively. Similarly, the function to transfer data to the area allocated on the GPU side is "cudaMemcpy()", and this corresponds to memcpy(), which performs equivalent processing on the host side. Therefore, we must write code to explicitly allocate and transfer the data (all four elements of vectors **a** and **b**) between the CPU and GPU prior to executing the kernel function at line 25 in Listing 5.6.

[3] Using the Unified Memory functions allows us to make it as if virtually shared introduced at the chapter "Unified Memory Programming" in CUDA document [8].

5.5.2 *Performing Different Calculations on CUDA Threads*

The CUDA thread can be executed once the data required to perform parallel computing on the GPU are prepared. Here, the four elements of the vector are calculated by four threads; thus, it is appropriate to configure the CUDA thread to begin with one grid and four blocks (lines 23 and 24 in Listing 5.6). Then, pass the memory address where the data of the vector element to be calculated is stored to the CUDA thread as a parameter, and start the kernel function "AddVector()" (line 25). After execution of the kernel function is completed, the vector calculation result is copied from GPU memory space to host memory space using cudaMemcpy(), and the result is shown on the screen (line 29). Here, it is necessary to **synchronize** all CUDA threads using the cudaDeviceSynchronize() function (Sect. 5.4); however, that is performed implicitly when copying to the host memory space using the cudaMemcpy() function.

Listing 5.7 Compilation, execution and result of "vector.cu"

```
$nvcc -o vector vector.cu
$./vector
host_s[4]={11, 22, 33, 44}
```

Listing 5.7 shows the compile command and execution result. While vectors $\mathbf{a} = \{1, 2, 3, 4\}$ and $\mathbf{b} = \{10, 20, 30, 40\}$ are defined at lines 11 and 12, respectively, in Listing 5.6, the expected result is $\mathbf{s} = \{(1 + 10), (2 + 20), (3 + 30), (4 + 40)\}$, which are the same as the execution result.

5.6 Parallel Reduction

The third example program deals with the parallelized computation of the sum of integers series a_k of length N, which is expressed in Eq. (5.1).

$$S = \sum_{k=1}^{N} a_k = (a_1 + a_2 + \cdots + a_N) \tag{5.1}$$

Listing 5.8 Sequential version of reduction code

```
int S = 0; //Initialize result S by zero.
for (int k=1; k<=N; k++) S += a[k]; //Add one by one sequentially.
```

This is generally referred to as a **reduction** algorithm that obtains a single result by performing operations on multiple elements. It is not so difficult to describe code snippet of this operation by a sequential algorithm shown in Listing 5.8.

The parallelized version of this reduction algorithm is described as follows. Here, even if the order of the addition is changed, the final result S does not change. In addition, for the number of elements N, the total number of additions required is $(N - 1)$. For efficient parallelization, it is desirable to execute as many partial additions simultaneously as possible. Generally, the value of N is very large; however, in this example, $N = 8$ is assumed in Eq. (5.2). The sequential calculations shown in Listing 5.8 are expressed in Eq. (5.3), which considers the order of addition. Here, the additions in the for loop are added one by one in order; thus, a total of seven addition operations are sequential from ①, ②, ..., and ⑦. Assuming the time required to execute a single addition operation is T, the total time required to obtain the result is $7T$. If multiple addition operations can be performed simultaneously, the results can be obtained in less time by performing the order shown in Eq. (5.4). The four addition operations in ①' are executed simultaneously during the first T time. Then, the obtained four partial sums ①' are paired, and the addition operations ②' are executed simultaneously in second T time. Finally, the pair of ②' can be added at ③' in third T time; thus, the total value S can be obtained in a total time of $3T$ with the parallel version, which is faster than the elapsed time of the sequential calculation, i.e., $7T$.

$$S = \sum_{k=1}^{8} a_k = (a_1 + a_2 + a_3 + a_4 + a_5 + a_6 + a_7) \tag{5.2}$$

$$= (((((((\underbrace{a_1 + a_2)}_{①} \underbrace{+a_3)}_{②} \underbrace{+a_4)}_{③} \underbrace{+a_5)}_{④} \underbrace{+a_6)}_{⑤} \underbrace{+a_7)}_{⑥} \underbrace{+a_8)}_{⑦} \tag{5.3}$$

$$= \underbrace{\underbrace{\big(\underbrace{(a_1 + a_2)}_{①'} + \underbrace{(a_3 + a_4)}_{①'}\big)}_{②'} + \underbrace{\big(\underbrace{(a_5 + a_6)}_{①'} + \underbrace{(a_7 + a_8)}_{①'}\big)}_{②'}}_{③'} \tag{5.4}$$

The CUDA source code for **parallel reduction** is shown in Listing 5.9 (corresponding to Eq. (5.4)). First, consider the main() function. The final total value is stored in the variable S defined at line 17. The eight integers a_k to be added are stored in the allocated memory region indicated by the pointer variable *a (lines 18–26). This part includes a CUDA API function "cudaMallocManaged()" for the first time. This function will be described in detail in a subsequent section. In the immediate context it can be considered a memory allocation function to secure a memory region that can be accessed from both the host function and the device function. At this point, we are ready to calculate the total value S. Since the calculation of the total sum is executed by the device function, the kernel function "reduce()" is called in lines 29–31. The remaining part of the main() function includes the cudaDeviceSynchroize(), which waits for the device function to finish, and cudaFree(), which releases the allocated memory. Both cudaDeviceSynchroize() and cudaFree() were described in the previous section. In line 36, the total value is obtained from a[0], which is a part of the original integers. Since this is closely related to the "reduce()" kernel algorithm, it will be explained next.

Listing 5.9 Sample program: "reduction.cu"

```
#include <cuda_runtime.h>
#include <stdio.h>
const int SIZE = 8;

__global__ void reduce(int *a)
{
  int tid = threadIdx.x; // Thread ID
  for (int i=1; i<SIZE; i*=2) {
    if (tid % (2 * i) == 0) {
      a[tid] += a[tid + i];
    }
    __syncthreads(); //-- synchronize all threads in this block.
  }
}
int main()
{
  int S; //--Result of total sum.
  int *a; //--Number array to be summed.
  //-- Allocate memory for both host and device.
  cudaMallocManaged(&a, sizeof(int)*SIZE);
  //-- Initialize allocated memory region.
  printf("a[]={ ");
  for (int k=0; k<SIZE; k++){
    a[k] = k + 1;
    printf("%d ", a[k]);
  }
  puts("}");
  //-- calculate sum by device(GPU).
  dim3 gridDim(1,1,1);
  dim3 blockDim(SIZE,1,1);
  reduce <<<gridDim, blockDim>>> (a);

  //-- Wait for finishing kernel function.
  cudaDeviceSynchronize();
  //-- Show the results.
  S = a[0];
  printf("S = %d\n", S);
  cudaFree(a);
  return 0;
}
```

Figure 5.7 shows the internal processing of the kernel function "reduce()" in Listing 5.9. The horizontal axis indicates how partial additions are performed in parallel using eight values in the array a[ ]. The vertical axis represents the operation of the eight CUDA threads to be launched. The first step comprises the four additions performed when index $i = 1$ in the for loop. These four additions are executed by each CUDA thread which have an even tid number, and their partial sums are overwritten on the even elements of a[ ]. At this time, threads with odd numbers that do nothing are masked by the conditional expression. The second step comprises the two additions performed when index $i = 2$ in the for loop. These two additions are executed by

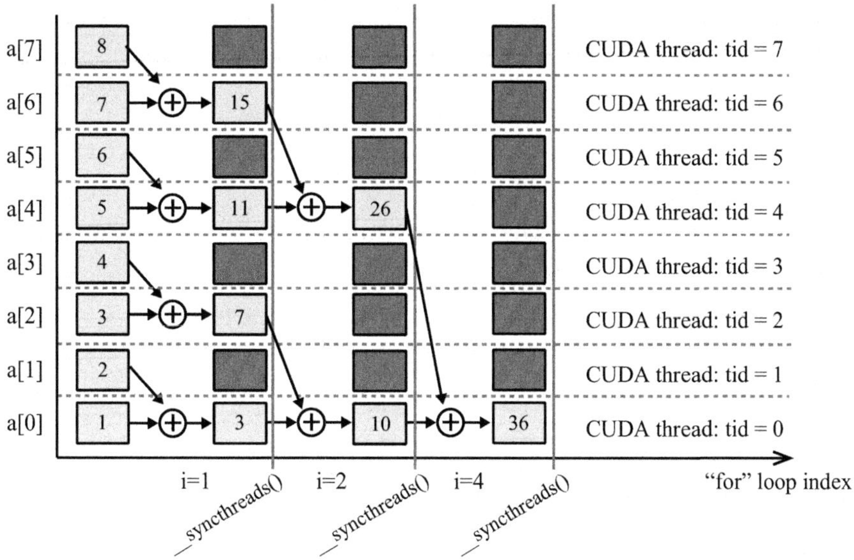

Fig. 5.7 Time chart of CUDA parallel reduction in the "reduct()" kernel function

the CUDA thread whose tid is 0 or 4, and similarly the partial sum overwrites a[0] and a[4]. In the final step, the thread with tid = 0 performs addition to obtain the final result, i.e., "36", overwrites the element in $a[0]$, and exit the kernel function.

A "__syncthreads()" function plays an important role in synchronizing the execution timings of multiple CUDA threads in same thread block. It is often assumed that CUDA threads are unconditionally guaranteed to proceed simultaneously, as represented in Fig. 5.7. In reality, there is no guarantee that multiple CUDA threads will run simultaneously. Therefore, the programmer needs to control the execution flow of multiple CUDA threads. When a particular CUDA thread reaches the __syncthread(), execution is suspended until other threads reach the same __syncthreads(). For example, if __syncthreads() is not defined, the second part may be forcibly calculated before the calculation of the first partial sum ends, which may lead to incorrect results.

Listing 5.10 shows the compilation commands and the execution result. Note, the sample code in Listing 5.9 was saved as filename "reduction.cu". Line 3 lists the eight elements to be added, and line 4 shows the result of sum 36, corresponding to Fig. 5.7. Listing 5.10 is relatively simple sample code. Sample code dealing with reduction is also included in the CUDA toolkit and uses more advanced techniques than this example. Examples of reduction algorithms can be found in the "samples/" directory in the CUDA toolkit.[4]

---

[4] This was found in directory named /usr/local/cuda/samples/6_Advanced/reduction/ in author's computer.

Listing 5.10 Compilation, execution and result of "reduction.cu"

```
$nvcc -o reduction reduction.cu
$./reduction
a[] = { 1 2 3 4 5 6 7 8 }
Sum = 36
```

Studies dealing with the detailed mechanics of CUDA and more advanced programming techniques can be found in the literature [8–11].

References

1. Nvidia Corporation, *CUDA Toolkit Archive*, URL:https://developer.nvidia.com/cuda-toolkit-archive, (Retrieved 2021)
2. Nvidia Corporation, *CUDA GPUs*, URL:https://developer.nvidia.com/cuda-gpus, (Retrieved 2021)
3. Nvidia Corporation, *CUDA Quick Start Guide*, CU-05347-301_v9.2, (2018) https://docs.nvidia.com/cuda/archive/9.2/cuda-quick-start-guide/index.html
4. Nvidia Corporation, *CUDA Installation Guide for Microsoft Windows*, https://docs.nvidia.com/cuda/archive/9.2/pdf/CUDA_Installation_Guide_Windows.pdf,DU-05349-001_v9.2, (August 2018)
5. Nvidia Corporation, *CUDA Installation Guide for Mac OS X*, DU-05349-001_v9.2, (2018) https://docs.nvidia.com/cuda/archive/9.2/pdf/CUDA_Installation_Guide_Mac.pdf
6. Nvidia Corporation, *CUDA Installation Guide for Linux*, DU-05347-001_v9.2, (2018) https://docs.nvidia.com/cuda/archive/9.2/pdf/CUDA_Installation_Guide_Linux.pdf
7. Nvidia Corporation, *CUDA Samples*, TRM-06704-001_v9.2, (August 2018) https://docs.nvidia.com/cuda/archive/9.2/pdf/CUDA_Samples.pdf
8. Nvidia Corporation, *CUDA C Programming Guide*, DU-02829-001_v9.2, (August 2018) https://docs.nvidia.com/cuda/archive/9.2/pdf/CUDA_C_Programming_Guide.pdf
9. Jason Sanders, Edward Kandrot, *CUDA by Example: An Introduction to General-Purpose GPU Programming*, ISBN 978-0-13-138768-3.
10. John Cheng, Max Grossman, Ty McKercher, *Professional CUDA C Programming*, Published by John Wiley & Sons,Inc, ISBN 978-1-11-873932-7.
11. David B.Kirk, Wen-mei W.Hwu, *Programming Massively Parallel Processors: A Hands-on Approach*, Morgan Kaufmann, Elsevier, ISBN 978-0-12-811986-0.

Chapter 6
Basics of OpenCL

Takashi Nishitsuji

Abstract Open Computing Language (OpenCL), which is generally called a heterogeneous computing system, is an open programming framework of parallel computing for a calculation system comprising different computers (e.g., CPU, GPU, DSP, FPGA). Although CUDA only applies to NVIDIA's GPU, OpenCL can drive the GPUs of different vendors (AMD, NVIDIA, Intel, Qualcomm), as well as the CPU or another computer, via the same OpenCL-written source code. Thus, OpenCL is more portable than CUDA. In this chapter, OpenCL, as well as the strategy for constructing a calculation environment, is briefly introduced employing a source code example for calculating a computer-generated hologram (CGH). Based on the contents of this chapter, holography calculations employing OpenCL can be attempted. Readers who wish to improve their OpenCL coding skills, programming guides that are published by chip vendors, etc., may be consulted.

6.1 General Introduction of OpenCL

Open Computing Language (**OpenCL**) is an open framework of parallel computing for many devices (GPU, CPU, FPGA); it is dissimilar to CUDA that only supports NVIDIA's GPU. The specification of OpenCL was developed by the Khronos group [1], which is an open consortium of software frameworks.

Although device vendors supply the Software Development Kits (SDKs) of OpenCL that comply with the specifications of the Khronos groups, the extension deviates from the approved specifications. They exhibit two types of application programming interfaces (APIs): one is a candidate for future specifications, and the

Supplementary Information The online version contains supplementary material available at https://doi.org/10.1007/978-981-99-1938-3_6.

T. Nishitsuji (✉)
Faculty of Science, Toho University, 2-1-1 Miyama, Funabashi-shi, Chiba, Japan
e-mail: takashi.nishitsuji@is.sci.toho-u.ac.jp

T. Shimobaba and T. Ito (eds.), *Hardware Acceleration of Computational Holography*,
https://doi.org/10.1007/978-981-99-1938-3_6

other is vendor dependent and can be distinguished by their names [2]. Thus, the consumers must consider the conformance of each API.

Although OpenCL is based on the C language, there are some wrappers for other languages, e.g., the official C++ wrapper [3] and PyOpenCL [4] for python (further information are descriptive on the website of STREAM HPC [5]), enabling many software engineers to utilize GPGPU. This chapter focuses on OpenCL based on C/C++.

The basic techniques for accelerating a program with OpenCL and CUDA are almost similar; thus, this chapter focuses on clarifying the technique for utilizing OpenCL on your devices, as well as its differences with CUDA. However, owing to the page limit, the details of OpenCL (the definition of APIs) cannot be discussed; thus, the programming guides, which are released by vendors of the computing device, can be referenced by readers who wish to learn OpenCL detailedly [6–8].

6.2 Setting Up an OpenCL Environment

Most vendors of OpenCL-supporting devices avail their SDKs for developers; these SDKs include the OpenCL library of their devices and standard headers (.h), as well as other headers for extended functions that support only their devices. Therefore, intending users of OpenCL must first download and install the SDKs of their devices.

Notably, a Windows 10 64-bit environment was employed in this chapter, although readers employing other environments, e.g., macOS and Linux, can substitute the filenames or extensions according to the available environment, e.g., OpenCL.dll -> OpenCL.so for Linux users. The static"OpenCL.lib" and dynamic link "OpenCL.dll" libraries are the required libraries for developing and executing the OpenCL program. "OpenCL.lib" is available in the directories of an SDK, while "OpenCL.dll" is preinstalled in the system directories of Windows, following the installation of the graphics driver. Further, a header file ("cl.h"), which is available in the directory of an SDK, should be included in the program.

6.3 Constructing an OpenCL Program

This section introduces the construction of an OpenCL program employing a simple computer-generated hologram (CGH) calculation source code as a "Hello, world" program of OpenCL, which is depicted on Listings 6.1 (host program) and 6.2 (device program). Readers who have already set up the OpenCL environment can attempt to execute the sample codes by copying Listing 6.1 (with an appropriate name for a C++ file) to your computer and Listing 6.2 with the name, "CGH_helloworld.cl," which should be placed in the same directory with an executable file of Listing 6.1. After executing the program, a kinoform-type CGH with a resolution of 1024 × 1024 in the "bfh_CGH" buffer can be obtained, as shown in Fig. 6.1.

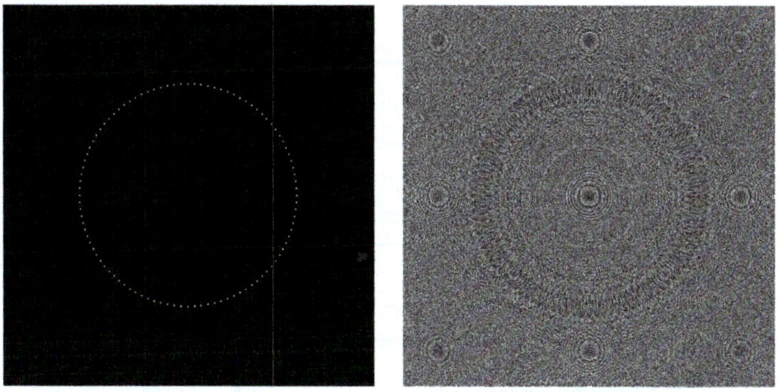

Fig. 6.1 Input and output of the example code: **a** a 3D model with 100 point clouds (input, generated in the program), **b** kinoform-type CGH (output)

An OpenCL program comprises two types of source codes, the host (.c or .cpp, .h) and device (.cl) codes. A standard OpenCL program adopts the online compile of the device code to improve its portability. Therefore, a C/C++ compiler, e.g., clang, gcc, and Visual C++, compiles the host code employing the OpenCL static library and creates the executable file, which will read and compile the device code according to specified devices for the program, following its execution. Noteworthily, OpenCL also supports offline compile.

The most significant differences between CUDA and OpenCL are the concepts of the platform and the devices. Since OpenCL supports many computing devices, an OpenCL program requires the availability of the available devices; users must specify the desired devices to execute the program. Every device must correspond to a platform. For example, when executing OpenCL on a CPU Intel Core–i7 8700K CPU employing an Intel OpenCL SDK environment, the platform would be "Intel OpenCL," and two devices (Integrated GPU, Intel UHD Graphics 630, and Intel Core i7-8700K CPU), which are available on the platform, would be utilized. The platforms and devices are specified by IDs; thus, many OpenCL APIs requests set the IDs in the arguments.

6.3.1 Creating OpenCL Objects That Are Not Required in CUDA

Dissimilar to CUDA, OpenCL defines many objects, e.g., the memory and kernel objects, to manage the device-related information, such as memory address and binary code of an executing program, since OpenCL is assumed to be executed on different platforms and devices. Thus, OpenCL requires the creation of such objects before the execution of a kernel. Table 6.1 and Fig. 6.2 exhibit the required

Table 6.1 Definition of the objects in OpenCL

Name of object	Role	Defined per
Context	Manages all the objects on a platform	Platform
Command queue	Manges all the commands to a device	Device
Program object	Manages the device program	Device source code
Kernel object	Compiles the kernel function of the device	Kernel function
Memory object	Manages the memory space on a device	Buffer

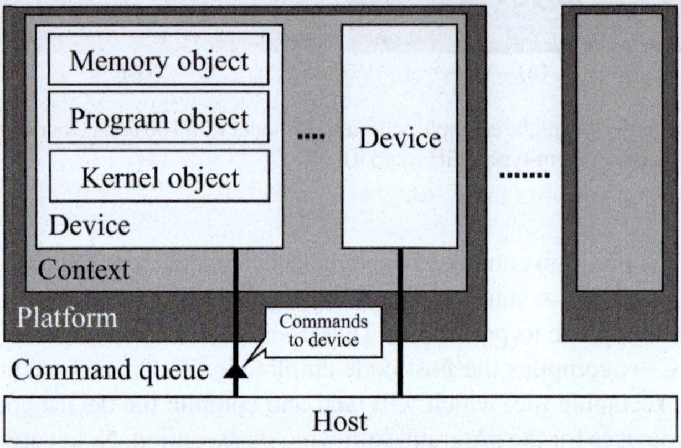

Fig. 6.2 Calculation model of OpenCL employing relation between the objects

object in a standard OpenCL program and the roles and relation between the objects, respectively. The OpenCL objects that are not required in CUDA are introduced in this subsection with reference to the sample code in Listing 6.1.

Context object is a fundamental object for managing all the objects on a platform; thus, it must be declared on the first line of an OpenCL program with the intended platform ID, as well as the number of devices on the platform. The available platforms and devices can be obtained by "clGetPlatformIDs()," which was employed on Lines 65 and 69 of Listing 6.1, and "clGetDeviceIDs()," which was used on Line 86 of Listing 6.1, for the platforms and devices, respectively. Detailed information on the platforms and devices can be obtained by "clGetPlatformInfo()" and "clGetDeviceInfo()," which employed utilized on Lines 77 and 91, respectively. Here, this program obtains the names of the platforms and devices. The context object is created by API "clCreateContext()," which was employed on Line 115 of the list.

command-queue object is an interface that manages all the commands, e.g., the execute-the-kernel and the transfer-the-data-in-a-buffer functions; thus, it must be declared per all to-be-utilized devices. A command-queue object is created by "clCreateCommandQueueWithProperties()" with a corresponding device ID, which is depicted on Line 118 of the list. The commands to a device are queued by the "clEnqueue***()" API via a command-queue object. For example, to copy data from the

Table 6.2 Corresponding names of memory

CUDA	OpenCL
Global memory	Global memory
Constant memory	Constant memory
Shared memory	Local memory
Register	Private memory
Local memory	

memory of a device to a host, "clEnqueueReadBuffer()," Line 165 of the list, is called employing the command-queue object in the first argument. Worthy, the commands are only enqueued; thus, the time of executing is unknown, and it depends on the preceding commands on the queue.

program object is an object that manages a raw (readable text) source code, as well as the compiled program of a device function. Thus, it must read a device source code as a text buffer before creating it. Lines 122–133 on the list show an example of reading the device source code from a file (CGH_helloworld.cl) to a char buffer (src), as well as creating a program object with "clCreateProgramWithSource()" on Line 128. After creating the program object, it can be built by "clBuildProgram()" employing a specified platform ID, as shown on Line 131 of the List.

kernel object is an object, which is created by the "clCreateKernel()" function employing program object and named the kernel function, that specifies the kernel function in a program object; thus, it must be created per device functions to be executed. On the List, only one device function is defined in the device code (Listing 6.2); therefore, only one kernel object is created on Line 136 of the Listing 6.1.

memory object is an object that manages the memory buffer on a device. It functions as a memory pointer. The memory object is created by "clCreateBuffer()" with context object and attributions that pertain to memory (size and writability), as obtainable in "cudaMalloc()" of CUDA. On Listing 6.1, four memory objects were created on Lines 139–142. Noteworthy, the hierarchical memory architectures of OpenCL and CUDA are almost the same (Table 6.2), and the memory buffer, which was created by "clCreateBuffer()," is assigned on the global memory.

The creations of the discussed objects indicate that the preparation for executing the kernel is almost completed. Further, the following section introduces the procedure for driving the OpenCL kernel.

6.3.2 Executing the Kernel Function

Dissimilar to CUDA, OpenCL requires a two-step setup before executing the enqueued kernel. The first step involves setting up the arguments of the kernel function via the "clSetKernelArg()" function (Lines 151–155 on Listing 6.1). Notably, all the arguments must be passed by a void* type pointer.

The second step involves the definition of the division unit for parallel execution; these units are called the grid, block, and thread in CUDA. However, the "grid," "block," and "thread" correspond to "NDRange," "workgroup," and "workitem," respectively. The sizes of NDRange and workgroup are specified by multidimensional size_t-type arrays, as exhibited on Lines 158 and 159 of the List. In the sample code, the size of NDRange was set to be equal to the size of the CGH, and the size of the workgroup was set to 256 $\times$ 1. The maximum number of workitems in a work group is defined by the specifications of hardware.

After the two-step preparation, the command for executing the kernel function can be enqueued by "clEnqueueNDRange()" employing the sizes of NDRange (globalSize), workgroup (localSize), and the queue object (Line 162 of the list).

Finally, the kernel function can be executed by transferring the buffer data from the device to the host. "clEnqueueReadBuffer()" is a transfer function; it is executed to transfer the buffer data from the device to the host (Line 165 of the list), and it is equivalent to "cudaMemcopy()" in CUDA. To ensure complete transfer, a call function for synchronizing the device to the host must be executed before subjecting the data to the host buffer (bfh_CGH). In the sample code, the "clFinish()" function, which was waiting to execute the last command that was enqueued in the command queue, was executed. Noteworthy, there are other functions, e.g., clWaitForEvents() with an event object, for achieving a finer synchronization; thus, those APIs can be referenced by readers who wish to construct a more complex OpenCL program.

This subsection only discusses the method for executing data-parallel-type computation. However, OpenCL comprises methods for parallelizing the calculation in a task unit, as obtainable in CUDA. Readers who wish to employ the task-parallel program may refer to the instruction manual of OpenCL, which is supplied by the vendors of devices.

To summarize the above introductions, the standard structure of the host program of OpenCL is, as follows:

1. Determine an available platform, as well as devices, and specify the appropriate devices.
2. Create a context object, which manages all the objects on a platform.
3. Create a command-queue object, which is connected to a device to manage the commands to be executed therein.
4. Read a device program as a text and build it, thereby treating it as a program object.
5. Create the kernel objects from a program object by specifying the name of the function that was written in the .cl file
6. Create the memory objects, which manage the memory space on a device.
7. Set the arguments and workgroup size, which are to be executed by the kernel.
8. Execute the kernel function.
9. Copy the result from the device memory.

Table 6.3 Corresponding names of the modifiers of the variables and memories

CUDA	OpenCL	Meaning
__device__	__global	On the global memory
__constant__	__constant	On the constant memory
__shared__	__local	On the shared memory

Table 6.4 Corresponding names of the modifier of the functions

CUDA	OpenCL	Meaning
__global__	__global	Kernel function
__device__	Not required	Inner function of the kernel

6.3.3 *Writing the Kernel Function*

The kernel function is one, which would be executed by a device. The grammars and syntaxes of the kernel functions of OpenCL and CUDA are almost the same, although the names of the modifiers of their variables, memories, and functions, as well as the methods for obtaining their index values, e.g., "gridDim" in CUDA, are different. Tables 6.3, 6.4, and 6.5 present the correlations of the modifiers and other basic functions of CUDA and OpenCL. N in Table 6.5 indicates that a dimension must be obtained employing the functions; thus, blockDim.x in CUDA is equivalent to get_num_groups(0);

The standard kernel function for calculating CGH is presented on Listing 6.2, which is a simplified version of the sample code of calculating CGH employing CUDA (Listing 10.2). For the readers who wish to execute an OpenCL program, the modification of Listing 6.2 is an easy technique for first building the OpenCL program. Here (Listing 6.2), three pre-processors are defined to substitute the constant

Table 6.5 Corresponding methods for obtaining the index values: N is a dimension

CUDA	OpenCL	Meaning
gridDim	get_num_groups(N)	Number of blocks per grid
blockDim	get_local_size(N)	Size of a block
blockIdx	get_group_id(N)	Index of a block
threadIdx	get_local_id(N)	Index of a thread
threadIdx + blockIdx * blockDim	get_global_id(N)	Global index of a thread
gridDim * blockDim	get_global_size(N)	Size of a grid

values. "CNS_255_DIV_2_PI" and "CNS_2_PI_DIV_LAMBDA" correspond to $\frac{255}{2\pi}$ and $\frac{2\pi}{\lambda}$, respectively ($\lambda = 532$ [nm] and "CNS_PITCH" represents the pixel pitch of a displaying device.

The calculation times for this execution are 95.2 ms with NVIDIA Quadro P1200 GPU and CUDA 11.0, 1738 ms with an Intel Core i7-8850H CPU, and 324 ms with an Intel UHD Graphics 630 GPU, all of them are evaluated with OpenCL. The kernel source code (Listing 6.2) is a very simple structure to understand; thus, applying the optimization techniques that are mentioned in Chapter 6 will be quite fast. Unfortunately, the techniques described in those sections are not within the scope of OpenCL, although readers who already briefly understand the differences and similarities of CUDA and OpenCL can easily apply those techniques in their OpenCL codes.

Moreover, only a few literature illustrate the fast calculation of CGH via OpenCL, although readers can refer to [9] as a practical example of implementing OpenCL to calculate CGH.

Listing 6.1 Simple CGH calculation employing OpenCL (host code)

```
#include <CL/cl.h>
#include <stdio.h>
#include <math.h>

#define MAX_CL_SOURCE_SIZE 10000
#define PI 3.14159265358979323846

int main()
{
  //Constants**********************************
  const char st_CLSrcName[1024] = "CGH_helloworld.cl";
  const int numPLS = 100; //number of PLS
  const int cgh_width = 1024; //width of CGH [pixel]
  const int cgh_height = 1024; //height of CGH [pixel]
  const float p = 0.000008; //pixel pitch for displaying device [m]

  //Classes************************************
  FILE* fp_CLSrc = fopen(st_CLSrcName, "rb");

  //Control variables for OpenCL
  cl_int status = 0;

  cl_platform_id v_SelectedPlatformID = 0;
  cl_platform_id* v_PlatformIDs;
  unsigned int v_SelectedPlatform;
  unsigned int v_NumPlatforms;

  cl_device_id v_SelectedDeviceID = 0;
  cl_device_id** v_DeviceIDs;
  unsigned int v_SelectedDevice;
  unsigned int v_NumDevices;

  cl_context context;
  cl_command_queue queue;
```

```
cl_program prog;
cl_kernel ker_CGH;

//Buffers***********************************
//(host)
cl_uchar* bfh_CGH = new cl_uchar[cgh_width * cgh_height];
cl_float* bfh_ox = new cl_float[numPLS];
cl_float* bfh_oy = new cl_float[numPLS];
cl_float* bfh_oz = new cl_float[numPLS];

//(device)
cl_mem bfd_CGH;
cl_mem bfd_ox;
cl_mem bfd_oy;
cl_mem bfd_oz;

//===Create Point cloud (circle)===
float r = 300 * p;   //radius of circle
float cx = cgh_width * 0.5 * p; //center of circle (x)
float cy = cgh_width * 0.5 * p; //center of circle (y)

for (int i = 0; i < numPLS; i++)
{
  bfh_ox[i] = r * cos(i / (float)numPLS * 2.0 * PI) + cx;
  bfh_oy[i] = r * sin(i / (float)numPLS * 2.0 * PI) + cy;
  bfh_oz[i] = 0.1 + 0.001*i;
}

//====Select the platform and devices to use====//
//Obtain the number of available platforms
status = clGetPlatformIDs(0, NULL, &v_NumPlatforms);
v_PlatformIDs = new cl_platform_id[v_NumPlatforms];

//Obtain the IDs of available platform
status = clGetPlatformIDs(v_NumPlatforms, v_PlatformIDs, &v_NumPlatforms);
v_DeviceIDs = new cl_device_id*[v_NumPlatforms];

//Show available platforms and device IDs
char msg[1024];
for (int i = 0; i < v_NumPlatforms; i++)
{
  //Obtain platform information (name of platform)
  clGetPlatformInfo(v_PlatformIDs[i], CL_PLATFORM_NAME, sizeof(msg), msg,
      NULL);
  printf("[%d] : %s\n", i, msg);

  //Obtain the number of available devices on the platform
  status = clGetDeviceIDs(v_PlatformIDs[i], CL_DEVICE_TYPE_ALL, NULL, NULL,
      &v_NumDevices);
  printf("Found %d devices\n", v_NumDevices);

  //Obtain the IDs of available platform
  v_DeviceIDs[i] = new cl_device_id[v_NumDevices];
```

```
    status = clGetDeviceIDs(v_PlatformIDs[i], CL_DEVICE_TYPE_ALL, v_NumDevices, v_DeviceIDs[i], &v_NumDevices);

    //Show the avaialble devices in the platform
    for (int j = 0; j < v_NumDevices; j++)
    {
      clGetDeviceInfo(v_DeviceIDs[i][j], CL_DEVICE_NAME, sizeof(msg), msg, NULL);
      printf("\t[%d][%d]%s\n", i, j, msg);
    }
  }

  //Select the platform and devices to use
  printf("Select platform ID to use: ");
  scanf_s("%d", &v_SelectedPlatform);

  v_SelectedPlatformID = v_PlatformIDs[v_SelectedPlatform];
  clGetPlatformInfo(v_SelectedPlatformID, CL_PLATFORM_NAME, sizeof(msg), msg, NULL);
  printf("Selected: %s\n\n", msg);

  printf("Select device ID to use: ");
  scanf_s("%d", &v_SelectedDevice);
  v_SelectedDeviceID = v_DeviceIDs[v_SelectedPlatform][v_SelectedDevice];
  clGetDeviceInfo(v_SelectedDeviceID, CL_DEVICE_NAME, sizeof(msg), msg, NULL);
  printf("Selected: %s\n\n", msg);

  //====Create a context====//
  //obtain the number of devices in the selected platform
  clGetDeviceIDs(v_SelectedPlatformID, CL_DEVICE_TYPE_ALL, NULL, NULL, &v_NumDevices);

  //Create a context for the selected platform
  context = clCreateContext(NULL, v_NumDevices, v_DeviceIDs[v_SelectedPlatform], NULL, NULL, &status);

  //====Create a command queue on the context====//
  queue = clCreateCommandQueueWithProperties(context, v_DeviceIDs[v_SelectedPlatform][v_SelectedDevice], NULL, &status);

  //====Build a program from a .cl source====//
  //Read .cl file to char buffer as text
  char* src;
  src = new char[MAX_CL_SOURCE_SIZE];
  size_t v_SizeOfSrc = fread(src, sizeof(char), MAX_CL_SOURCE_SIZE - 1, fp_CLSrc);
  src[v_SizeOfSrc] = '\0';

  //Create program object with the .cl source file
  prog = clCreateProgramWithSource(context, 1, (const char**)&src, NULL, &status);

  //Build program
  status = clBuildProgram(prog, v_NumDevices, v_DeviceIDs[v_SelectedPlatform], NULL, NULL, NULL);

```

```
delete[] src;

//====Create kernels to execute====//
ker_CGH = clCreateKernel(prog, "simpleCGH", &status);

//====Create memory objects====//
bfd_CGH = clCreateBuffer(context, CL_MEM_READ_WRITE, sizeof(cl_uchar)*
    cgh_width*cgh_height, NULL, &status);
bfd_ox = clCreateBuffer(context, CL_MEM_READ_WRITE, sizeof(float) * numPLS,
    NULL, &status);
bfd_oy = clCreateBuffer(context, CL_MEM_READ_WRITE, sizeof(float) * numPLS,
    NULL, &status);
bfd_oz = clCreateBuffer(context, CL_MEM_READ_WRITE, sizeof(float) * numPLS,
    NULL, &status);

//====Transfer the PLS data from the host ====//
status = clEnqueueWriteBuffer(queue, bfd_ox, CL_TRUE, 0, sizeof(cl_float) * numPLS,
    bfh_ox, 0, NULL, NULL);
status = clEnqueueWriteBuffer(queue, bfd_oy, CL_TRUE, 0, sizeof(cl_float) * numPLS,
    bfh_oy, 0, NULL, NULL);
status = clEnqueueWriteBuffer(queue, bfd_oz, CL_TRUE, 0, sizeof(cl_float) * numPLS,
    bfh_oz, 0, NULL, NULL);

//====Execute kernels====//
//Set arguments of the kernel
status = clSetKernelArg(ker_CGH, 0, sizeof(cl_mem), (void*)&bfd_CGH);
status = clSetKernelArg(ker_CGH, 1, sizeof(int), (void*)&numPLS);
status = clSetKernelArg(ker_CGH, 2, sizeof(cl_mem), (void*)&bfd_ox);
status = clSetKernelArg(ker_CGH, 3, sizeof(cl_mem), (void*)&bfd_oy);
status = clSetKernelArg(ker_CGH, 4, sizeof(cl_mem), (void*)&bfd_oz);

//Set the division unit for parallel execution
size_t globalSize[] = { (size_t)cgh_width, (size_t)cgh_height };
size_t localSize[] = { 256, 1 };

//Execute the kernel
status = clEnqueueNDRangeKernel(queue, ker_CGH, 2, NULL, globalSize, localSize, 0,
    NULL, NULL);

//====Transfer the CGH data from the device====//
status = clEnqueueReadBuffer(queue, bfd_CGH, CL_TRUE, 0, sizeof(cl_char)*
    cgh_width*cgh_height, bfh_CGH, 0, NULL, NULL);

//Wait for finish the last enqueued command
clFinish(queue);

//====Termination (Freeing memory)====//
fclose(fp_CLSrc);
clReleaseMemObject(bfd_CGH);
clReleaseMemObject(bfd_ox);
clReleaseMemObject(bfd_oy);
clReleaseMemObject(bfd_oz);

```

```
    delete[] bfh_CGH;
    delete[] bfh_ox;
    delete[] bfh_oy;
    delete[] bfh_oz;

    return 0;
}
```

Listing 6.2 Simple CGH calculation employing OpenCL (device code; CGHspshelloworld.cl)

```
#define CNS_255_DIV_2_PI 40.58451049
#define CNS_2_PI_DIV_LAMBDA 11810498.7
#define CNS_PITCH 0.000008

__kernel void simpleCGH(__global uchar* dbf_CGH, const int numPLS, __global float*
        ox, __global float* oy, __global float*oz)
{
    float x = get_global_id(0) * CNS_PITCH;
    float y = get_global_id(1) * CNS_PITCH;
    int width = get_global_size(0);
    int dst_addr = get_global_id(0) + get_global_size(0) * get_global_id(1);

    float2 c = (float2)(0.0, 0.0);

    for (int i = 0; i < numPLS; i++)
    {
        float phase = CNS_2_PI_DIV_LAMBDA * sqrt(pow(ox[i]-x, 2) + pow(oy[i]-y, 2) +
            pow(oz[i], 2));
        c += (float2)(cos(phase), sin(phase));
    }

    float arg = CNS_255_DIV_2_PI * atan2(c.y, c.x);
    dbf_CGH[dst_addr] = convert_uchar((int)arg);
}
```

Fundings This work was supported by JSPS KAKENHI Grant Number 22H03616.

References

1. Official OpenCL website of khronos group https://www.khronos.org/opencl/ Cited 30 Oct. 2019.
2. Khonos group, The OpenCL Extension Specification https://www.khronos.org/registry/OpenCL/specs/2.2/html/OpenCL_Ext.html. Cited 30 Oct. 2019
3. Khronos's github repository for OpenCL C++ bindings https://github.khronos.org/OpenCL-CLHPP/ Cited 30 Oct. 2019
4. A. Klöckener, PyOpenCL https://mathema.tician.de/software/pyopencl/. Cited 30 Oct. 2019
5. STREAM High Peformance Computing, OpenCL Wrappers https://streamhpc.com/knowledge/for-developers/opencl-wrappers/. Cited 30 Oct. 2019
6. NVIDIA, OpenCL Programming Guide for the CUDA Architecture Ver 4.2 http://developer.download.nvidia.com/compute/DevZone/docs/html/OpenCL/doc/OpenCL_Programming_Guide.pdf. Cited 30 Oct. 2019

7. Intel, Intel FPGA SDK for OpenCL Pro Edition Programming Guide 19.3 https://www.intel.com/content/dam/www/programmable/us/en/pdfs/literature/hb/opencl-sdk/aocl_programming_guide.pdf. Cited 30 Oct. 2019
8. Qualcomm, Snapdragon Mobile Platform OpenCL General Programming and Optimization, Nov. 3, 2017 https://developer.qualcomm.com/software/adreno-gpu-sdk/tools. Cited 30 Oct. 2019
9. Shimobaba, T., Ito. T., Masuda, N., Ichihashi. Y., Takada, N.: Fast calculation of computer-generated-hologram on AMD HD5000 series GPU and OpenCL, Opt. Express **18**, 10, 9955–9960 (2010).

Chapter 7
Basics of Field-Programmable Gate Array

Yota Yamamoto

Abstract A field-programmable gate array (FPGA) is a large-scale integration (LSI) that enables the user to modify the internal circuit structure. Central processing units (CPUs) are also implemented on LSIs and have one or more arithmetic logic units (ALUs). FPGAs, however, can have tens or hundreds of thousands of ALU-equivalent arithmetic cores using their on-board logic resources. CPU ALUs can operate as fast as 3 GHz, whereas FPGAs are nearly an order of magnitude slower at around 500 MHz. To build a high-speed special-purpose computer using FPGAs, we must select suitable algorithms that have less dependence on data, employ low precision, and are easily parallelizable. Effective parallel computation can be attained by taking advantage of the FPGAs' plentiful arithmetic units.

7.1 Structure of Field-Programmable Gate Array

Field-programmable gate arrays (**FPGA**s) are large-scale integrations (LSIs) that enable the user to modify the internal circuit structure. Figure 7.1 reveals a typical FPGA structure. Their internal circuits, unlike CPUs and graphics processing units (GPUs), are not functionally connected, and they work by loading precise circuit configuration data upon launch. The circuit configuration data configure the different blocks in the FPGA, like the programmable logic blocks (LBs) that implement logic circuits, programmable input-output blocks (IOBs) that offer the interface to external circuits, and programmable routing blocks [connection blocks (CBs) and switching blocks (SBs)], which connect each block. Inside the FPGA, these elements are arranged in a grid.

Supplementary Information The online version contains supplementary material available at https://doi.org/10.1007/978-981-99-1938-3_7.

Y. Yamamoto (✉)
Faculty of Engineering, Tokyo University of Science, 6-3-1 Niijuku, Katsushika-ku, Tokyo 125-8585, Japan
e-mail: yy-yamamoto@rs.tus.ac.jp

T. Shimobaba and T. Ito (eds.), *Hardware Acceleration of Computational Holography*,
https://doi.org/10.1007/978-981-99-1938-3_7

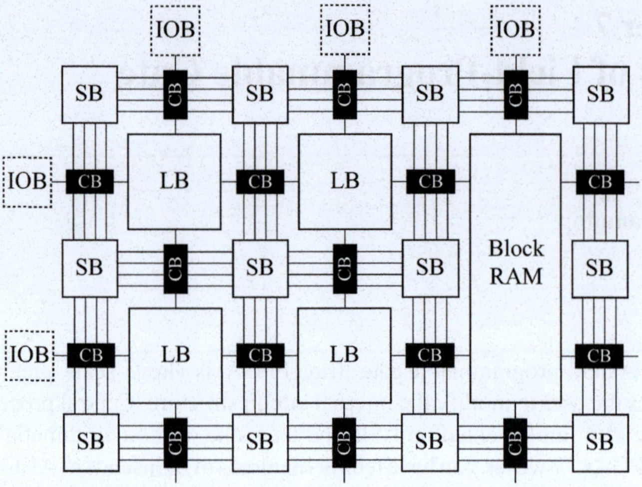

Fig. 7.1 Structure of FPGA

LBs are based on the lookup table (LUT) and multiplexer (MUX) cells to implement certain logic functions. The LB's name differs among FPGA vendors: Xilinx calls it a configurable LB (CLB) [1], and Intel calls it a logic array block (LAB) [2]. Furthermore, even if it is from the same vendor, the internal structure of the LB varies depending on the family.

There are two primary kinds of LBs: hard logic and soft logic. Hard logic includes a **digital signal processor** (**DSP**) [3] and block RAM [4]. Although it lacks the flexibility of LUT-based soft blocks, it can run predetermined logic functions at a high speed. Soft logic, which comprises LUTs and so on, fulfills any logic functions that are unavailable in hard logic.

The primary difference between CPUs and FPGAs is the arithmetic units' number. CPUs have one or more arithmetic logic units (ALUs), whereas FPGAs can have tens or hundreds of thousands of ALU-equivalent arithmetic cores using their on-board logic resources. CPU ALUs can perform as fast as 3 GHz, while FPGA ALUs are nearly an order of magnitude slower at around 500 MHz.

7.2 Hardware Description Language (HDL)

The circuit configuration data are produced by compiling the source code written by **hardware description languages** (**HDLs**) using a tool offered by FPGA vendors (Fig. 7.2). First, the HDL is transformed into an intermediate code called a netlist using a process called **logic synthesis**. The netlist is then mapped to the physical pin assignments and LBs of the actual device by a process called implementation, and the circuit configuration data are produced [5].

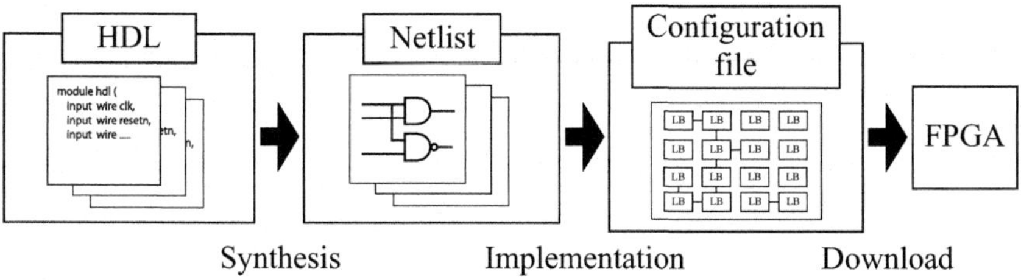

Fig. 7.2 Programming flow of FPGA

Fig. 7.3 Block diagram of the adder circuit

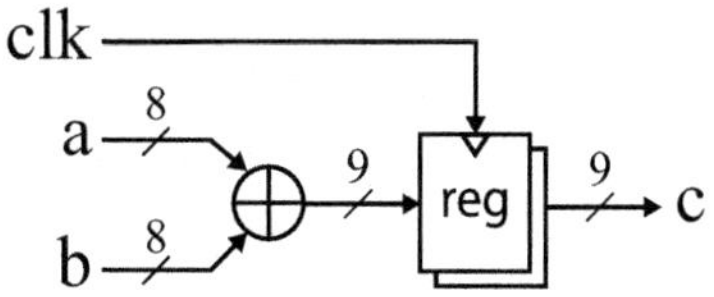

There are different types of HDLs, and there are compilers called high-level synthesis tools that produce HDL from abstract descriptions in C language [6]. In this section, we shortly present the **VHDL** [7] and **SystemVerilog** [8]. Listings 7.1 and 7.2 reveals the source code of an adder circuit using VHDL and SystemVerilog, and Fig. 7.3 reveals the block diagram.

In HDL, it is possible to describe the logic circuit to be implemented in FPGA using addition and subtraction of variables, conditional branching, and so on, just as in C programming. However, it is crucial to note that the operations of CPUs and FPGAs are very different. Software programming, including C programming, describes how the CPU operates, and the process is run sequentially. However, FPGA hardware programming explains the logic circuits' structure. All the illustrated logic circuits operate simultaneously.

The defining of input and output signals is the first step in both VHDL and SystemVerilog. The circuit is synchronized with "clk," which is a clock signal that repeats "0" and "1," The circuit conducts the addition of the input values of "a" and "b." The bit width of the CPU and GPU is fixed, whereas that of the FPGA can be freely determined by the user.

Listing 7.1 Source code for adder circuit using VHDL

```
library IEEE;
use IEEE.std_logic_1164.all;
use ieee.std_logic_arith.all;
use ieee.std_logic_unsigned.all;

entity adder_vh is
  generic (
    INPUT_WIDTH : integer := 8
  );
  port (
    clk : in std_logic;
```

```
        a : in std_logic_vector(INPUT_WIDTH−1 downto 0);
        b : in std_logic_vector(INPUT_WIDTH−1 downto 0);
        c : out std_logic_vector(INPUT_WIDTH−1+1 downto 0)
    );
end adder_vh ;

architecture rtl of adder_vh is
    signal add : std_logic_vector(INPUT_WIDTH−1+1 downto 0);
begin
    c <= add;

    process (clk)
    begin
        if clk'event and clk = '1' then
            add <= ('0'&a) + ('0'&b);
        end if;
    end process;

end architecture;
```

Listing 7.2 Source code for adder circuit using SystemVerilog

```
module adder_sv #(
    parameter int INPUT_WIDTH = 8
)(
    input wire clk,
    input wire [INPUT_WIDTH−1:0] a,
    input wire [INPUT_WIDTH−1:0] b,
    output wire [INPUT_WIDTH−1+1:0]c
);

    logic [INPUT_WIDTH−1+1:0] add;

    assign c = add;

    always_ff @(posedge clk) begin
            add <= a + b;
    end

endmodule
```

7.3 Special-Purpose Computation Circuit Using FPGA

To build a high-speed special-purpose computer using FPGAs, it is a must to use tens to hundreds of thousands of arithmetic units. However, FPGAs are about one order of magnitude slower than CPUs in terms of operating frequency, and it is crucial to consider effective data flow to build a high-speed special-purpose computer using FPGAs.

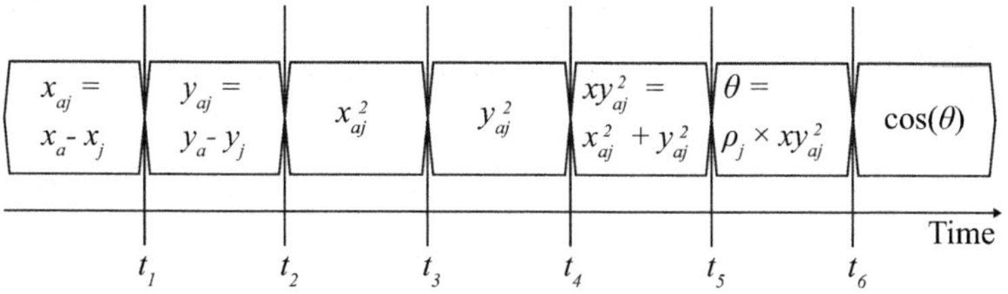

Fig. 7.4 Sequential execution of Eq. 7.1

There are two important factors for efficient computation: throughput and latency. **Throughput** is the processing capacity per unit of time. **Latency** is the delay time needed for each process. By enhancing throughput and latency, faster computation becomes possible.

To enhance throughput and latency, pipeline and data parallelization can be employed. To explain this, we consider the implementation of Eq. 7.1 to compute a computer-generated hologram (CGH):

$$I\left(x_a, y_a\right) = \sum_{j=0}^{M-1} \cos\left[\rho_j\left\{\left(x_a - x_j\right)^2 + \left(y_a - y_j\right)^2\right\}\right], \tag{7.1}$$

where(x_a, y_a) represents a coordinate on the CGH plane, $\rho_j = \pi/2\lambda z_j$, (x_j, y_j, z_j) are the coordinates of the 3D object's point cloud, M denotes the point-cloud number, and λ represents the reference light's wavelength.

For a sequential computation on a CPU, the computation inside Σ in Eq. 7.1 is shown in Fig. 7.4.

For example, we consider the computation time t [s] for $1{,}024 \times 1{,}024$-pixel CGH from $M = 100$ object points at the latency shown in Fig. 7.4. Assuming that each operation is run at 250 MHz (4 ns), the computation time is

$$t = \frac{1}{250\text{ MHz}} \times 7 \times 100 \times 1{,}024 \times 1{,}024 = 2.94\text{ s}. \tag{7.2}$$

Since x_{aj} and y_{aj} are independent of each other, the computations for them can be parallelized as illustrated in Fig. 7.5. Here, the latency is reduced from 7 to 5, and the computation time can be lowered to

$$t = \frac{1}{250\text{ MHz}} \times 5 \times 100 \times 1{,}024 \times 1{,}024 = 2.10\text{ s}. \tag{7.3}$$

Although we have focused on the computation of only a single CGH pixel, the CGH computation can be parallelized for each CGH pixel. Figure 7.6 reveals the five-step computation in Fig. 7.5 parallelized for two CGH pixels:

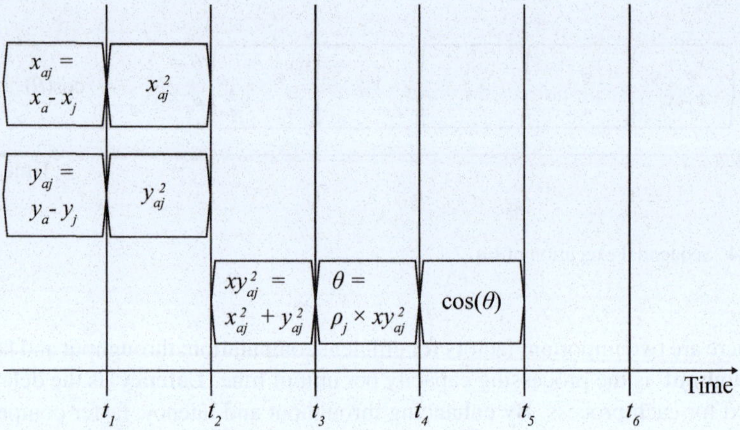

Fig. 7.5 Parallelization of x and y calculations

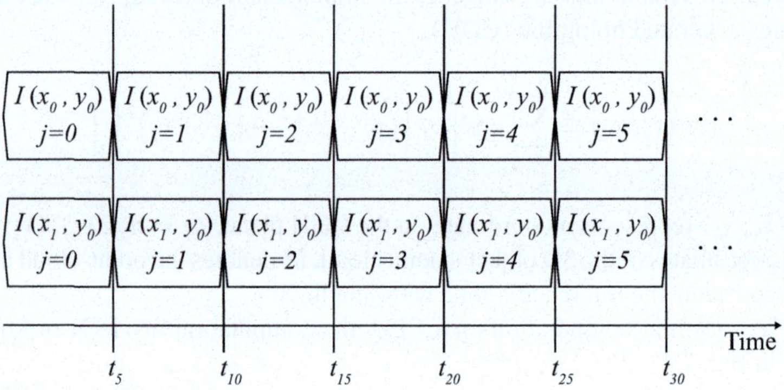

Fig. 7.6 Pixel-by-pixel parallelization

$$t = \frac{1}{250\text{MHz}} \times 5 \times 100 \times 1{,}024 \times 1{,}024 \div 2 = 1.05 \text{ s.} \tag{7.4}$$

This parallelization approach, which takes advantage of the lack of dependency between data and performs operations in parallel, is called **data parallelization**.

The computation time can be further accelerated using **pipeline parallelization**. Data parallelization is user-controllable not only in FPGAs but also in CPUs and GPUs, whereas pipeline parallelization is a user-controllable parallelization approach only in FPGAs. Here, we denote $x_a - x_j$ and $y_a - y_j$ operations, ${x_{aj}}^2$ and ${y_{aj}}^2$ operations, ${xy_{aj}}^2$ operation, θ operation, and $\cos(\theta)$ operation in Fig. 7.4 as OP0_j, OP1_j, OP2_j, OP3_j, and OP4_j, respectively. In pipeline parallelization, the amount of arithmetic units needed for the entire computation is arranged as illustrated in Fig. 7.7 for Fig. 7.4. Additionally, it is parallelization at the operator level. Here, the latency is the same as that in data parallelization. However, the throughput is enhanced. In

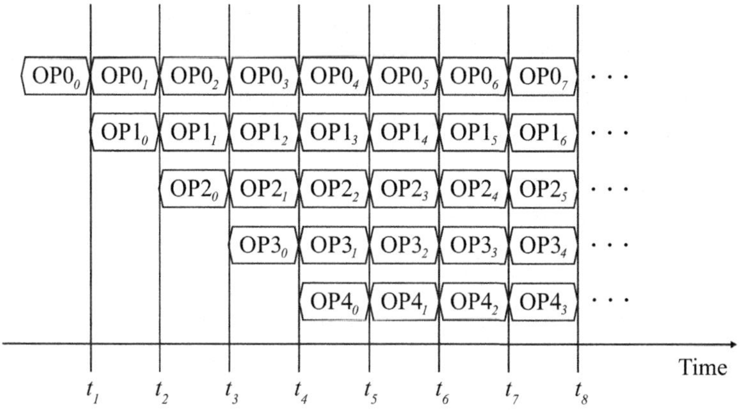

Fig. 7.7 Pipeline parallelization

the case of data parallelization only, the next data cannot be input to the circuit until all five operations are completed. However, in the case of pipeline parallelization, the following object point data can be input immediately. The computation time can be expressed as follows:

$$t = \frac{1}{250\,\text{MHz}} \times (5 + 100 - 1) \times 1024 \times 1024 = 0.44\,s. \tag{7.5}$$

The computation is nearly five times faster than when neither data parallelization nor pipeline parallelization is employed. Combining pipeline parallelization and data parallelization is also possible. When the two are combined, the computation time in Eq. 7.5 is reduced by the number of parallels. If 10 CGH pixels can be data-parallelized, for example, the computation time can be further accelerated from Eq. 7.5 as

$$t = \frac{1}{250\,\text{MHz}} \times (5 + 100 - 1) \times 1024 \times 1024 \div 10 = 0.044\text{ s}. \tag{7.6}$$

FPGAs can attain high-speed computation by employing data parallelization and pipeline parallelization, as well as an efficient parallel computation that takes advantage of the reconfigurable resources inside the FPGA. To attain high-speed computation, the algorithm should have less resilience on data, should be able to compute with as low accuracy as feasible, and should be readily parallelized.

7.4 Fixed-Point and Floating-Point Arithmetic

Floating-point arithmetic are frequently employed in CPUs. Floating-point arithmetic employs exponential representation to denote numerical values, and the IEEE754 [9] standard defines the data format. Although floating-point arithmetic can handle a wide range of values, exponentiation operations are required. However, fixed-point arithmetic is frequently employed for numerical computations in FPGAs. In fixed-point arithmetic, the user places the decimal point's position arbitrarily. Compared with floating-point arithmetic, fixed-point arithmetic has a smaller range of values, but they do not need exponentiation operations and can be computed with simple hardware.

Equation 7.7 is the phase computation part of Eq. 7.1, and we describe how to compute it using fixed-point arithmetic.

$$\theta = \rho_j \left\{ \left(x_a - x_j\right)^2 + \left(y_a - y_j\right)^2 \right\}. \tag{7.7}$$

FPGAs can use any data width, whereas floating-point arithmetic use 32-bit or 64-bit data widths. The smaller the data width, the more resources (gates or transistors) can be employed to construct the arithmetic unit and the more parallelism can be realized.

If x_a, x_j, y_a, y_j in Eq. 7.7 are normalized by the sampling interval of CGH, they are integer values. The normalized values' data width is determined from the minimum and maximum values. Here, (x_a, y_a) is the coordinate on the CGH plane and x_j, y_j is the point cloud's coordinate. These coordinates range from -2,048 to 2,047 when using a CGH with 4,096 $\times$ 4,096 pixels; therefore, x_a, x_j, y_a, y_j are denoted by 12 bits. Figure 7.8 reveals the data widths and decimal point positions of fixed-point integer arithmetic. In the fixed-point arithmetic's addition and subtraction between integers, no decimal point change occurs. However, the data width is extended by 1 bit in addition. Also, in multiplication, the sum of the data widths of both operands is extended.

Fixed-point arithmetic in binary numbers is each digit weights units of powers of two as shown in Fig. 7.9. Figure 7.9 reveals an example of an unsigned binary number; in a signed binary number represented in two's complement, the most significant bit's weight is -2^3 as in the case of Fig. 7.9.

Figure 7.10 shows the data width of two fixed-point arithmetics and how multiplication moves the decimal point. The multiplication of integer and decimal fraction fixed-point arithmetics can also be computed in a straightforward manner. However, the decimal point is shifted, and the decimal point's position in the computation finding θ becomes the 32nd bit position.

Here, θ is represented as a fixed-point number with a 32-bit decimal part. Here, the decimal part's minimum value is 0.000000000232 (2^{-32}). In other words, we must treat θ as estimated values with some error. This error is known as quantization error. The quantization error may have a large influence on some computations, so it is necessary to assess the effect of the quantization error in advance by simulation.

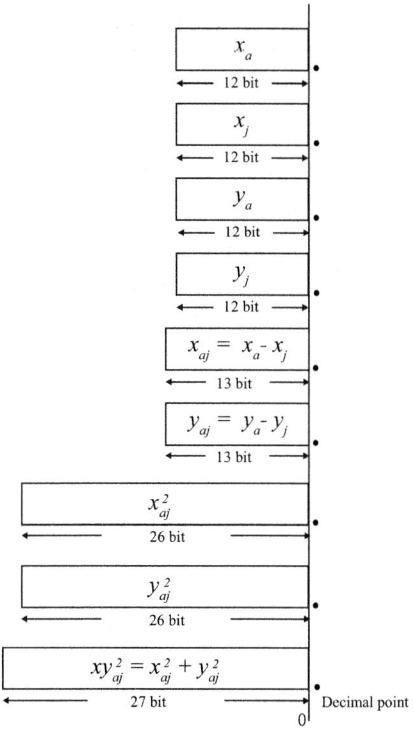

Fig. 7.8 Integer fixed-point arithmetic

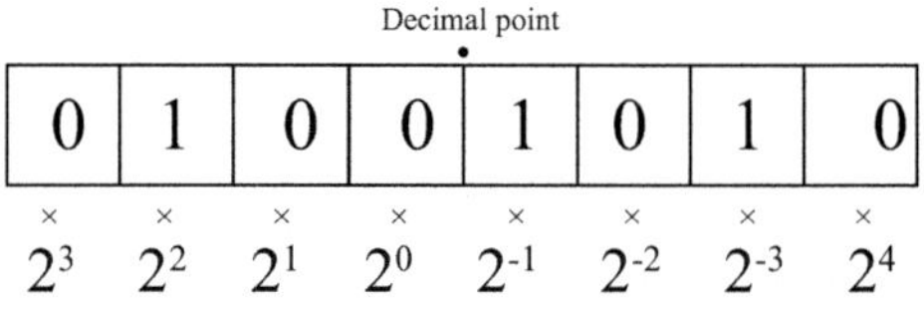

Fig. 7.9 Fixed-point arithmetic representation. Here, the decimal number is 8.625

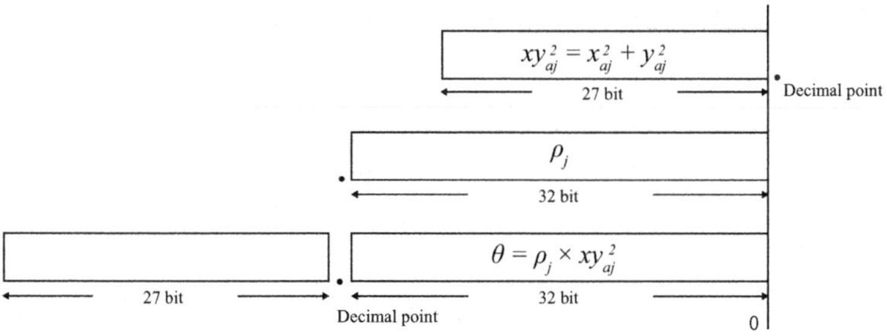

Fig. 7.10 Fixed-point arithmetic of decimals

7.5 Communication Between FPGAs and CPUs

A special-purpose computer using FPGAs is not employed alone but is connected to CPUs (FPGA embedded or on a PC) that send and pre- and post-process the data. There are different communication protocols, including Universal Serial Bus (USB) and PCI Express. In Xilinx FPGAs, the **advanced extensible interface** (**AXI**) [10] is employed for communication between a CPU (hard logic embedded on an FPGA) and programmable logic (Fig. 7.11) [11]. Even if the FPGA is communicated with a host PC through PCI Express or Zynq [12, 13] with a built-in CPU, we can employ AXI communication by developing auxiliary circuits.

7.6 AXI Communication

AXI is an inter-module communication protocol created by ARM Ltd [10]. There are three AXI communications: AXI(-Full), AXI-Lite, and AXI-Stream. AXI Lite is employed for small-scale data communication (e.g., control signals), whereas AXI(-Full) and AXI-Stream are used for large-scale data communication.

A circuit that requests data is called a "requester," and a circuit that sends and receives data in response to the request is called a "responder." The requester retains complete control over the data's transmission and reception. In AXI(-Full) and AXI-Lite, the data may be sent from the requester to the responder or from the responder to the requester in this chapter. AXI-Stream always sends data from the requester to the responder. Figure 7.12 reveals a diagram of the basic communication.

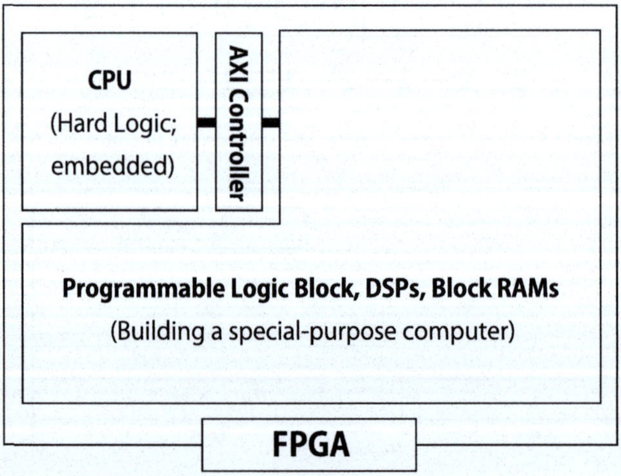

Fig. 7.11 Outline of the circuit connected by AXI

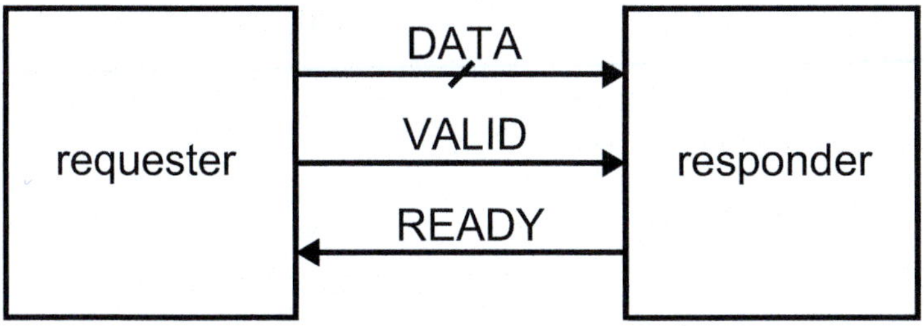

Fig. 7.12 AXI basic communication

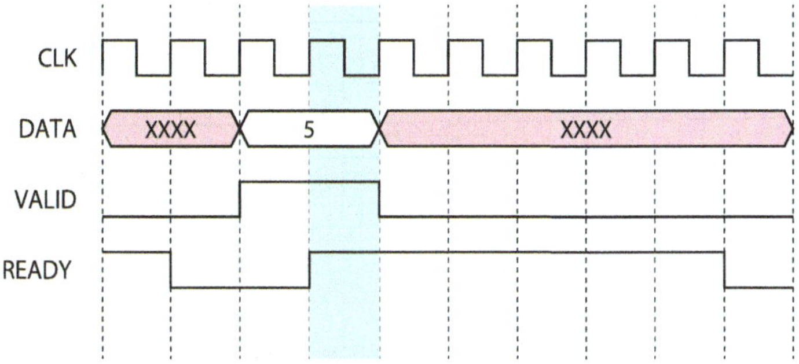

Fig. 7.13 The VALID-READY communication

When both VALID and READY of AXI signals are set to 1, the transfer is complete. The form of communication in which VALID shows that valid data are being introduced and READY illustrates that the data can be received is known as VALID-READY communication. The AXI protocol can employ several channels based on VALID-READY communication to improve communication capacity (Fig. 7.13).

Figure 7.14 shows a block diagram of a communication circuit using the AXI protocol (the signal's description can be found in Tables 7.1 and 7.2). In Fig. 7.14, regulating to write and read data are implemented as state machines, which transition the internal state depending on the input and current state. Figure 7.15 reveals the state transition diagrams for reading and writing data. Listing 7.3 indicates the implementation of Fig. 7.13 written by SystemVerilog.

The READ state machine, which is the data read from a CPU to an FPGA, comprises R_IDLE (read wait state) and R_READ (read response). After the start (e.g., assertion of the reset signal), the circuit begins in the R_IDLE state. A transition is made to the R_READ state when the signal S_AXI_ARVALID, which shows that a valid address is an output from the requestor (CPU), becomes 1. In the R_READ state, the FPGA maintains the signal S_AXI_RVALID as 1, indicating that it is outputting valid data, and returns to the R_ILDE state after receiving the read response (S_AXI_BVALID set as 1).

The WRITE state machine, which is the data written from the CPU to the FPGA, comprises the W_IDLE state, which is the write wait state, and the W_RESP state,

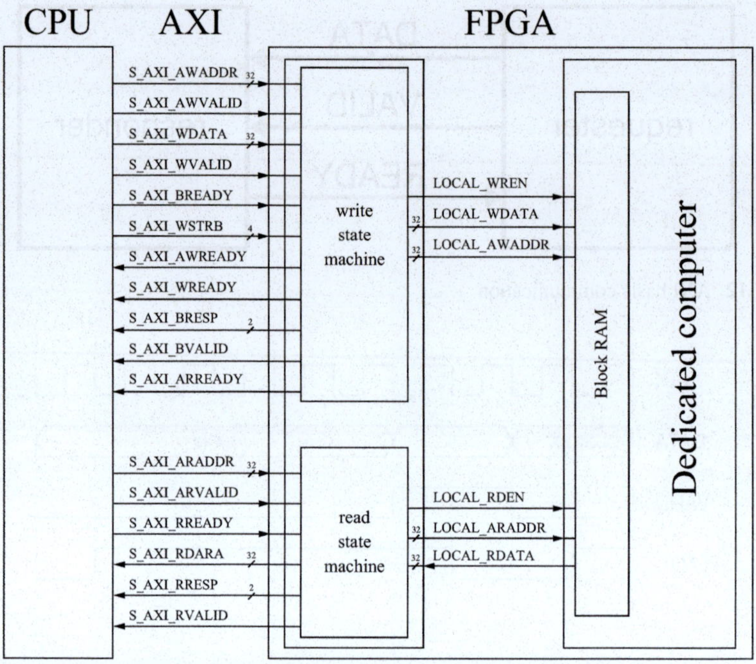

Fig. 7.14 Block diagram of the communication circuit (excluding clock and reset signals). The shaded lines on the signal lines in the figure show the bit width

Table 7.1 Description of AXI signals

Signal name	Description
S_AXI_AWADDR	Write start address
S_AXI_AWVALID	Write address valid
S_AXI_WDATA	Write data
S_AXI_WVALID	Write data valid
S_AXI_BREADY	Acceptable
S_AXI_WSTRB	Byte enable
S_AXI_AWREADY	Write address can be accepted
S_AXI_WREADY	Writing can be accepted
S_AXI_BRESP	Write response
S_AXI_BVALID	Write response enabled
S_AXI_ARREADY	Readable address can be accepted
S_AXI_ARADDR	Read start address
S_AXI_ARVALID	Read address valid
S_AXI_RREADY	Read data can be accepted
S_AXI_RDATA	Read data
S_AXI_RRESP	Read response
S_AXI_RVALID	Read data valid

Table 7.2 Description of local signals. These signals are defined by the author

Signal name	Description
LOCAL_WREN	Write data and address valid
LOCAL_WDATA	Write data
LOCAL_AWADDR	Write start address
LOCAL_RDEN	Read data and address response
LOCAL_ARADDR	Read start address
LOCAL_RDATA	Read data

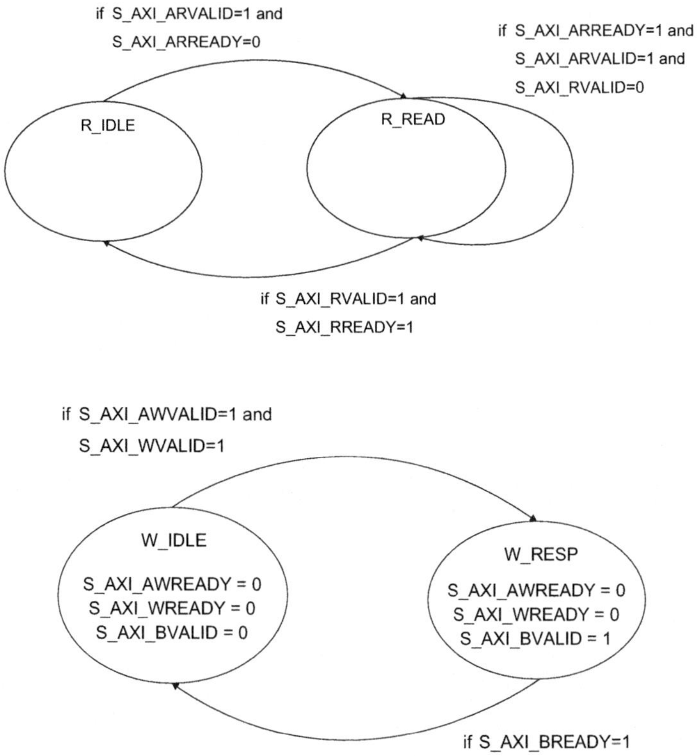

Fig. 7.15 State transition diagram for AXI Lite. The upper and bottom figures show READ and WRITE state machines, respectively

which is the write response state. The state machine starts in W_IDLE after initialization. When both the signal S_AXI_AWVALID, showing that the address is generating a valid value, and the signal S_AXI_WVALID, indicating that the data are valid, are set to 1 by the requestor (CPU), a transition to the W_RESP state happens. The FPGA returns to the W_IDLE state after a successful read response in W_RESP.

Listing 7.3 reveals the sample source code for the AXI Lite response side.

Listing 7.3 Source code for AXI Lite

```
module axi_lite_s # (
  parameter integer C_S_AXI_DATA_WIDTH = 32,
  parameter integer C_S_AXI_ADDR_WIDTH = 32
) (
  // Users to add ports here
  output wire        local_wren,
  output wire [C_S_AXI_DATA_WIDTH−1 : 0]  local_wdata,
  output wire [C_S_AXI_ADDR_WIDTH−1 : 0]  local_awaddr,
  output wire        local_rden,
  input wire [C_S_AXI_DATA_WIDTH−1 : 0]  local_rdata,
  output wire [C_S_AXI_ADDR_WIDTH−1 : 0]  local_araddr,
  output wire [(C_S_AXI_DATA_WIDTH/8)−1 : 0]  local_wstrb,

  // Ports of Axi S Bus Interface S_AXI
  input  wire        s_axi_aclk,
  input  wire        s_axi_aresetn,
  input  wire [C_S_AXI_ADDR_WIDTH−1 : 0]  s_axi_awaddr,
  input  wire [2 : 0]      s_axi_awprot,
  input  wire        s_axi_awvalid,
  output  wire        s_axi_awready,
  input  wire [C_S_AXI_DATA_WIDTH−1 : 0]  s_axi_wdata,
  input  wire [(C_S_AXI_DATA_WIDTH/8)−1 : 0] s_axi_wstrb,
  input  wire        s_axi_wvalid,
  output  wire        s_axi_wready,
  output  wire [1 : 0]      s_axi_bresp,
  output  wire        s_axi_bvalid,
  input  wire        s_axi_bready,
  input  wire [C_S_AXI_ADDR_WIDTH−1 : 0]  s_axi_araddr,
  input  wire [2 : 0]      s_axi_arprot,
  input  wire        s_axi_arvalid,
  output  wire        s_axi_arready,
  output  wire [C_S_AXI_DATA_WIDTH−1 : 0]  s_axi_rdata,
  output  wire [1 : 0]      s_axi_rresp,
  output  wire        s_axi_rvalid,
  input  wire        s_axi_rready
);

  localparam  W_IDLE = 2'd0, W_RESP = 2'd1;
  localparam  R_IDLE = 2'd0, R_READ = 2'd1;

  logic [C_S_AXI_ADDR_WIDTH−1 : 0]  axi_awaddr;
  logic      axi_awready;
  logic [C_S_AXI_DATA_WIDTH−1 : 0]  axi_wdata;
  logic      axi_wready;
  logic      axi_bvalid;
  logic [C_S_AXI_ADDR_WIDTH−1 : 0]  axi_araddr;
  logic      axi_arready;
  logic      axi_rvalid;
  logic [(C_S_AXI_DATA_WIDTH/8)−1 : 0] axi_wstrb;

```

```
logic [1:0] w_state, r_state;

// I/O Connections assignments
assign s_axi_awready = axi_awready;
assign s_axi_wready  = axi_wready;
assign s_axi_bresp  = 2'b0; // 'OKAY' response
assign s_axi_bvalid  = axi_bvalid;
assign s_axi_arready = axi_arready;
assign s_axi_rresp  = 2'b0; // 'OKAY' response
assign s_axi_rvalid  = axi_rvalid;

always_ff @( posedge s_axi_aclk ) begin
  if ( s_axi_aresetn == 1'b0 ) begin
    w_state  <= W_IDLE;
    axi_awready <= 1'b0;
    axi_awaddr  <= 0;
    axi_wready  <= 1'b0;
    axi_wdata  <= 0;
    axi_wstrb <= 0;
    axi_bvalid <= 1'b0;
  end else begin
    case ( w_state )
      W_IDLE: begin
        if ( ~axi_awready && ~axi_wready && s_axi_awvalid && s_axi_wvalid ) begin
          axi_awready <= 1'b1;
          axi_awaddr  <= s_axi_awaddr;
          axi_wdata  <= s_axi_wdata;
          axi_wstrb <= s_axi_wstrb;
          axi_wready  <= 1'b1;
          w_state  <= W_RESP;
        end else begin
          axi_awready <= 1'b0;
          axi_wready  <= 1'b0;
          axi_bvalid  <= 1'b0;
        end
      end
      W_RESP: begin
        if ( s_axi_bready && axi_bvalid ) begin
          axi_bvalid  <= 1'b0;
          w_state  <= W_IDLE;
        end else begin
          axi_awready <= 1'b0;
          axi_wready  <= 1'b0;
          axi_bvalid  <= 1'b1;
        end
      end
      default: begin
        w_state <= W_IDLE;
      end
    endcase
  end
end

```

```
  always_ff @( posedge s_axi_aclk ) begin
    if ( s_axi_aresetn == 1'b0 ) begin
      axi_arready <= 1'b0;
        axi_araddr <= 0;
      axi_rvalid <= 1'b0;
      r_state <= R_IDLE;
    end else begin
      case ( r_state )
        R_IDLE: begin
          if ( ~axi_arready && s_axi_arvalid ) begin
              axi_arready <= 1'b1;
              axi_araddr <= s_axi_araddr;
            r_state <= R_READ;
            end else begin
              axi_arready <= 1'b0;
            axi_rvalid <= 1'b0;
            end
        end
        R_READ: begin
            if ( axi_arready && s_axi_arvalid && ~axi_rvalid ) begin
              axi_rvalid <= 1'b1;
            axi_arready <= 1'b0;
            end else if ( axi_rvalid && s_axi_rready ) begin
            axi_rvalid <= 1'b0;
            axi_arready <= 1'b0;
            r_state <= R_IDLE;
          end
        end
        default: begin
          r_state <= R_IDLE;
        end
      endcase
    end
  end

  assign local_wren = axi_wready && s_axi_wvalid && axi_awready && s_axi_awvalid;
  assign local_rden = axi_arready && s_axi_arvalid && ~axi_rvalid;
  assign local_araddr = axi_araddr;
  assign local_awaddr = axi_awaddr;
  assign local_wdata = axi_wdata;
  assign s_axi_rdata = local_rdata;
  assign local_wstrb = axi_wstrb;

endmodule
```

7.7 Communication Program Between CPU and FPGA

Figure 7.15 shows that the CPU and FPGA are connected to communicate data, and it is crucial to create a dedicated driver. However, since creating a driver is outside the scope of this book, we will present an approach using /dev/mem [15], which is slow but easy to read and write data from/to an FPGA. As a prerequisite, we examine a situation where a Linux OS, including Petalinux (Linux manufactured by Xilinx) [14], is operating on the CPU embedded in Zynq. In Linux, reading and writing data to a device (here, an FPGA) can be replaced by reading and writing to a special file called /dev/mem. Listing 7.4 shows a sample program.

Listing 7.4 Source code for AXI Lite

```
#include <stdio.h>
#include <stdlib.h>
#include <unistd.h>
#include <fcntl.h>
#include <sys/mman.h>

FPGA_ADDR_START=0xA0000000;
FPGA_ADDR_SIZE=0x8000;

int main()
{
    uint32_t *uio;
    int fd;

    // open "/dev/mem"
    fd = open("/dev/mem", O_RDWR | O_SYNC);
    if (fd < 1) {
        perror("Failed to open devfile");
        return -1;
    }

    // map FPGA physical address into user space
    uio = (uint32_t *)mmap(NULL, FPGA_ADDR_SIZE, PROT_READ|PROT_WRITE,
          MAP_SHARED, fd, FPGA_ADDR_START);

    // write "5" to FPGA
    uio[0] = 0x5;

    // cleanup
    munmap((void*)address, 0x1000);
    close(fd);

    return 0;
}
```

Listing 7.4 reveals an example where the start address for writing data to the FPGA is 0xA0000000. This address is determined by vendors (refer to the datasheet for details). The Linux system commands (C language functions) “write” and “read” are employed to send data as if reading and writing to a file.

Using the /dev/mem technique removes the need for building a driver, but it should be noted that this is not a permanent approach from the viewpoint of security and speed. This is only for confirmation purposes. To improve the communication speed, device drivers need to be created.

7.8 Discussion

In this study, we presented a communication scheme between CPU and FPGA abstracted by AXI in Xilinx FPGAs. Data communication is a barrier to parallelization in FPGAs and CPUs and GPUs: in the CGH computation example, if the data size that can be sent by the communication circuit is 128 bits, the number of pixels that can be sent at a time (assumed to be 8 bits) is 16. Here, there will be a delay in data transmission if more than 16 are parallelized. If data transmission and reception are slow, the communication time becomes a barrier that lowers the circuit's arithmetic efficiency and makes it impossible to produce an arithmetic speed commensurate with parallelization.

A possible countermeasure is to create high-speed communication circuits that can transmit and receive numerous data at high speeds using direct memory access (DMA); DMA allows asynchronous communication so that the computation circuit can operate while sending and receiving data. Pipeline parallelization is completed at the operator level, so if all needed data can be stored in the FPGA, there are no communication constraints during computation. The longer the pipeline, the higher the pipeline parallelization's speed-up rate. By integrating pipeline parallelization with data parallelization, special-purpose computers that are unaffected by communication barriers can be built.

Funding This work was supported by JSPS KAKENHI Grant Number JP21K21294.

References

1. Xilinx 7 Series FPGAs: The Logical Advantage, Xilinx. https://docs.xilinx.com/v/u/en-US/wp405-7Series-Logical-Advantage
2. Intel MAX 10 FPGA Device Architecture, intel. https://www.intel.com/content/www/us/en/docs/programmable/683105/current/logic-array-block.html
3. DSP Solution, Xilinx. https://www.xilinx.com/products/technology/dsp.html
4. Memory Solution, Xilinx. https://www.xilinx.com/products/technology/memory.html
5. Implementation, Xilinx. https://www.xilinx.com/products/design-tools/vivado/implementation.html
6. Introduction to FPGA Design with Vivado High-Level Synthesis, Xilinx. https://docs.xilinx.com/v/u/en-US/ug998-vivado-intro-fpga-design-hls
7. IEEE Standard for VHDL Language Reference Manual, IEEE 1076-2019. https://standards.ieee.org/ieee/1076/5179/
8. IEEE Standard for SystemVerilog–Unified Hardware Design, Specification, and Verification Language, IEEE 1800-2017. https://standards.ieee.org/ieee/1800/6700/
9. ISO/IEC/IEEE 60559:2011. https://www.iso.org/standard/57469.html
10. AMBA AXI and ACE Protocol Specification Version E, ARM. https://developer.arm.com/documentation/ihi0022/e/
11. AXI Interconnect, Xilinx. https://www.xilinx.com/products/intellectual-property/axi_interconnect.html#overview
12. SoCs with Hardware and Software Programmability, Xilinx. https://japan.xilinx.com/products/silicon-devices/soc/zynq-7000.html

13. Zynq UltraScale+ MPSoC, Xilinx. https://www.xilinx.com/products/silicon-devices/soc/zynq-ultrascale-mpsoc.html
14. PetaLinux Tools, Xilinx. https://www.xilinx.com/products/design-tools/embedded-software/petalinux-sdk.html
15. The Linux Programming Interface, man7.org. https://man7.org/linux/man-pages/man4/mem.4.html

Part III
Acceleration and Advanced Techniques in Computational Holography

Part III consists of 11 chapters. Each chapter describes the implementation of the algorithms in computational holography with specific source code. Diffraction calculations play an important role in computational holography. CPU and GPU implementations of the main diffraction calculations are shown. The point-cloud method, polygon method, layer method and light field method of hologram computation algorithms will be described, and GPU cluster acceleration will be presented. Compressed sensing, hologram image quality evaluation, and recent digital holography acceleration will also be presented.

Chapter 8
CPU and GPU Implementations of Diffraction Calculations

Soma Fujimori

Abstract In this chapter, we describe a set of C++ and CUDA programs to perform diffraction calculations using a Fourier transform. The programs implement the Fresnel diffraction calculation and angular spectrum method. We provide sample source code for both central processing unit (CPU) and graphics processing unit (GPU) hardware because the execution of the computations can be accelerated on the latter.

8.1 Implementation of the Fresnel Diffraction Calculation

As explained in Chap. 1, the **Fresnel diffraction** described as follows.

$$\begin{aligned} u_2(x_2, y_2) &= \frac{\exp\left(i\frac{2\pi z}{\lambda}\right)}{i\lambda z} \times u_1(x_1, y_1) \otimes \exp\left(i\frac{\pi}{\lambda z}(x_1^2 + y_1^2)\right) \\ &= \frac{\exp\left(i\frac{2\pi z}{\lambda}\right)}{i\lambda z} \times u_1(x_1, y_1) \otimes h(x_1, y_1), \end{aligned} \tag{8.1}$$

where $\otimes$ is **convolution** operator, $u_1(x_1, y_1)$ and $u_2(x_2, y_2)$ are, respectively, the complex amplitude of the source and destination planes, z is a propagation distance, and λ is wavelength of light.

Let us rewrite Eq. (8.1) for computation on standard computer hardware. The first term in Eq. (8.1) can be ignored when only the light intensity is considered. By the convolution theorem [1], we replace the convolution operation with the Fourier transform $\mathcal{F}$ and express Eq. (8.1) as follows.

Supplementary Information The online version contains supplementary material available at https://doi.org/10.1007/978-981-99-1938-3_8.

S. Fujimori (✉)
Graduate School of Science and Engineering, Chiba University, 1-33 Yayoi-cho, Inage-ku, Chiba-shi, Chiba, Japan
e-mail: soma.fjmr@chiba-u.jp

T. Shimobaba and T. Ito (eds.), *Hardware Acceleration of Computational Holography*,
https://doi.org/10.1007/978-981-99-1938-3_8

$$u_2(x_2, y_2) = \mathcal{F}^{-1}[\mathcal{F}[u_1(x_1, y_1)]\mathcal{F}[h(x_1, y_1)]]. \tag{8.2}$$

Subsequently, we obtain $x_1 = pm_1, x_2 = pm_2, y_1 = pn_1, y_2 = pn_2$ by discretizing x_1, x_2, y_1, y_2 with a sampling interval p. Hence,

$$u_2(m_2, n_2) = \mathcal{F}^{-1}[\mathcal{F}[u_1(m_1, n_1)]\mathcal{F}[h(m_1, n_1)]], \tag{8.3}$$

where m_1, m_2, n_1, n_2 are discretized coordinates.

The Fourier transform is commonly computed using the fast Fourier transform (FFT). Therefore, the Fresnel diffraction calculation can be formulated as given below.

$$u_2(m_2, n_2) = \mathrm{FFT}^{-1}[\mathrm{FFT}[u_1(m_1, n_1)]\mathrm{FFT}[h(m_1, n_1)]]. \tag{8.4}$$

In developing a computer program to calculate Eq. (8.4), we focused on the following points.

(a) Linear and circular convolution
We considered both linear and circular convolution in this work. For example, the program computes the two types of convolution operations as shown in Fig. 8.1. The **linear convolution** of Eq. (8.1) used only the data within a given region. By contrast, the convolution computed by the FFTs in Eq. (8.4) is a **circular convolution** that interprets the given data $u_1(m_1)$ as periodic due to the properties of FFT, which causes a **wraparound effect**. Therefore, to obtain the same result as the linear convolution with the circular convolution, **zero padding** was used, as shown in Fig. 8.2, to avoid the wraparound effect. Finally, the same result as linear convolution was obtained by cropping only the necessary parts. For 2D data, zero padding was implemented as shown in Fig. 8.3.

(b) Arrangement of the frequency spectrum obtained by FFT
As shown in Fig. 8.4, the spectrum obtained by the 2D FFT exhibits low-frequency components in the periphery and increases in frequency toward the center. The DC component is shown at the lower left of the image. If this arrangement is inconvenient, the quadrants are exchanged such that the low-frequency components are in the center of the image. This operation is called an **FFT shift**.

(c) Origin position in spatial and frequency domains
In FFT libraries, the origin (DC component) is set at the edge of the image in both the spatial and frequency domains. Therefore, if the origin of data input to the FFT algorithm is placed in the center of the image, the quadrants need to be appropriately exchanged by the FFT shift to agree with the origin of the FFT as shown in Fig. 8.5.

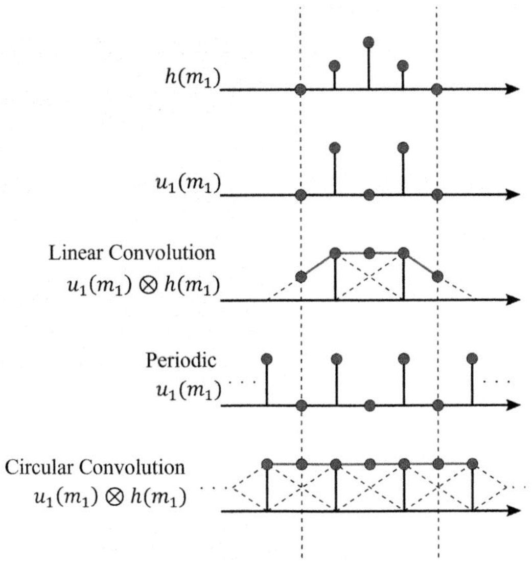

Fig. 8.1 Linear and circular convolution (Based on [2])

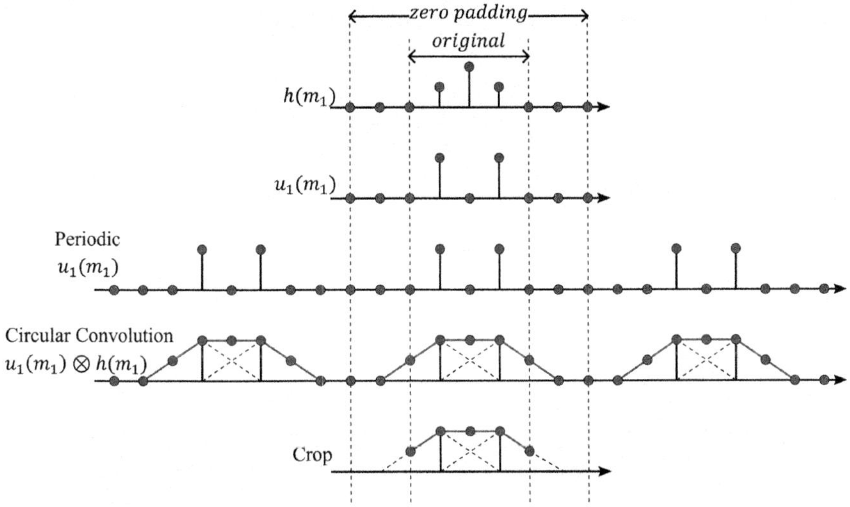

Fig. 8.2 Avoiding the wraparound caused by circular convolution with zero padding (Based on [2])

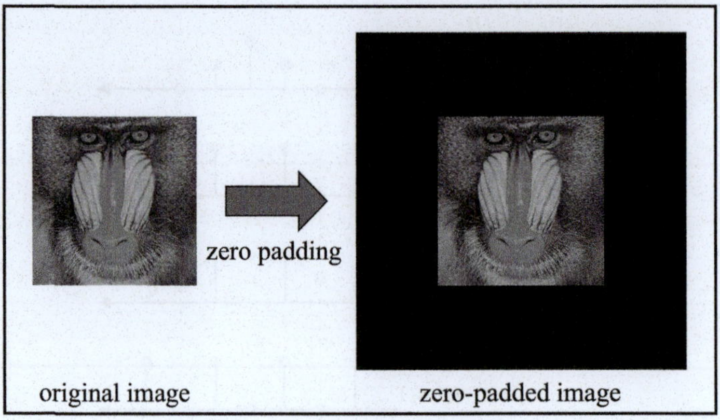

Fig. 8.3 Zero padding for 2D data

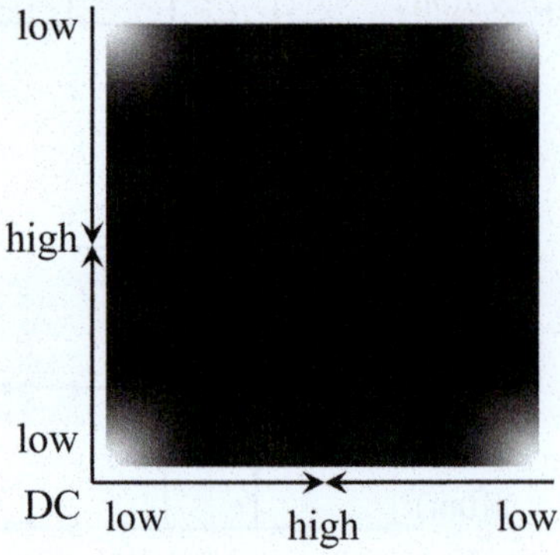

Fig. 8.4 Frequency spectrum obtained via FFT

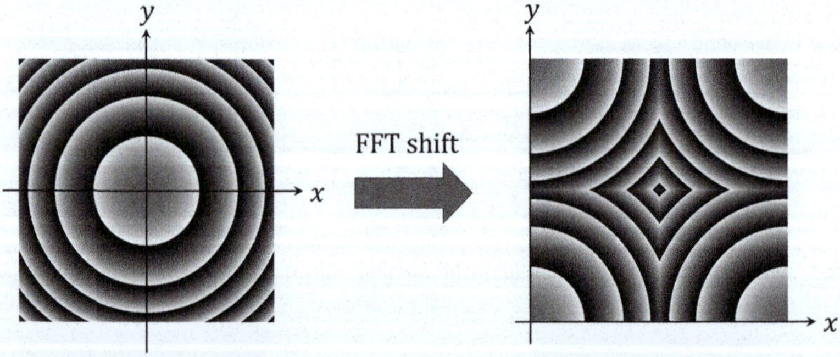

Fig. 8.5 Moving the origin by FFT shift

8.1.1 Process Flow of the Fresnel Diffraction Calculation

Considering the above, the program developed to perform the Fresnel diffraction calculation executes the following steps.

Step 1. Apply zero padding to the complex amplitude of the source u_{src} to avoid the wraparound effect caused by circular convolution and obtain the zero-padded complex amplitude u_{pad}.
Step 2. Apply FFT to u_{pad} to obtain the Fourier spectrum U_{pad}.
Step 3. Compute the impulse response h (we assume that the origin is set at the center of the computational windows) and move the position of the origin to the edge of the image by using FFT shift, as shown in Fig. 8.5.
Step 4. Apply FFT to h to obtain the transfer function H.
Step 5. Multiply the spectrum U_{pad} by the transfer function H.
Step 6. Apply inverse FFT to the result of the multiplication.
Step 7. Crop only the necessary calculation result and obtain the complex amplitude of the propagation destination, u_{dst}.

8.1.2 CPU Implementation

Let us consider implementing the Fresnel diffraction calculation on a CPU. We used the **FFTW** [3, 4] FFT library. Listing 8.1 shows functions for FFT and inverse FFT using the FFTW. Both the functions "fft" and "ifft" can be implemented by (1) creating a **plan** (fftwf_plan_dft2d), (2) executing the calculation (fftwf_execute), and (3) destroying the plan (fftwf_destroy_plan). **Multi-thread** calculations are also possible by specifying the number of threads before creating a plan. In that case, the **OpenMP** [5] interface is required.

Listing 8.1 Functions for FFT and inverse FFT using FFTW

```
void fft(std::complex<float>* src, std::complex<float>* dst, int32_t ny, int32_t nx)
{
  fftwf_complex* _usrc = reinterpret_cast<fftwf_complex*>(src);
  fftwf_complex* _udst = reinterpret_cast<fftwf_complex*>(dst);
  fftwf_init_threads();
  fftwf_plan_with_nthreads(omp_get_max_threads());
  fftwf_plan p = fftwf_plan_dft_2d(ny, nx, _usrc, _udst, FFTW_FORWARD,
        FFTW_ESTIMATE);
  fftwf_execute(p);
  fftwf_destroy_plan(p);
}

void ifft(std::complex<float>* src, std::complex<float>* dst, int32_t ny, int32_t nx)
{
  fftwf_complex* _usrc = reinterpret_cast<fftwf_complex*>(src);
  fftwf_complex* _udst = reinterpret_cast<fftwf_complex*>(dst);
  fftwf_init_threads();
```

```
  fftwf_plan_with_nthreads(omp_get_max_threads());
  fftwf_plan p = fftwf_plan_dft_2d(ny, nx, _usrc, _udst, FFTW_BACKWARD,
        FFTW_ESTIMATE);
  fftwf_execute(p);
  fftwf_destroy_plan(p);
}
```

Listing 8.2 shows an example of the CPU implementation of the Fresnel diffraction calculation. Each function call in the "FresnelProp" function corresponds to a step in the flow of the Fresnel diffraction calculation described in Sect. 8.1.1. "FresnelResponse" is a function that computes the impulse response $h(m_1, n_1)$.

Listing 8.2 Example of CPU implementation of the Fresnel diffraction calculation

```
void FresnelResponse(std::complex<float>* h, int32_t ny, int32_t nx, float dy, float dx, float
        lambda, float z)
{
  float tmp = 1 / (lambda * z);
  int32_t hny = ny / 2;
  int32_t hnx = nx / 2;
  for(int32_t n = 0;n < ny;n++){
    float y = ((float) (n − hny)) * dy;
    for(int32_t m = 0;m < nx;m++){
      int32_t idx = n * nx + m;
      float x = ((float) (m − hnx)) * dx;
      float phase = M_PI * (x * x + y * y) * tmp ;
      h[idx] = std::complex<float>(cos(phase),sin(phase));
    }
  }
}

void FresnelProp(std::complex<float>* u,int32_t ny, int32_t nx,float dy, float dx, float lambda,
        float z)
{
  int32_t ny2 = 2 * ny;
  int32_t nx2 = 2 * nx;
  std::complex<float>* upad = new std::complex<float>[ny2 * nx2];
  auto h = new std::complex<float>[ny2 * nx2];
  // Step1
  zeropadding(u, upad, ny, nx);
  // Step2
  fft(upad, upad, ny2, nx2);
  multscalar(upad, 1.0 / (ny2 * nx2), ny2, nx2);
  // Step3
  FresnelResponse(h, ny2, nx2, dy, dx, lambda, z);
  fftshift(h, ny2, nx2);
  // Step4
  fft(h, h, ny2, nx2);
  multscalar(h, 1.0 / (ny2 * nx2), ny2, nx2);
  // Step5
  mult(upad, h, upad, ny2, nx2);
  // Step6
  ifft(upad, upad, ny2, nx2);
  // Step7
```

```
  crop(upad, u, ny2, nx2);

  delete[] upad;
  delete[] h;
}
```

Listing 8.3 shows the implementation of each function used in Listing 8.2.

Listing 8.3 Functions of the Fresnel diffraction calculation on a CPU

```
void mult(std::complex<float>* src1, std::complex<float>* src2, std::complex<float>* dst,
      int32_t ny, int32_t nx)
{
  for(int32_t n = 0; n < ny;n++){
    for(int32_t m = 0;m < nx;m++){
      dst[m + n * nx] = src1[m + n * nx] * src2[m + n * nx];
    }
  }
}

void multscalar(std::complex<float>* u, float c, int32_t ny, int32_t nx)
{
  for(int32_t n = 0;n < ny;n++){
    for(int32_t m = 0;m < nx;m++){
      u[m + n * nx] *= c;
    }
  }
}

void zeropadding(std::complex<float>* src, std::complex<float>* dst, int32_t ny, int32_t nx)
{
  int32_t nx2 = nx * 2;
  int32_t ny2 = ny * 2;
  for(int32_t n = 0;n < ny2;n++){
    for(int32_t m = 0;m < nx2;m++){
      if(ny / 2 <= n && n < ny * 3 / 2 && nx / 2 <= m && m < nx * 3 / 2)
      {
        dst[m + n * nx2] = src[m - nx / 2 + (n - ny / 2) * nx];
      }
      else{
        dst[m + n * nx2] = 0;
      }
    }
  }
}

void crop(std::complex<float>* src, std::complex<float>* dst, int32_t ny, int32_t nx)
{
  for(int32_t n = 0;n < ny;n++){
    for(int32_t m = 0;m < nx;m++){
      if(ny / 4 <= n && n < ny * 3 / 4 && nx / 4 <= m && m < nx * 3 / 4)
      {
        dst[m - nx / 4 + (n - ny / 4) * nx / 2] = src[m + n * nx];
      }
```

```
    }
  }
}

void fftshift(std::complex<float>* u, int32_t ny, int32_t nx)
{
  int32_t hny = ny / 2;
  int32_t hnx = nx / 2;
  for(int32_t n = 0;n < hny;n++){
    for(int32_t m = 0;m < hnx;m++){
      int32_t idx1, idx2;
      std::complex<float> tmp;
      idx1 = n * nx + m;
      idx2 = (n + hny) * nx + (m + hnx);
      tmp = u[idx1];
      u[idx1] = u[idx2];
      u[idx2] = tmp;
      idx1 = n * nx + (m + hnx);
      idx2 = (n + hny) * nx + m;
      tmp = u[idx1];
      u[idx1] = u[idx2];
      u[idx2] = tmp;
    }
  }
}
```

8.1.3 GPU Implementation

By contrast, Fresnel diffraction can be calculated at high speed by implementing the procedure on a GPU using the **CUDA** toolkit, which provides **cuFFT** [6] as an FFT library. FFT and inverse FFT functions using cuFFT are shown in Listing 8.4. Both the functions "fft" and "ifft" can be implemented by (1) creating a plan (cufftPlan2d), (2) executing the calculation (cufftExecC2C), and (3) destroying the plan (cufftDestroy). Because creating a plan with cuFFT takes a relatively long time, creating a plan only once and using it for multiple FFTs is appropriate. For this reason, Listing 8.4 provides a function "set" to create a plan.

Listing 8.4 FFT and inverse FFT functions using cuFFT

```
class gFFT{
  public:
  cufftHandle fftplan;
  bool flag = false;
  ~gFFT(){
    if (flag == true)
      cufftDestroy(fftplan);
  }
  void set(int32_t ny, int32_t nx){
    cufftPlan2d(&fftplan, ny, nx, CUFFT_C2C);
```

```
    flag = true;
  }
  void fft(cufftComplex* src,cufftComplex* dst){
    if (flag == true){
      cufftExecC2C(fftplan, src, dst, CUFFT_FORWARD);
      cudaDeviceSynchronize();
    }
  }
  void ifft(cufftComplex* src,cufftComplex* dst){
    if (flag == true){
      cufftExecC2C(fftplan, src, dst, CUFFT_INVERSE);
      cudaDeviceSynchronize();
    }
  }
};
```

Listing 8.5 shows an example of a GPU implementation of the Fresnel diffraction calculation. Each function call in the "prop" function in the C++ class "gFresnelProp" corresponds to a step in the flow of the Fresnel diffraction calculation described in Sect. 8.1.1. "gFresnelResponse" is a function that computes the impulse response $h(m_1, n_1)$.

Listing 8.5 Example of the GPU implementation of the Fresnel diffraction calculation

```
__global__ void gFresnelResponseKernel(cufftComplex* u, int32_t ny, int32_t nx,float dy,
    float dx, float lambda, float z){
  int32_t m = blockIdx.x * blockDim.x + threadIdx.x;
  int32_t n = blockIdx.y * blockDim.y + threadIdx.y;
  int32_t idx = n * nx + m;
  if ( (m < nx) && (n < ny) ){
    int32_t hnx = nx / 2;
    int32_t hny = ny / 2;
    float x = (m - hnx) * dx;
    float y = (n - hny) * dy;
    float tmp = M_PI * (x * x + y * y) / (lambda * z);
    u[idx] = make_cuComplex(cos(tmp),sin(tmp));
  }
}

void gFresnelResponse(cufftComplex* u,int32_t ny, int32_t nx, float dv,float du, float lambda,
    float z)
{
  dim3 block(16, 16, 1);
  dim3 grid(ceil((float)nx / block.x), ceil((float)ny / block.y), 1);
  gFresnelResponseKernel<<<grid,block>>>(u, ny, nx, dv, du, lambda, z);
  cudaDeviceSynchronize();
}

class gFresnelProp{
  private:
    cufftComplex* buf1, *buf2;
    gFFT fft;
  public:
    gFresnelProp(int32_t ny, int32_t nx){
```

```
    int32_t ny2 = ny * 2;
    int32_t nx2 = nx * 2;
    size_t mem_size = sizeof(cufftComplex) * ny2 * nx2;
    cudaMalloc((void**)&buf1, mem_size);
    cudaMalloc((void**)&buf2, mem_size);
    fft.set(ny2,nx2);
  }
  ~gFresnelProp(){
    cudaFree(buf1);
    cudaFree(buf2);
  }
  void prop(cufftComplex* u, int32_t ny, int32_t nx,float dy, float dx, float lambda, float z){
    int32_t ny2 = ny * 2;
    int32_t nx2 = nx * 2;
    // Step1
    gzeropadding(u, buf1, ny, nx);
    // Step2
    fft.fft(buf1,buf1);
    gmultscalar(buf1, 1.0f / (ny2 * nx2), ny2, nx2);
    // Step3
    gFresnelResponse(buf2, ny2, nx2, dy, dx, lambda, z);
    gfftshift(buf2, ny2, nx2);
    // Step4
    fft.fft(buf2, buf2);
    gmultscalar(buf2, 1.0f / (ny2 * nx2), ny2, nx2);
    // Step5
    gmult(buf1, buf2, buf1, ny2, nx2);
    // Step6
    fft.ifft(buf1, buf1);
    // Step7
    gcrop(buf1, u, ny2, nx2);
  }
};
```

Listing 8.6 shows the implementation of each function used in Listing 8.5.

Listing 8.6 Functions of the Fresnel diffraction calculation on a GPU

```
__global__ void gmultKernel(cufftComplex* src1, cufftComplex* src2, cufftComplex* dst,
    int32_t ny, int32_t nx)
{
  int32_t m = blockIdx.x * blockDim.x + threadIdx.x;
  int32_t n = blockIdx.y * blockDim.y + threadIdx.y;
  uint32_t idx = m + n * nx;
  if ( m < nx && n < ny){
    dst[idx] = cuCmulf(src1[idx], src2[idx]);
  }
}

void gmult(cufftComplex* src1, cufftComplex* src2,cufftComplex* dst,
  int32_t ny, int32_t nx)
{
  dim3 block(16, 16, 1);
  dim3 grid(ceil((float)nx / block.x), ceil((float)ny / block.y), 1);
```

```
  gmultKernel<<<grid,block>>>(src1, src2, dst, ny, nx);
  cudaDeviceSynchronize();
}

__global__ void gmultscalarKernel(cufftComplex* u, float c, int32_t ny, int32_t nx)
{
  int32_t m = blockIdx.x * blockDim.x + threadIdx.x;
  int32_t n = blockIdx.y * blockDim.y + threadIdx.y;
  size_t idx = (m + n * nx);
  if ( m < nx && n < ny){
    u[idx].x *= c;
    u[idx].y *= c;
  }
}

void gmultscalar(cufftComplex* u,float c, int32_t ny, int32_t nx)
{
  dim3 block(16, 16, 1);
  dim3 grid(ceil((float)nx / block.x), ceil((float)ny / block.y), 1);
  gmultscalarKernel<<<grid,block>>>(u, c, ny, nx);
  cudaDeviceSynchronize();
}

__global__ void gzeropaddingKernel(cufftComplex* src,cufftComplex* dst, int32_t ny,int32_t nx)
{
  int32_t m = blockIdx.x*blockDim.x + threadIdx.x;
  int32_t n = blockIdx.y*blockDim.y + threadIdx.y;
  if (m < 2 * nx && n < 2 * ny){
    if(ny / 2 <= n && n < ny * 3 / 2 && nx / 2 <= m && m < nx * 3 / 2)
    {
      dst[m + n * 2 * nx] = src[m - nx / 2+(n - ny / 2) * nx];
    }
    else{
      dst[m + n * 2 * nx] = make_cuComplex(0.0f, 0.0f);
    }
  }
}

void gzeropadding(cufftComplex* src, cufftComplex* dst, int32_t ny, int32_t nx){
  int32_t ny2 = ny * 2;
  int32_t nx2 = nx * 2;
  dim3 block(16, 16, 1);
  dim3 grid(ceil((float)nx2 / block.x), ceil((float)ny2 / block.y), 1);
  gzeropaddingKernel<<<grid, block>>>(src, dst, ny, nx);
  cudaDeviceSynchronize();
}

__global__ void gcropKernel(cufftComplex* src,cufftComplex* dst, int32_t ny, int32_t nx)
{
  int32_t m = blockIdx.x * blockDim.x + threadIdx.x;
  int32_t n = blockIdx.y * blockDim.y + threadIdx.y;
  if ( m < nx && n < ny){
```

```
    if(ny / 4 <= n && n < ny * 3 / 4 && nx / 4 <= m && m < nx * 3 / 4)
    {
      dst[m - nx / 4 + (n - ny / 4) * nx / 2] = src[m + n * nx];
    }
  }
}

void gcrop(cufftComplex* src, cufftComplex* dst, int32_t ny, int32_t nx){
  dim3 block(16, 16, 1);
  dim3 grid(ceil((float)nx / block.x), ceil((float)ny / block.y), 1);
  gcropKernel<<<grid,block>>>(src, dst, ny, nx);
  cudaDeviceSynchronize();
}

__global__ void gfftshiftKernel(cufftComplex* u, int32_t ny, int32_t nx){
  int32_t m = blockIdx.x * blockDim.x + threadIdx.x;
  int32_t n = blockIdx.y * blockDim.y + threadIdx.y;
  int32_t hnx = nx / 2; int32_t hny = ny / 2;
  if (m >= hnx || n >= hny){
    return;
  }
  int32_t idx1, idx2;
  cufftComplex tmp;
  idx1 = n * nx + m;
  idx2 = (n + hny) * nx + (m + hnx);
  tmp = u[idx1];
  u[idx1] = u[idx2];
  u[idx2] = tmp;
  idx1 = n * nx + (m + hnx);
  idx2 = (n + hny) * nx + m;
  tmp = u[idx1];
  u[idx1] = u[idx2];
  u[idx2] = tmp;
}

void gfftshift(cufftComplex* u, int32_t ny, int32_t nx){
  dim3 block(16, 16, 1);
  int32_t hny = ny / 2;
  int32_t hnx = nx / 2;
  dim3 grid(ceil((float) hnx / block.x), ceil((float)hny / block.y), 1);
  gfftshiftKernel<<<grid, block>>>(u, ny, nx);
  cudaDeviceSynchronize();
}
```

8.2 Implementation of the Angular Spectrum Method

As explained in Chap. 1, the **angular spectrum method** is described as follows.

$$\begin{aligned} u_2(x_2, y_2) &= \mathcal{F}^{-1}\left[\mathcal{F}[u_1(x_1, y_1)]\exp\left(i2\pi z\sqrt{\frac{1}{\lambda^2} - f_x^2 - f_y^2}\right)\right] \\ &= \mathcal{F}^{-1}[\mathcal{F}[u_1(x_1, y_1)]H(f_x, f_y)]. \end{aligned} \tag{8.5}$$

By discretizing Eq. (8.5) in the same manner as the Fresnel diffraction calculation with sampling intervals p_{xy} in the spatial domain (x_1, y_1) and (x_2, y_2) and p_f in the frequency domain (f_x, f_y), $x_1 = p_{xy}m_1, x_2 = p_{xy}m_2, y_1 = p_{xy}n_1, y_2 = p_{xy}n_2, f_x = p_f u, f_y = p_f v$ can be obtained. Equation (8.5) can be expressed as follows using the FFTs.

$$u_2(m_2, n_2) = \mathrm{FFT}^{-1}[\mathrm{FFT}[u_1(m_1, n_1)]H(u, v)]. \tag{8.6}$$

8.2.1 *Process Flow of the Angular Spectrum Method*

As in the Fresnel diffraction calculation, the program to calculate the angular spectrum method is implemented in the following steps, focusing on the wraparound due to circular convolution and quadrant exchange.

1. Apply zero padding to the complex amplitude of the source u_{src} to avoid the wraparound caused by the circular convolution and obtain the zero-padded complex amplitude u_{pad}.
2. Apply FFT to u_{pad} to obtain the Fourier spectrum U_{pad}.
3. Compute the transfer function H and move the position of the origin to the edge of the image by FFT shift.
4. Multiply the spectrum U_{pad} by the transfer function H in the angular spectrum method.
5. Apply inverse FFT to the result of the multiplication.
6. Crop only the necessary calculation result and obtain the complex amplitude of the propagation destination u_{dst}.

8.2.2 *CPU Implementation*

Listing 8.7 shows an example of CPU implementation of the angular spectrum method. Each function call in the "AsmProp" function corresponds to a step in the flow of the angular spectrum method described in Sect. 8.2.1. The function

"AsmTransferF" computes the transfer function $H(u, v)$. The implementation of the other functions is as shown in Listing 8.3.

Listing 8.7 Example of the CPU implementation of the angular spectrum method

```
void AsmTransferF(std::complex<float>* H, int32_t ny, int32_t nx, float dv, float du, float lambda, float z)
{
  float tmp = 1 / (lambda * lambda);
  for (int32_t n = 0;n < ny;n++){
    float v = (n - ny / 2) * dv;
    for (int32_t m = 0;m < nx;m++){
      int32_t idx = m + n * nx;
      float u = (m - nx / 2) * du;
      float w = sqrt(tmp - u * u - v * v);
      float phase = 2 * M_PI * w * z;
      H[idx] = std::complex<float>(cos(phase),sin(phase));
    }
  }
}

void AsmProp(std::complex<float>* u, int32_t ny, int32_t nx, float dy, float dx, float lambda, float z)
{
  int32_t ny2 = ny * 2;
  int32_t nx2 = nx * 2;
  float du = 1 / (dx * nx2);
  float dv = 1 / (dy * ny2);
  auto upad = new std::complex<float>[ny2 * nx2];
  auto H = new std::complex<float>[ny2 * nx2];
  // Step1
  zeropadding(u, upad, ny, nx);
  // Step2
  fft(upad, upad, ny2, nx2);
  multscalar(upad, 1.0 / (ny2 * nx2), ny2, nx2);
  // Step3
  AsmTransferF(H, ny2, nx2, dv, du, lambda, z);
  fftshift(H, ny2, nx2);
  // Step4
  mult(upad, H, upad, ny2, nx2);
  // Step5
  ifft(upad, upad, ny2, nx2);
  // Step6
  crop(upad, u, ny2, nx2);

  delete[] upad;
  delete[] H;
}
```

8.2.3 GPU Implementation

Listing 8.8 shows an example of GPU implementation of the angular spectrum method. Each function call in the "prop" function in the C++ class "gAsmProp" corresponds to a step in the flow described in Sect. 8.2.1. The function "gAsmTransferF" computes the transfer function $H(u, v)$. The implementation of the other functions is the same as that in Listing 8.6.

Listing 8.8 Example of the GPU implementation of the angular spectrum method

```
__global__ void gAsmTransferFKernel(cufftComplex* H, int32_t ny, int32_t nx, float dv, float
     du, float lambda, float z)
{
  int32_t m = blockIdx.x * blockDim.x + threadIdx.x;
  int32_t n = blockIdx.y * blockDim.y + threadIdx.y;
  if ( (m < nx) && (n < ny) ){
    int32_t idx = n * nx + m;
    int32_t hnx = nx / 2;
    int32_t hny = ny / 2;
    float u = (m − hnx) * du;
    float v = (n − hny) * dv;
    float w = sqrt(1 / (lambda * lambda) − u * u − v * v);
    float phase = 2.0f * M_PI * w * z;
    H[idx] = make_cuComplex(cos(phase), sin(phase));
  }
}

void gAsmTransferF(cufftComplex* H, int32_t ny, int32_t nx, float dv, float du, float lambda,
     float z)
{
  dim3 block(16, 16, 1);
  dim3 grid(ceil((float)nx / block.x), ceil((float)ny / block.y), 1);
  gAsmTransferFKernel<<<grid, block>>>(H, ny, nx, dv, du, lambda, z);
  cudaDeviceSynchronize();
}

class gAsmProp{
  private:
    cufftComplex* buf1, *buf2;
    gFFT fft;
  public:
    gAsmProp(int32_t ny, int32_t nx){
      int32_t ny2 = ny * 2;
      int32_t nx2 = nx * 2;
      size_t mem_size = sizeof(cufftComplex) * ny2 * nx2;
      cudaMalloc((void**)&buf1, mem_size);
      cudaMalloc((void**)&buf2, mem_size);
      fft.set(ny2, nx2);
    }
    ~gAsmProp(){
      cudaFree(buf1);
      cudaFree(buf2);
    }
```

```
    void prop(cufftComplex* u, int32_t ny, int32_t nx,float dy, float dx, float lambda, float z){
        int32_t ny2 = ny * 2;
        int32_t nx2 = nx * 2;
        float du = 1 / (dx * nx2);
        float dv = 1 / (dy * ny2);
        // Step1
        gzeropadding(u, buf1, ny, nx);
        // Step2
        fft.fft(buf1, buf1);
        gmultscalar(buf1, 1.0f / (ny2 * nx2), ny2, nx2);
        // Step3
        gAsmTransferF(buf2, ny2, nx2, dv, du, lambda, z);
        gfftshift(buf2, ny2, nx2);
        // Step4
        gmult(buf1, buf2, buf1, ny2, nx2);
        // Step5
        fft.ifft(buf1, buf1);
        // Step6
        gcrop(buf1, u, ny2, nx2);
    }
};
```

8.3 Results

Figure 8.6a and b shows an original image (1024 × 1024 pixels) and its diffracted result obtained using the angular spectrum method, respectively. The calculation conditions included a sampling interval of 8 μm, a wavelength of 633 nm, and a propagation distance of 0.02 m.

Finally, we compared the computation speed of the Fresnel diffraction calculation for three different cases, including a single-thread CPU, a multithreaded CPU, and

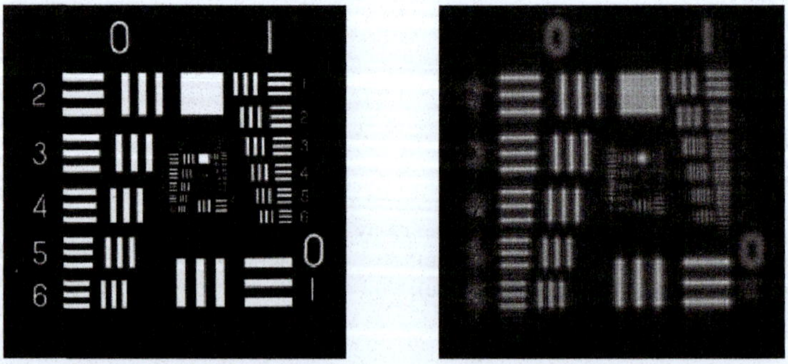

(a) Original image (b) Diffracted result

Fig. 8.6 Result of the angular spectrum method

Table 8.1 Speed of the Fresnel diffraction calculation

Number of pixels	CPU (1 thread) (ms)	CPU (16 threads) (ms)	GPU (ms)
2048 × 2048	2362	806	37
4096 × 4096	6109	1808	140
8192 × 8192	40031	13272	487

the GPU hardware. The computational environment used was a system with an Intel Corei9-11900K CPU and an NVIDIA GeForce RTX 3060 GPU. Table 8.1 shows the results of this comparison. Note that the measurement time for the GPU includes the time required to send and receive data between the CPU and the GPU. As shown in Table 8.1, the GPU had the lowest computation time for all numbers of pixels.

References

1. J. W. Goodman, "Introduction to Fourier optics (3rd.)" Roberts and Company Publishers (2005).
2. T. Shimobaba, T. Ito "Computer Holography" CRC Press (2019).
3. Matteo Frigo, Steven G. Johnson, "The Design and Implementation of FFTW3," Proceedings of the IEEE 13(5), 216–231 (2005).
4. FFTW, "https://www.fftw.org/"
5. E. Ayguade et al., "The Design of OpenMP Tasks," IEEE Transactions on Parallel and Distributed Systems 20(3), 404–418 (2009).
6. cuFFT, "https://docs.nvidia.com/cuda/cufft/index.html"

Chapter 9
Acceleration of CGH Computing from Point Cloud for CPU

Takashige Sugie

Abstract We practically accelerate the calculation for computer-generated hologram (CGH) by using seven types of programs as examples. The source code is improved step by step, finally achieving a speed increase of approximately 140 times. First, we show the importance of carefully choosing variable types. Next, we use OpenMP to implement multithreading and SIMD instruction parallel processing. We speed up the calculation by applying an algorithm to the calculation formula that is suitable for the CPU. We shorten high latency instructions using the table look-up method. It is important to supply data to the CPU at high speed to achieve high-speed computation. We discuss the reduction of communication bandwidth between CPU and system memory. Finally, we show how to calculate decimals using integer variables.

9.1 Introduction

In this chapter, specific programs for **computer-generated hologram (CGH)** are created and explained in tutorial format. Seven types of programs have been prepared and are gradually improved to allow high-speed calculations to be performed. We evaluated the amount of processing time and memory required to convert the coordinate data of object points into CGH pixels. The personal computer (PC) shown in Table 9.1 is used as a test machine, and we used C programming language. We used Linux [1] as the operating system and GNU Compiler Collection (GCC) [2] as the compiler. Another major compiler is Intel C/C++ Compiler (ICC) [3]. Both compiler optimizations are excellent, with the little difference in the performance of the executable program. The author occasionally encounters people who claim that

Supplementary Information The online version contains supplementary material available at https://doi.org/10.1007/978-981-99-1938-3_9.

T. Sugie (✉)
Chiba University, 1-33 Yayoi-cho, Inege-ku, Chiba 263-8522, Japan
e-mail: myrhrk@gmail.com

T. Shimobaba and T. Ito (eds.), *Hardware Acceleration of Computational Holography*,
https://doi.org/10.1007/978-981-99-1938-3_9

Table 9.1 Test machine

CPU	i7-7700 K (4 cores, 4.50 GHz)
System memory	32 GiB
OS	linux-5.4.2 (x86_64), glibc-2.30
Compiler	gcc-9.2.0

changing the compiler increases the speed. This is because they wrote a program that could be improved, and in most cases, the problem is with the source code rather than the compiler.

The calculation time depends on the number of object points and the resolution of the hologram plane. If the calculation time varies significantly, we should check the execution environment. We should check with the "top" or "ps" command whether any heavy program is running in the background, such as daemons. If it exists, we perform appropriate action such as terminating the process. If the computation time still varies, we must check the functions and settings of the kernel. Depending on the setting of the preemption model for interactive events and the governor for the CPU operating frequency, the performance may change owing to the influence of the runtime situation. This workaround requires recompiling the kernel and we can ignore it if we are unsure. It is important to note that kernel tuning can affect computation time.

9.2 Sample Source Code Package

9.2.1 How to Download and Build the Sample Source Code Package

A sample source code package is available on the book site. We decompress the downloaded sample program package and move to "sample" directory. Directories corresponding to the contents of each subsection can be found in Sect. 9.3. Their directory names start with a number from one to seven. The "appendix" directory contains auxiliary source files for explanation. We can use the "make" command to build all sample programs by inputting the following command on a terminal software:

```
% make
```

An execution command "cgh" is generated in each directory. This command can be executed by specifying the object point cloud data file as an argument. The "cgh" command calculates the CGH pixels, normalizes the result to 256 gradations, and outputs it as a BMP format file [4]. To simplify the program, the "cgh" command can read an object point cloud data file of 3df format only. The 3df format is described in Sect. 9.2.3. Each directory contains several source files. The principal program is written in cgh.c. We only need to read cgh.c.

9.2.2 Compile Optimization Options

The current CPU has a highly parallel architecture. As mentioned in Chap. 4, the CPU has parallelism such as pipeline processing of instruction, throughput and latency of instruction, superscalar, SIMD processing, and multi-core. It is very difficult to schedule combinations of instructions that can be calculated efficiently and correctly while considering all of these parallel architectures. The compiler solves this problem better than trial and error in the source code. Therefore, the compiler option that specifies the strength of optimization to the compiler is important. The best known optimization option is "-O". This option is usually used with an optimization level. Optimization levels range from zero to seven, with up to three currently implemented. The optimization option for level three is "-O3" and strong optimization is applied, whereas, "-O0" means no optimization.

The safest level is two, which generates a stable executable file. Because level three is an optimization that assumes a numerical computation algorithm, the compiler misinterprets program algorithms in extremely rare cases. Compiler optimization is weak for algorithms that are difficult to predict program flow. For example, an algorithm in which the loop condition changes depend on the calculation result in the loop, or jump from inside a loop into another loop depending on the condition.

However, "-Ofast" is a stronger optimization option that is not a numerical level specification. The -Ofast option attempts to optimize ignoring strict standard compliance. We can expect even faster than the "-O3" option. It is worth using "-Ofast" for numerical calculation programs. We must not forget that optimization options stronger than "-O2" may generate instructions that are not what we intended. If the behavior is strange using "-Ofast", we change to "-O2" and check the operation. Alternatively, we use "-Ofast" after confirming that the correct result is obtained with "-O2". We use "-Ofast" to compile the sample programs in this chapter.

Even with the "-O" option alone, we can obtain a good optimization effect. In addition, if we notify the compiler the hardware environment to run, it may be faster. The compiler optimizes using all the technologies and functions allowed in that environment. The compiler options are "-march=native" and "-mtune=native". Because the execution code depending on a specific machine is generated, the compatibility is low. It does not guarantee that the program functions correctly even on machines with the same CPU series. To improve compatibility, we change the parameter from "native" to "generic". Three compiler options for optimization are sufficient: "-Ofast -march=native -mtune=native".

9.2.3 Main Program Flow

Here is a brief explanation of the program flow in the main function. The main function is described in cgh.c.

Listing 9.1 Main function

```c
int main(int argc, char **argv)
{
  FILE *obj_file;
  int32_t n; // number of object points
  float (*obj_xyz)[3];
  struct timespec ts_start, ts_diff;
  char *bmp_fn;

  if (argc != 2) return (-EINVAL);

  obj_file = fopen(argv[1], "r");
  if (!obj_file) return (-ENOENT);
  fread(&n, sizeof (n), 1, obj_file);
  if (n <= 0) {
    fclose(obj_file);
    printf("Too few object points\n");
    return (-EINVAL);
  }
  if (MAX_N_POINTS < n) {
    fclose(obj_file);
    printf("Too many object points\n");
    return (-EINVAL);
  }
  obj_xyz = (float (*)[3])malloc(sizeof (float) * 3 * n);
  if (obj_xyz) fread(obj_xyz, sizeof (float) * 3 * n, 1, obj_file);
  fclose(obj_file);
  if (!obj_xyz) return (-ENOMEM);

  stopwatch_get_time(&ts_start);
  xyz_to_psi(n, obj_xyz);
  stopwatch_diff_from(&ts_start, &ts_diff);
  printf("xyz(%'d) => psi(%'dx%'d): %'41d.%091d sec.\n", n,
      LCD_WIDTH, LCD_HEIGHT, ts_diff.tv_sec, ts_diff.tv_nsec);

  free(obj_xyz);

  bmp_fn = malloc(strlen(argv[1]) + sizeof (".bmp"));
  if (!bmp_fn) return (-ENOMEM);
  strcpy(bmp_fn, argv[1]);
  strcat(bmp_fn, ".bmp");
  save_bitmap(bmp_fn);
  free(bmp_fn);

  return (0);
}
```

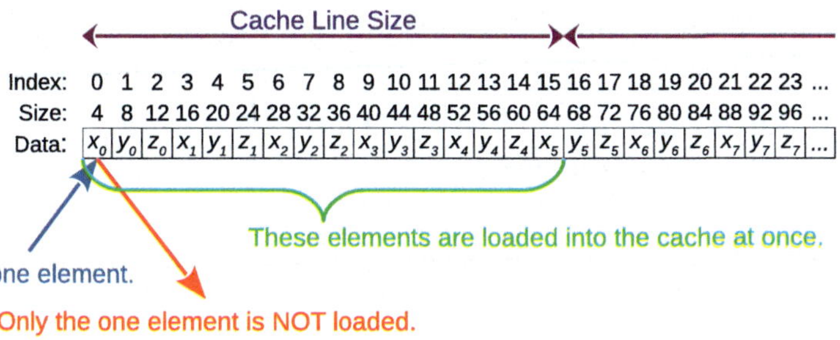

Fig. 9.1 Image diagram of the coordinate data format of the object point considering the cache line

Only the minimum necessary processing is implemented to keep the program for becoming highly complex. The processing is roughly divided into three.

1. Read the coordinate data of the object points from a file.
2. Convert from coordinate data to CGH pixels.
3. Save it to a bitmap format file.

The file format of the file in which the coordinate data of the object point is recorded is simple. The number of object points is recorded as a 32-bit integer at the beginning. Subsequently, float type (32-bit) coordinate data is recorded in the order of x, y, and z. Then, the x, y, and z coordinate data are repeated for the number of object points. Therefore, we read 32 bits first to obtain the number of object points. Next, we allocate memory to hold the coordinate data and read the coordinate data from the file.

A program that converts coordinate data into CGH pixels is written in the xyz_to_psi function. In the next section, we will improve the processing in this function and increase the calculation speed. We will evaluate the memory size and processing time required to execute this function.

Finally, we normalize the CGH pixels to 256 gradations and save it as a bitmap format file. The output file name is the input file name with ".bmp" appended.

We consider the coordinate data format of an object point. The coordinate data has three variables, x, y, and z, and each variable type is float type (4 bytes). CGH calculation can be performed by the coordinate data of an object point. The calculation is possible with loading only 12 bytes of 3D coordinate data. We need to focus on data cache misses. When the instruction to load the 0th data is executed, 64 bytes including it are read into the cache memory. If the subsequent data is used immediately, the probability of being in the cache increases, and we can expect an improvement in performance. If the data is not used immediately, the probability of being deleted from the cache memory increases. Because the calculations are performed using the variables x, y, and z in order, it is recommended to prepare the coordinate data of the object point as shown in Fig. 9.1.

9.3 Implementation of the xyz_to_psi Function

9.3.1 Direct Calculation Using Double-Precision Floating-Point Type

The CGH pixels $\psi(x_\alpha, y_\alpha)$ at the hologram plane (x_α, y_α) is obtained by

$$\psi(x_\alpha, y_\alpha) = \sum_{j}^{N_{obj}} \cos\left(2\pi \frac{p}{\lambda} \frac{(x_\alpha - x_j)^2 + (y_\alpha - y_j)^2}{2|z_j|}\right), \tag{9.1}$$

where (x_j, y_j, z_j) are the discretized coordinates of the j-th point consisting the object, p is the sampling rate of the hologram plane (corresponding to the pixel pitch of the spatial light modulator (SLM)), λ is the wavelength of the reference light, and N_{obj} is the total number of points consisting the object. We can obtain the hologram by repeatedly calculating Eq. (9.1) for all the pixels consisting the hologram. Listing 9.2 is a program that describes Eq. (9.1) in C programming language.

Listing 9.2 xyz_to_psi function using the double-precision floating-point type

```c
float psi[LCD_HEIGHT][LCD_WIDTH];

void xyz_to_psi(int32_t obj_n, float (*obj_xyz)[3])
{
  int xa, ya, n;
  float xj, yj, zj;

  for (ya = 0; ya < LCD_HEIGHT; ya++) {
    for (xa = 0; xa < LCD_WIDTH; xa++) {
      psi[ya][xa] = 0.0;
    }
  }

  for (ya = 0; ya < LCD_HEIGHT; ya++) {
    for (xa = 0; xa < LCD_WIDTH; xa++) {
      for (n = 0; n < obj_n; n++) {
        xj = obj_xyz[n][0] * XYZ_EXP + X_OFFSET;
        yj = obj_xyz[n][1] * XYZ_EXP + Y_OFFSET;
        zj = obj_xyz[n][2] * XYZ_EXP + Z_OFFSET;
        psi[ya][xa] += cos(2.0 * M_PI * (LCD_DOT_PITCH / LAMBDA) * ((xa − xj) * (xa − xj) + (ya − yj) * (ya − yj)) / (2.0 * fabs(zj)));
      }
    }
  }
}
```

First, we initialize the two-dimensional array psi with zeros. Next, we prepare a triple for-loop to calculate all hologram pixels using all object points. Because Eq. (9.1) normalizes the coordinates with the pixel pitch of the hologram, we can simply increment x_α and y_α. In the triple for-loop, we calculate CGH pixels using Eq. (9.1).

Table 9.2 Performance of the xyz_to_psi function using the double-precision floating-point type

N_{obj}	Calculation time [s]	Used memory [MiB]	$(12N_{obj} + 4N_{hol})$
710	28.42	7.92	$(0.01 + 7.91)$
44,647	1,828.25	8.42	$(0.51 + 7.91)$
978,416	40,804.55	19.11	$(11.20 + 7.91)$

We evaluate the calculation time for obtaining $1{,}920 \times 1{,}080$ hologram from the coordinate data of the points consisting of the object, and the amount of memory used at that time. At the beginning of cgh.c, the hologram size and the wavelength of the reference light are defined. We observe three kinds of objects consisting of approximately 1,000 points, 40,000 points, and 1,000,000 points. The computer used for the evaluation is the test PC shown in Table 9.1. Table 9.2 shows the performance of Listing 9.2, which is the standard program in this subsection. N_{hol} is the number of pixels in the hologram. The calculated hologram image is shown in Sect. 9.4.

We can calculate the hologram by using a buffer that stores the coordinate data of the object point and the calculation result. The amount of memory used is as follows.

Point cloud data size: size of float type $\times$ 3D coordinates $\times$ number of the object points = $12N_{obj}$
Hologram data size: size of float type $\times$ number of pixels of the hologram = $4N_{hol}$

Therefore, the amount of memory used in xyz_to_psi function is $12N_{obj} + 4N_{hol}$.

Listing 9.2 is the most straightforward program and also requires a very long calculation time. We will investigate the problem, considering why this program is slow, based on the functions and features of the CPU introduced in Chap. 4 or the behavior of the compiler and programming language characteristic. From the next subsection, we will improve the xyz_to_psi function step by step.

9.3.2 Direct Calculation Using Single-Precision Floating-Point Type

We improve Listing 9.2 in the previous subsection. The improvement is simple and we only need to use float types instead of using double types. We may think that the program does not originally have a double type. In addition, the double type is used. Even though the constant 0.0 in the initialization of the two-dimensional array psi is not a problem, constants such as 2.0 and the actual value of M_PI are the double type. The coordinate data is a float type, although when the calculation progresses and a double type is used, it changes to processing using the double type to keep accuracy. Furthermore, we use the cos function and the fabs function of the math library. Because both functions perform calculations with double precision, we also use the double type here. We may use different types unconsciously if we are not careful.

Listing 9.3 is a program that rewrites Listing 9.2 to allow it to calculate only with the float type. We add suffix to constants to clarify the type. If the value is a float type constant, we append suffix 'f'. For variables, we convert the type using the cast operator. The cast operator is a prefix of parentheses. We write the converted type in parentheses. We should use the cosf function instead of the cos function and the fabsf function instead of the fabs function.

Listing 9.3 xyz_to_psi function using the single-precision floating-point type

```
float psi[LCD_HEIGHT][LCD_WIDTH];

void xyz_to_psi(int32_t obj_n, float (*obj_xyz)[3])
{
  int xa, ya, n;
  float xj, yj, zj;

  for (ya = 0; ya < LCD_HEIGHT; ya++) {
    for (xa = 0; xa < LCD_WIDTH; xa++) {
      psi[ya][xa] = 0.0f;
    }
  }

  for (ya = 0; ya < LCD_HEIGHT; ya++) {
    for (xa = 0; xa < LCD_WIDTH; xa++) {
      for (n = 0; n < obj_n; n++) {
        xj = obj_xyz[n][0] * XYZ_EXP + X_OFFSET;
        yj = obj_xyz[n][1] * XYZ_EXP + Y_OFFSET;
        zj = obj_xyz[n][2] * XYZ_EXP + Z_OFFSET;
        psi[ya][xa] += cosf(2.0f * (float)M_PI * (float)(LCD_DOT_PITCH / LAMBDA) * ((xa - xj) * (xa - xj) + (ya - yj) * (ya - yj)) / (2.0f * fabsf(zj)));
      }
    }
  }
}
```

With only this modification, we can improve the calculation speed. The measurement results are shown in Table 9.3. The acceleration rate is a magnification that is accelerated from Listing 9.2 shown in Sect. 9.3.1. The value in parentheses is a magnification that is accelerated from the program shown in previous subsection. From the results, we can observe that we can obtain about 1.6 times faster by focusing on the type for data. Because most instructions have a lower latency for the 32-bit float type than for the 64-bit double type, we can accelerate the computation. We should always attempt to program while focusing on the word length of variables, rather than randomly choosing a variable type. The word length of variables is related to the data pack for executing SIMD operations, and directly affects the performance of parallel processing. Therefore, we should choose the variable type carefully.

Type conversion is not possible without computational cost. Computational cost is incurred each time a type conversion is performed. The problem is that using

Table 9.3 Performance of the xyz_to_psi function using the single-precision floating-point type. The value in parentheses is a magnification that is accelerated from the program shown in previous subsection

N_{obj}	Calculation time [s]	Acceleration rate	Used memory [MiB]	$(12N_{obj} + 4N_{hol})$
710	17.40	1.63 (1.63)	7.92	(0.01 + 7.91)
44,647	1,148.72	1.59 (1.59)	8.42	(0.51 + 7.91)
978,416	26,663.88	1.53 (1.53)	19.11	(11.20 + 7.91)

type conversion in an iterative process such as for-loop significantly increases the overhead, which cannot be ignored. For example, in Listing 9.3, the computational cost of type conversion is $O(N_{obj} \times N_{hol})$. We should avoid using cast operators in programs.

Furthermore, the cast operator is used in Listing 9.3, although the timing at which the type conversion is performed is different. A cast operator is attached to a constant in Listing 9.3. Constants do not need to be typecast at runtime. This type conversion does not affect the calculation speed because it can be calculated when compiling.

9.3.3 Multi-thread and SIMD Programming Using OpenMP

To enable a multi-core CPU to perform parallel processing, we create multiple threads and assign them to each CPU core for processing. In parallel processing using SIMD, we write programs using CPU-dependent instructions such as AVX [5]. Multithreaded programming is difficult to control threads. Because a program using SIMD instructions is not similar to mathematical expressions, the readability of the source code is poor. Therefore, in this chapter, we perform parallel processing using **OpenMP** [6].

OpenMP is a tool that can implement parallel processing relatively easily even with little specialized knowledge of hardware. If we inform the compiler that we use OpenMP, **SIMD** instructions may be generated without writing a SIMD instruction program in the source code. The author assumes that the compiler understands the program well and generates SIMD instructions. Therefore, even if we do not intentionally program SIMD processing, the performance of the program is nearly the same. Therefore, this subsection focuses on the implementation of multi-thread processing using OpenMP. This subsection explains the preferred data for SIMD instructions, not programs for generating SIMD instructions.

When using OpenMP, we prepare in two simple ways. We add "-fopenmp" to the compile options and include "omp.h" in the source file. The programming method is a way to parallelize using the pragma directive. The program part immediately following the pragma directive is converted into a parallel program by OpenMP. The pragma directive itself is supported by C programming language. It is a preprocessor instruction for passing specific information to the compiler, not a function provided by OpenMP. From here, we understand the pragma directives in Listing 9.4.

Listing 9.4 xyz_to_psi function using OpenMP

```
#define SIMD_ALIGN     (256 / 8)     // avx2: 256 bits

float psi[LCD_HEIGHT][LCD_WIDTH] __attribute__((aligned(SIMD_ALIGN)));

void xyz_to_psi(int32_t obj_n, float (*obj_xyz)[3])
{
  int32_t n;

#pragma omp parallel for // (1st)
  for (int32_t ya = 0; ya < LCD_HEIGHT; ya++) {
    for (int32_t xa = 0; xa < LCD_WIDTH; xa++) {
      psi[ya][xa] = 0.0f;
    }
  }

#pragma omp parallel for simd private(n) // (2nd)
  for (n = 0; n < obj_n; n++) {
    obj_xyz[n][0] = obj_xyz[n][0] * XYZ_EXP + X_OFFSET;
    obj_xyz[n][1] = obj_xyz[n][1] * XYZ_EXP + Y_OFFSET;
    obj_xyz[n][2] = obj_xyz[n][2] * XYZ_EXP + Z_OFFSET;
  }

#pragma omp barrier // (3rd)

#pragma omp parallel for private(n) reduction(+:psi) // (4th)
  for (int32_t ya = 0; ya < LCD_HEIGHT; ya++) {
    for (int32_t xa = 0; xa < LCD_WIDTH; xa++) {
      float xaj, yaj;
      for (n = 0; n < obj_n; n++) {
        xaj = xa - obj_xyz[n][0];
        yaj = ya - obj_xyz[n][1];
        psi[ya][xa] += cosf(2.0f * (float)M_PI * (float)(LCD_DOT_PITCH / LAMBDA) * (
      xaj * xaj + yaj * yaj) / (2.0f * fabsf(obj_xyz[n][2])));
      }
    }
  }
}
```

The first **pragma directive** (#pragma omp parallel for) converts the following for-loop to the parallel processing using **multithread**. We use the parallel clause to create multi-threads using OpenMP. The first pragma directive also specifies the parallelization target, which is the for clause that follows the parallel clause. OpenMP divides the for-loop processing that immediately follows the pragma directive and assigns it to each thread. By default, OpenMP adjusts the number of threads to use all CPU logical cores. If we know an efficient load balancing method from the characteristics of the algorithm, we can specify a parallel schedule. We refer to

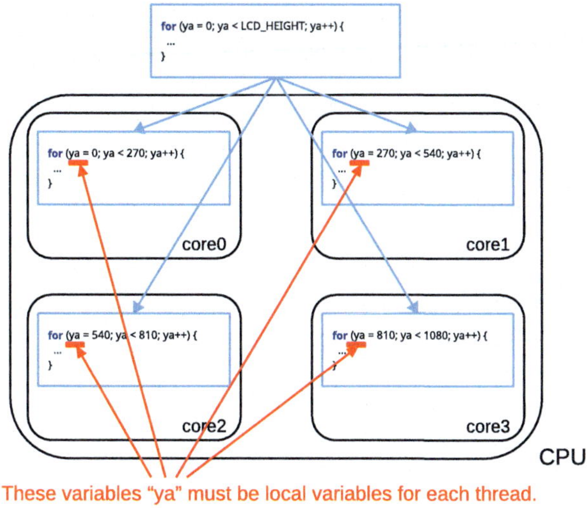

Fig. 9.2 Thread local variables

the OpenMP reference guide [7] for details. In this way, we can perform parallel processing by inserting only one line of pragma directives.

We must design programs while always considering the situation where multiple programs are executed simultaneously. For example, we consider how the variable ya, which is used in the for-loop is accessed. Figure 9.2 shows the image when the processing of the for-loop statement of the loop counter ya is equally allocated to the 4-core CPU. Because each CPU core processes different parts of the for-loop, the loop counter ya must be a local variable for each thread. If we declare the variable ya at the beginning of the xyz_to_psi function, as in Listing 9.3, all threads use that single variable ya, preventing the correct processing. Therefore, we declare the variable ya in the for statement to allow it to become a local variable in the thread. The same applies to the variable xa, and we rewrite the program to allow the variable xa to become a local variable of the thread. OpenMP simply rewrites the source code part to be parallelized into a program that uses threads. OpenMP does not understand exactly how the variables are used after parallelization. We must declare variables by assuming exactly what happens when the program is processed in parallel.

The second pragma directive (#pragma omp parallel for simd private (n)) contains nearly the same contents as the first and is intentionally changed for a supplementary explanation. We can also create thread local variables using the OpenMP private clause. The variable n should be prepared for each thread, although it is declared at the beginning of the xyz_to_psi function. The second pragma directive uses a private clause with the variable n. OpenMP replaces the variable n with the local variable of each thread. The difference to the first pragma directive is whether C programming language grammar is used to localize variable scope or OpenMP functionality is used.

We can use either one. The simd clause analyzes whether the for-loop is replaced by a SIMD instruction. OpenMP generates a SIMD instruction if possible.

Owing to parallel processing, different degrees of progress exist. Even if we distribute the computational load evenly to the threads, the progress speed is different in each thread. Processing of the xyz_to_psi function includes initialization of the two-dimensional array psi, correction of coordinate data, and calculation of the CGH pixels. Figure 9.3 shows an example of the flow in which four threads process the xyz_to_psi function. Figure 9.3 has two figures at the top and bottom. The top figure shows an example of an error that can occur when the program is simply executed. The bottom figure shows the state of appropriate processing for the error. The arrows in the figure indicate the processing progress position. In the top figure, Thread0 and Thread1 are in the process of initializing the two-dimensional array psi, and Thread2 is performing the correction calculation of coordinate data. Because there is no dependency between the initialization of the two-dimensional array psi and the correction calculation of coordinate data, we can obtain the correct result regardless of the order of processing. However, we must not start the calculation of the CGH pixels similar to Thread3. To calculate the CGH pixels correctly, the two-dimensional array psi must be initialized to zero, and the coordinate data must be converted to the correct corrected value. Even if we distribute the computational load equally among the threads, the degrees of processing progress are extremely different. One of the main reasons is that the CPU usage right is deprived by other processes because the OS is multitasking. Therefore, to enter the for-loop that calculates the CGH pixels, we must ensure that all threads have reached that point. The **barrier** clause of the third pragma directive (#pragma omp barrier) prevents execution of the program until all threads reach there. It acts as a barrier as shown in Fig. 9.3. After all threads are

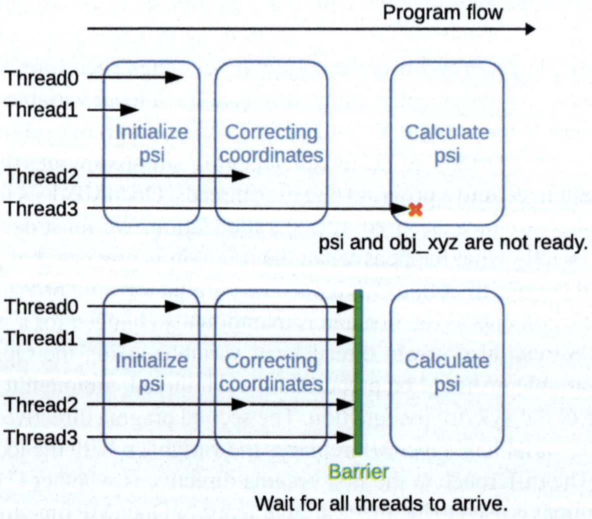

Fig. 9.3 Processing synchronization using barrier

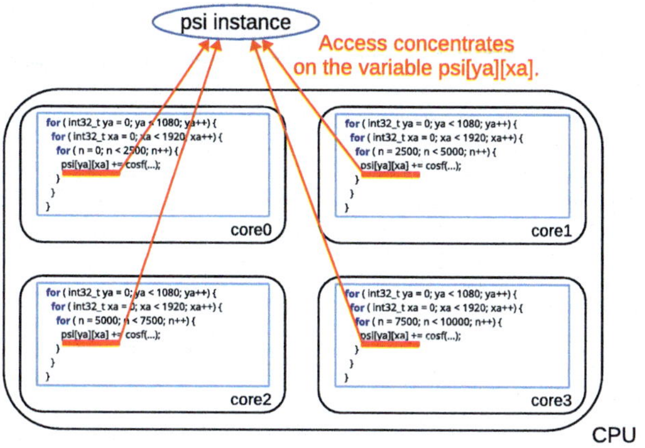

Fig. 9.4 Image diagram of a situation in which multiple threads access the same address

completed the processing up to that point, parallel processing by the fourth pragma directive is started.

We can parallelize the for-loop that calculates the CGH pixels with one line of pragma directive. The variable n is declared as a local variable, as shown on line 25 of Listing 9.4. In other words, in this for-loop, the processing is faster by parallelizing for the variable n. However, in contrast to the first and second directives, inefficient processing is included. Figure 9.4 shows an example of that processing. It shows a situation using a 4-core CPU. Each CPU core calculates the cosine and summates the result to the element of the two-dimensional array psi. The problem is that multiple CPU cores simultaneously access the exact same element in the two-dimensional array psi depending on the timing of processing. The program part written about variables accessed from multiple threads is called a critical section. Critical sections require exclusive processing to maintain the integrity of shared variables. Exclusive processing is processing that restricts access to shared variables by only one thread. Although it is necessary for correct calculation, calculation efficiency decreases because the time between calculations increases. Therefore, we process the superposition for the element of the two-dimensional array psi in two steps as shown in Fig. 9.5. We prepare one local variable for each thread. Each thread performs a summation operation on its local variable. Because each thread is always ready to perform a summation operation, the CPU does not need wait for acquiring the access right and can execute effective processing continuously. When the calculation of the CGH pixels of the object point assigned to each thread is completed, the local variables of each thread are summated as the next step. Therefore, we obtain a complete value that is summated from all object points. Finally, we only need to write it once for the element of the two-dimensional array psi.

OpenMP can also generate programs such complicated processing. As in the fourth pragma directive (#pragma omp parallel for private (n) reduction (+:psi)), we can inform OpenMP using the **reduction** clause. First, we specify an operator that

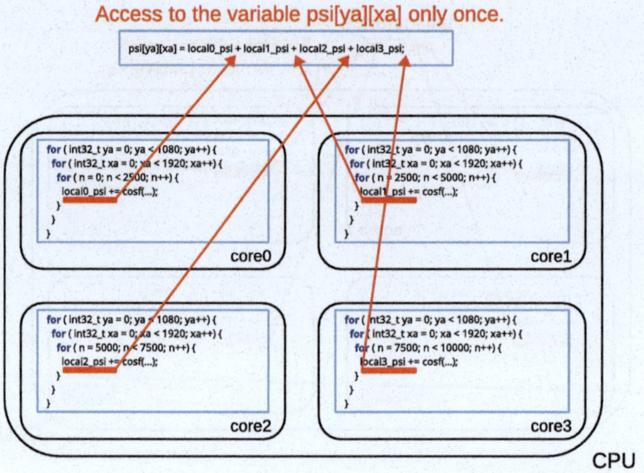

Fig. 9.5 Image diagram of reducing access to shared variables using the reduction clause

calculates the local variables of each thread. In the calculation of the two-dimensional array psi, we specify a plus operator because it is a summation calculation. OpenMP prepares a thread local variable and modifies the program to prevent it from becoming a critical section. Because the thread local variable is one variable in this study, the probability that it can exist in the low-level cache of each CPU core increases. Therefore, we can also expect the effects of reducing access to the system memory and cache misses. It is very important to use the reduction clause when there are many accesses to shared variables in parallel processing.

SIMD instructions can perform multiple operations of the same type with a single instruction. To generate SIMD instructions, we need a program that performs the same process repeatedly. Compilers can find efficiently such program parts. However, if we do not prepare for the SIMD calculation, smooth operation may be prevented. Operands used in SIMD instructions should be adjusted to an address where the CPU can easily process. Because SIMD instructions require packed data for operands, the CPU always accesses block data of several bytes. For AVX instructions, at least 32 bytes must be read and written. In addition, a 64-bit CPU operates using 64 bits data as a basic block. When we allocate memory in a normal procedure, the starting address is a position where at least 8 bytes can be read and written in one access. However, if the SIMD register size is the basic block, the address granularity increases. We should allocate memory from addresses that can simultaneously access the large size block. This is because a boundary exists in the memory device. When accessing across the boundary, the access occurs twice before and after the boundary. Regarding the existence of boundaries, we can easily understand that the system memory is composed of multiple memory chips rather than one large memory device. We should align addresses with SIMD register size to avoid performance degradation. Address alignment can be performed by two methods: using C programming language attribute to inform the compiler, and using a memory allocation function that

Table 9.4 Performance of the xyz_to_psi function using the OpenMP. The value in parentheses is a magnification that is accelerated from the program shown in previous subsection

N_{obj}	Calculation time [s]	Acceleration rate	Used memory [MiB]	$(12N_{obj} + 4N_{hol})$
710	0.38	74.14 (45.38)	7.92	$(0.01 + 7.91)$
44,647	22.04	82.97 (52.13)	8.42	$(0.51 + 7.91)$
978,416	1,892.76	21.56 (14.09)	19.11	$(11.20 + 7.91)$

can perform address alignment. The variable psi on line 3 of Listing 9.4 performs address alignment by adding attributes at the time of declaration. The value assigned to "aligned" is the amount of address to be aligned, that is, the amount of bytes in the SIMD register size. In the main function, the memory allocation of the coordinate data of the object point is rewritten to use the aligned_alloc function, which is the C11 standard [8]. For SIMD instruction implementations using OpenMP, we only focus on memory allocation. Although the compiler may not generate the SIMD instructions as expected, it is usually generated well.

By implementing parallel processing, we can expect significant performance improvements. The CPU of the test machine has four physical cores, and we can use AVX2 for SIMD instructions. The AVX2 instruction can process data packed with 8 float type data. Therefore, we can expect a speedup of approximately 32 times. In addition, in Listing 9.4, we consider coordinate correction calculations out of the triple for-loop and eliminate unnecessary calculations. The results are shown in Table 9.4. We have achieved higher speed than expected except for an object consisting of one million points. Thus, we can expect a significant improvement in computation speed by parallel processing. Implementing parallel processing is indispensable to fully exploit CPU performance.

9.3.4 Recurrence Algorithm Implementation

In this subsection, we further increase the processing speed by improving the calculation algorithm. We have calculated the distances between all object points and pixels of the hologram. From here, we adopt an algorithm that can calculate nearly all of the hologram plane by a **recurrence formula** [9, 10]. Chapter 2 describes the detail of the recurrence algorithm. This algorithm eliminates many distance calculations and can reduce the amount of calculation. The recurrence formula is a sequential process. It is highly suitable for a computer architecture because it can continuously process the same instruction. However, we must be careful because the recurrence formula is a typical formula that cannot be processed in parallel.

We consider applying the recurrence formula toward the x-axis direction of the hologram plane. We calculate θ, Δ, and Γ at the first CGH point $(x_\alpha|_{\alpha=0}, y_\alpha)$ of each row that is the base point, using Eqs. (9.2), (9.3), and (9.4).

$$\theta_{\alpha j} = \frac{p}{\lambda}\frac{1}{2|z_j|}(x_{\alpha j}^2 + y_{\alpha j}^2) \tag{9.2}$$

$$\Delta_{\alpha j} = \frac{p}{2\lambda|z_j|}(2x_{\alpha j} + 1) \tag{9.3}$$

$$\Gamma_j = \frac{p}{\lambda|z_j|}, \tag{9.4}$$

where $x_{\alpha j}$ is $x_\alpha - x_j$. After the base points θ, Δ, and Γ are obtained, we can calculate the θ and Δ of the neighboring pixels by the recurrence formulas of Eqs. (9.5) and (9.6).

$$\theta_{(\alpha+n)j} = \theta_{(\alpha+n-1)j} + \Delta_{(\alpha+n-1)j} \tag{9.5}$$

$$\Delta_{(\alpha+n)j} = \Delta_{(\alpha+n-1)j} + \Gamma_j. \tag{9.6}$$

We can obtain the CGH pixels ψ by using θ as shown in Eq. (9.7).

$$\psi(x_\alpha, y_\alpha) = \sum_{j}^{N_{obj}} \cos(2\pi\theta_{\alpha j}) \tag{9.7}$$

Listing 9.5 describes these mathematical expressions in C programming language. We allocate memory to store gamma, delta, and theta for each object point necessary for the calculation. Because the calculations for delta and theta depend on the loop counters ya and xa, they must be thread local. By contrast, the calculation of gamma depends only on z_j (obj_xyz[n][2]) and can be performed in common for all threads. Therefore, we need to prepare only one variable as a buffer for gamma. However, it is possible that the same address is referenced from multiple threads and contends for access rights. In the sample program, we prepare gamma for each object point as array to avoid conflict. Although the amount of memory used increases, it becomes possible to process the recurrence formulas in parallel as will be described in the next paragraph. Then we calculate gamma, delta, and theta of the base point for all object points. Finally, we calculate the CGH pixels using the recurrence formula toward the x-axis.

Listing 9.5 xyz_to_psi function using the recurrence algorithm

```
#define SIMD_ALIGN     (256 / 8)     // avx2: 256 bits

float psi[LCD_HEIGHT][LCD_WIDTH] __attribute__((aligned(SIMD_ALIGN)));

void xyz_to_psi(int32_t obj_n, float (*obj_xyz)[3])
{
  const float ppl = LCD_DOT_PITCH / LAMBDA;
```

```

#pragma omp parallel for simd
  for (int32_t n = 0; n < obj_n; n++) {
    obj_xyz[n][0] = obj_xyz[n][0] * XYZ_EXP + X_OFFSET;
    obj_xyz[n][1] = obj_xyz[n][1] * XYZ_EXP + Y_OFFSET;
    obj_xyz[n][2] = obj_xyz[n][2] * XYZ_EXP + Z_OFFSET;
  }

  memset(psi, 0, sizeof (psi));

#pragma omp barrier

#pragma omp parallel for reduction(+:psi)
  for (int32_t ya = 0; ya < LCD_HEIGHT; ya++) {
    int32_t n, xa;
    float *gamma, *delta, *theta;
    float rho;

    gamma = aligned_alloc(SIMD_ALIGN, sizeof (float) * obj_n);
    delta = aligned_alloc(SIMD_ALIGN, sizeof (float) * obj_n);
    theta = aligned_alloc(SIMD_ALIGN, sizeof (float) * obj_n);

    for (n = 0; n < obj_n; n++) {
      rho = ppl / (2.0f * obj_xyz[n][2]);
      gamma[n] = rho * 2.0f;
      delta[n] = rho * (2.0f * (0.0f - obj_xyz[n][0]) + 1.0f);
      theta[n] = rho * ((0.0f - obj_xyz[n][0]) * (0.0f - obj_xyz[n][0]) + (ya - obj_xyz[n][1])
        * (ya - obj_xyz[n][1]));
    }

    for (xa = 0; xa < LCD_WIDTH; xa++) {
      for (n = 0; n < obj_n; n++) {
        psi[ya][xa] += cosf(2.0f * (float)M_PI * theta[n]);
        theta[n] += delta[n];
        delta[n] += gamma[n];
      }
    }

    free(theta);
    free(delta);
    free(gamma);
  }
}
```

At the beginning of this subsection, we mentioned that the recurrence formula is a mathematical formula that cannot be processed in parallel. Furthermore, no dependencies exist between object points in CGH calculations. In the for-loop for the loop counter n, psi[ya][xa] is a thread local variable by the reduction clause, and the executable condition of the program depends only on the object points. We can execute parallel processing by dividing the object point instead of dividing the hologram plane. If the gamma, delta, and theta of the base point for each object point are available, we can calculate in parallel. Therefore, we prepared as many gamma-rays as the number of object points.

The memory required for the calculation requires more buffers to store gamma, delta, and theta. The test machine had eight logical CPUs. The buffer sizes were as follows:

The amount of memory used increases by $96N_{obj}$ bytes. The amount of memory required is highly dependent on the number of object points. This is a major problem for high-definition objects. In this case, the calculation is impossible, regardless of the calculation speed. We solve this problem in Sect. 9.3.6.

Buffer size of gamma: number of threads × size of float × number of the object points = $32N_{obj}$
Buffer size of delta: number of threads × size of float × number of the object points = $32N_{obj}$
Buffer size of theta: number of threads × size of float × number of the object points = $32N_{obj}$

Table 9.5 lists the performances. By adopting an algorithm suitable for the CPU, the calculation speed nearly doubled.

9.3.5 Latency Reduction with Look-Up Table

We can achieve a speedup of more than 100 times compared to the program using the double type first introduced in Listing 9.2. Furthermore, to effectively speed up the calculation, we must understand the computational complexity in the program. We consider whether processing can be reduced by a heavily loaded program. The recurrence formula consists of addition operations with simple computational complexity. Thus, we focus on the cosine calculation. The calculation of the trigonometric functions is extremely complicated, therefore, the CPU is equipped with a vector-type cosine arithmetic circuit. Furthermore, the cosine function in glibc [11] (in math.h) simply passes arguments to a vector-type cosine arithmetic circuit. Despite using a vector-type circuit, it has a **latency** of over 150 [12]. This calculation is time-

Table 9.5 Performance of the xyz_to_psi function using the recurrence algorithm. The value in parentheses is a magnification that is accelerated from the program shown in previous subsection

N_{obj}	Calculation time [s]	Acceleration rate	Used memory [MiB]	$(12N_{obj} + 4N_{hol} + 96N_{obj})$
710	0.21	136.96 (1.85)	7.98	(0.01 + 7.91 + 0.07)
44,647	11.20	163.28 (1.97)	12.51	(0.51 + 7.91 + 4.09)
978,416	1,908.04	21.39 (0.99)	108.68	(11.20 + 7.91 + 89.58)

consuming. Because this is in the deepest for-loop, we need to improve it in some way.

We reduce the latency by allowing cosines to be calculated using memory. The basic operation of memory is that when an address is input, the corresponding data are output. We regard the memory address as an argument for the cosine function. If we write the cosine value corresponding to the argument as data in the memory, the memory reference is equivalent to calculating the cosine. A data array for data conversion is called a **look-up table** (**LUT**). For example, if data are output at the next clock given an address to the memory, the latency is one, and the calculation speed is remarkable. In practice, the latency is greater than one because of the hierarchical structure of the cache memory and pipeline processing of instructions. Because the LUT must be able to be referenced at high speed, it must be sufficiently large to be stored in cache memory. Moreover, the size must be sufficiently small that LUT data are not frequently cached out by other data. Therefore, the table size is limited, which affects the size of the address space and the accuracy of the data. Nevertheless, if we design to meet these requirements, latency can be reduced. We can expect a sufficient speedup.

The sample program uses an LUT that uses 8 bits for the number of elements and stores float type data (32 bits). The number of elements empirically determines the accuracy with which the original object can be observed when we study the reconstructed image. We sample 2π into 256 parts and store the corresponding cosine values as LUT data. The LUT size is $256 \times 4 = 1,024$ bytes. This is a practical size, considering that the L1 data cache size is 32 KiB. Listing 9.6 shows the make_cos_table function that creates an LUT. We create the LUT by storing the result of the cosf function in a one-dimensional array that has 256 float type data. We executed this function before executing the xyz_to_psi function. In the xyz_to_psi function, we rewrite the program to refer to the one-dimensional array cos_table, instead of calling the cosf function, as shown in Listing 9.7. The variable theta is considered as an array index, not as a function argument. Because the index must be an integer that does not exceed the number of elements in an array, we only need to extract the lower 8 bits of the variable theta. It is extremely easy to process because it is a calculation of the bitwise AND of variables theta and 0xff. This operation is no problem because trigonometric functions are periodic functions.

Listing 9.6 make_cos_table function to create LUT of float type data

```
#define COS_DEPTH     8 // [bit]
#define COS_N        (0x1 << COS_DEPTH)
#define COS_MASK      (COS_N - 1)

float cos_table[COS_N];

void make_cos_table()
{
  int i;

  for (i = 0; i < COS_N; i++) cos_table[i] = cosf(2.0 * M_PI * i / COS_N);
}
```

Listing 9.7 xyz_to_psi function using the Look-Up Table

```
#define SIMD_ALIGN      (256 / 8)    // avx2: 256 bits
#define FtoI(n)         (((n) < 0.0f) ? (int32_t)((n) - 0.5f) : (int32_t)((n) + 0.5f)) // float to int32_t

float psi[LCD_HEIGHT][LCD_WIDTH] __attribute__((aligned(SIMD_ALIGN)));

void xyz_to_psi(int32_t obj_n, float (*obj_xyz)[3])
{
  const float ppl = LCD_DOT_PITCH / LAMBDA;

#pragma omp parallel for simd
  for (int32_t n = 0; n < obj_n; n++) {
    obj_xyz[n][0] = obj_xyz[n][0] * XYZ_EXP + X_OFFSET;
    obj_xyz[n][1] = obj_xyz[n][1] * XYZ_EXP + Y_OFFSET;
    obj_xyz[n][2] = obj_xyz[n][2] * XYZ_EXP + Z_OFFSET;
  }

  memset(psi, 0, sizeof (psi));

#pragma omp barrier

#pragma omp parallel for reduction(+:psi)
  for (int32_t ya = 0; ya < LCD_HEIGHT; ya++) {
    int32_t n, xa;
    float *gamma, *delta, *theta;
    float rho;

    gamma = aligned_alloc(SIMD_ALIGN, sizeof (float) * obj_n);
    delta = aligned_alloc(SIMD_ALIGN, sizeof (float) * obj_n);
    theta = aligned_alloc(SIMD_ALIGN, sizeof (float) * obj_n);

    for (n = 0; n < obj_n; n++) {
      rho = ppl / (2.0f * obj_xyz[n][2]);
      gamma[n] = rho * 2.0f;
      delta[n] = rho * (2.0f * (0.0f - obj_xyz[n][0]) + 1.0f);
      theta[n] = rho * ((0.0f - obj_xyz[n][0]) * (0.0f - obj_xyz[n][0]) + (ya - obj_xyz[n][1])
        * (ya - obj_xyz[n][1]));
    }

    for (xa = 0; xa < LCD_WIDTH; xa++) {
      for (n = 0; n < obj_n; n++) {
        psi[ya][xa] += cos_table[FtoI(theta[n] * COS_N) & COS_MASK];
        theta[n] += delta[n];
        delta[n] += gamma[n];
      }
    }

    free(theta);
    free(delta);
    free(gamma);
  }
}
```

Table 9.6 Performance of the xyz_to_psi function using the look-up table. The value in parentheses is a magnification that is accelerated from the program shown in previous subsection

N_{obj}	Calculation time [s]	Acceleration rate	Used memory [MiB]	$(12N_{obj} + 4N_{hol} + 96N_{obj})$
710	0.11	260.90 (1.90)	7.98	(0.01 + 7.91 + 0.07)
44,647	6.02	303.81 (1.86)	12.51	(0.51 + 7.91 + 4.09)
978,416	1,573.55	25.93 (1.21)	108.68	(11.20 + 7.91 + 89.58)

The performance is presented in Table 9.6. The amount of memory used did not include the LUT size. We succeeded in achieving higher speeds.

9.3.6 Memory Reduction by Rearranging Program Procedure

The expected performance cannot be obtained when the number of object points increases. As we are aware, this problem is caused by the large amount of data required for the calculation. Because the amount of data exceeds the CPU cache size, calculations are always performed using system memory, which has a slow processing speed. Therefore, the calculation speed is limited by the speed of communication with system memory. If the object consists of tens of thousands of points, the data are all stored in the cache of the CPU, and thus the problem does not surface. Because the CPU is not equipped with a large **cache memory**, as we increase the number of object points, the calculation speed decreases significantly. In the case of the test machine (Table 9.1), as described in Sect. 9.3.4, the recurrence algorithm requires approximately 100 times more memory as the number of object points increases.

It is important to reduce the amount of data used as well as numerical calculations. It is also important to understand the data characteristics. One of the characteristics of the data is whether they must be prepared for each CPU core or whether they can be shared by multiple CPU cores. Moreover, we should better understand the characteristics and trends of the memory-access frequency, continuity, and range. Devices with different performances, such as cache and system memory, have a hierarchical structure and are involved in a complicated manner. What data can exist in what level of cache memory, or whether we need to devise to make it exist, can be hints for data design.

One of the easiest ways to increase the cache hit rate is by cache blocking. Cache blocking is a method of executing processing in small increments such that the amount of data used for calculation is less than the cache size. A sample source code file using cache blocking is provided as an appendix/5-recurrence_float_lut/xyz_to_psi.c in the sample program package. The size of the one-dimensional arrays gamma, delta, and theta exceed the cache size of the CPU. We delimit the processing to the extent that

Table 9.7 Performance of the xyz_to_psi function using the cache blocking. The value in parentheses is a magnification that is accelerated from the program shown in previous subsection.

N_{obj}	Calculation time [s]	Acceleration rate	Used memory [MiB]	$(12N_{obj} + 4N_{hol} + 96N_{obj})$
710	0.65	44.01 (0.17)	7.98	$(0.01 + 7.91 + 0.07)$
44,647	39.24	46.59 (0.15)	12.51	$(0.51 + 7.91 + 4.09)$
978,416	871.73	46.81 (1.81)	108.68	$(11.20 + 7.91 + 89.58)$

access to them is less than or equal to the CPU cache size. In the sample source code, we restrict the extent of access using the macro name BLOCK. The performance is shown in Table 9.7. We obtained approximately twice the improvement in computing objects consisting of one million points that exceeded the CPU cache size. The effect of cache blocking was certain. However, a better improvement way exists for Listing 9.5 and Listing 9.7.

Listing 9.8 xyz_to_psi function using the cache blocking

```c
#define SIMD_ALIGN     (256 / 8)     // avx2: 256 bits
#define FtoI(n)        (((n) < 0.0f) ? (int32_t)((n) - 0.5f) : (int32_t)((n) + 0.5f)) // float to int32_t
#define BLOCK          (87000)       // < 8 MiB (cache size) / 8 threads / 3 arrays / size of float

float psi[LCD_HEIGHT][LCD_WIDTH] __attribute__((aligned(SIMD_ALIGN)));

void xyz_to_psi(int32_t obj_n, float (*obj_xyz)[3])
{
  const float ppl = LCD_DOT_PITCH / LAMBDA;

#pragma omp parallel for simd
  for (int32_t n = 0; n < obj_n; n++) {
    obj_xyz[n][0] = obj_xyz[n][0] * XYZ_EXP + X_OFFSET;
    obj_xyz[n][1] = obj_xyz[n][1] * XYZ_EXP + Y_OFFSET;
    obj_xyz[n][2] = obj_xyz[n][2] * XYZ_EXP + Z_OFFSET;
  }

  memset(psi, 0, sizeof (psi));

#pragma omp barrier

#pragma omp parallel for reduction(+:psi)
  for (int32_t ya = 0; ya < LCD_HEIGHT; ya++) {
    int32_t n, xa, m;
    float *gamma, *delta, *theta;
    float rho;

    gamma = aligned_alloc(SIMD_ALIGN, sizeof (float) * obj_n);
```

```
    delta = aligned_alloc(SIMD_ALIGN, sizeof (float) * obj_n);
    theta = aligned_alloc(SIMD_ALIGN, sizeof (float) * obj_n);

    for (n = 0; n < obj_n; n++) {
      rho = ppl / (2.0f * obj_xyz[n][2]);
      gamma[n] = rho * 2.0f;
      delta[n] = rho * (2.0f * (0.0f − obj_xyz[n][0]) + 1.0f);
      theta[n] = rho * ((0.0f − obj_xyz[n][0]) * (0.0f − obj_xyz[n][0]) + (ya − obj_xyz[n][1])
        * (ya − obj_xyz[n][1]));
    }

    for (n = 0; n < obj_n; n += BLOCK) {
      for (xa = 0; xa < LCD_WIDTH; xa++) {
        for (m = 0; m < BLOCK && (n + m) < obj_n; m++) {
          psi[ya][xa] += cos_table[FtoI(theta[n + m] * COS_N) & COS_MASK];
          theta[n + m] += delta[n + m];
          delta[n + m] += gamma[n + m];
        }
      }
    }

    free(theta);
    free(delta);
    free(gamma);
  }
}
```

We already understood that the buffer sizes of the one-dimensional arrays gamma, delta, and theta are large. Although a large size is a problem, a more serious problem is that a wide bandwidth is required for communication with the system memory. Moreover, gamma, delta, and theta are closely related to both the calculation of the base point and the recurrence formula. For example, in the calculation of a recurrence formula in a triple for-loop, theta, cos_table, psi, delta, theta, gamma, and delta are stored. As the ratio of memory-access instructions to arithmetic instructions increases, we cannot hide the latency of the memory-access instructions. Hence, the communication time with the system memory accounts for most of the total processing time. It is best to reduce the memory-access instructions. If this is not possible, we should devise, if possible, such that the latency can be hidden well by balancing the calculation and memory-access instructions. Even if the data size used for the calculation is much larger than the cache memory size, the calculation time can be sufficiently long for the communication time, we can avoid the rate limiting by the system memory. In other words, we should reduce communication bandwidth rather than data size.

We solve the problem by reducing both the data size and the communication bandwidth. Listing 9.5 and Listing 9.7 complete the calculation for one pixel of the hologram plane before starting the calculation for the next pixel. Although one pixel of the hologram plane is calculated intensively, we change the program to calculate one line of the hologram plane using one object point. Because CGH pixels can be obtained by superposition calculation, they can be summated in any order or timing.

The deepest for-loop changes to a process of repeating the movement of pixels in the x-axis direction of the hologram plane instead of repeating for the object point. After the gamma, delta, and theta of the next pixel are obtained, the gamma, delta, and theta of the current pixel become unnecessary. We no longer need to store gamma, delta, and theta for all object points. We can calculate with only a few registers that can hold the current gamma, delta, and theta values.

The program appears similar to Listing 9.9. It just swaps the for-loop for the loop counter xa and the for-loop for the object point. Because arrays gamma, delta, and theta have each been reduced to a single variable, we can significantly reduce the amount of memory used. Communication with the system memory owing to access to the one-dimensional arrays gamma, delta, and theta, which occurred during the calculation, has been eliminated. The cosine table is a size that can be stored in the **L1 cache**. In the for-loop that computes a recurrence formula, the only access to system memory is to the two-dimensional array psi. The advantage of this is that all the operations except for the instruction to save data in the two-dimensional array psi can be performed only with the registers and the L1 cache. During communication of psi[ya][xa] by the store instruction, the CPU can process arithmetic instructions. Because we can hide the latency of store instructions, the apparent latency approaches the throughput. In addition to achieving a reduction in data volume and communication bandwidth, we have also achieved faster calculations.

Listing 9.9 xyz_to_psi function using the memory reduction by one line calculation of hologram plane

```
#define SIMD_ALIGN     (256 / 8)     // avx2: 256 bits
#define FtoI(n)        (((n) < 0.0f) ? (int32_t)((n) − 0.5f) : (int32_t)((n) + 0.5f)) // float
    to int32_t

float psi[LCD_HEIGHT][LCD_WIDTH] __attribute__((aligned(SIMD_ALIGN)));

void xyz_to_psi(int32_t obj_n, float (*obj_xyz)[3])
{
  const float ppl = LCD_DOT_PITCH / LAMBDA;

#pragma omp parallel for simd
  for (int32_t n = 0; n < obj_n; n++) {
    obj_xyz[n][0] = obj_xyz[n][0] ∗ XYZ_EXP + X_OFFSET;
    obj_xyz[n][1] = obj_xyz[n][1] ∗ XYZ_EXP + Y_OFFSET;
    obj_xyz[n][2] = obj_xyz[n][2] ∗ XYZ_EXP + Z_OFFSET;
  }

  memset(psi, 0, sizeof (psi));

#pragma omp barrier

#pragma omp parallel for // reduction(+:psi)
  for (int32_t ya = 0; ya < LCD_HEIGHT; ya++) {
    float xaj, yaj, rho, gamma, delta, theta;

    for (int32_t n = 0; n < obj_n; n++) {
      xaj = 0.0f − obj_xyz[n][0];
```

```
        yaj = (float)ya − obj_xyz[n][1];
        rho = ppl / (2.0f ∗ obj_xyz[n][2]);
        gamma = rho ∗ 2.0f;
        delta = rho ∗ (2.0f ∗ xaj + 1.0f);
        theta = rho ∗ (xaj ∗ xaj + yaj ∗ yaj);

        for (int32_t xa = 0; xa < LCD_WIDTH; xa++) {
          psi[ya][xa] += cos_table[FtoI(theta ∗ COS_N) & COS_MASK];
          theta += delta;
          delta += gamma;
        }
      }
    }
}
```

For the coordinate data of the object point, we only need to be able to load the coordinate data of one point during calculation of one line of the hologram plane. If we divide the loop of the loop counter ya evenly and provide it to the threads, all the threads calculate using the coordinate data of the object point in the same order. The coordinate data of the object point is data that can be shared by all CPU cores. Therefore, all CPU cores other than the CPU core that first accessed the coordinate data can load it from the L3 cache with high probability. The communication with the system memory is only one coordinate data per one line of the hologram plane. We can understand that it is even more efficient considering the cache line size that can have coordinate data of five object points as shown Fig. 9.1.

The for-loop for the loop counter xa is the deepest. The value of xa changes whenever the for-loop is processed even once. Multiple threads do not access the same psi[ya][xa]. Whether we use the OpenMP "reduction" clause or not, it does not affect the performance.

The results changed as shown in Table 9.8. The computation time for an object consists of one million points, where access to the system memory was a bottleneck, has been reduced by less than one-third. By contrast, the performance of other objects was worse. OpenMP analyzes for-loop iteration instructions and distributes the load to threads. In addition, OpenMP converts instructions repeated in for-loop into SIMD instructions. Therefore, for OpenMP, the loop counter is an important indicator for scheduling parallel processing. For example, Listing 9.7 is described by a program that accesses an array using the loop counter n. In Listing 9.9, the program changed

Table 9.8 Performance of the xyz_to_psi function using the memory reduction by one line calculation of hologram plane. The value in parentheses is a magnification that is accelerated from the program shown in previous subsection.

N_{obj}	Calculation time [s]	Acceleration rate	Used memory [MiB]	$(12N_{obj} + 4N_{hol})$
710	0.36	79.85 (0.31)	7.92	(0.01 + 7.91)
44,647	22.08	82.79 (0.27)	8.42	(0.51 + 7.91)
978,416	481.81	84.69 (3.27)	19.11	(11.20 + 7.91)

to an instruction with the loop counter removed. Although OpenMP can generate a parallel program using threads even if it loses the index, OpenMP does not seem to be able to generate packed data for SIMD instructions.

We calculate the recurrence formula toward the x-axis direction of the hologram plane and obtain the CGH pixels for one line. At this time, we can also calculate other lines of the hologram plane. We should be able to use the SIMD instruction because multiple rows can be computed simultaneously using the same formula. We can use 256-bit packed data for AVX. We can calculate eight lines of the hologram plane using SIMD instructions if the calculation precision is 32 bits. Therefore, the loop counter ya moves eight lines ahead. A sample program is prepared to help the compiler generate SIMD instructions using packed data. The source file is "appendix/6-recurrence_float_lut_line/xyz_to_psi.c" in the sample program package. However, it is far from the expected performance and the performance is only slightly improved. We use the Intel Intrinsics [13] to generate SIMD instructions without OpenMP.

9.3.7 Decimal Fraction Calculation Using Integer Type

The first improvement was to speed up the calculation by reducing the excessive data precision to an appropriate word length. This time, we accelerate the calculation by changing the data format to **integer type**. The data format changes from floating point to fixed point. In fixed-point arithmetic, we can process binary numbers as they are. Arithmetic circuits are simplified, and the throughput and latency of instructions are shorter than those of arithmetic instructions using floating point. Another advantage is the ability to perform operations using the nature of binary numbers. For example, we can calculate faster by shifting the value 1 bit to the right than by dividing it by two.

We cannot calculate correctly by simply changing the variable type. The integer type is a simple binary number. An n-bit register can represent zero to $(2^n - 1)$ if the variable type is an unsigned integer. However, the floating-point format uses a mantissa and an exponent to represent values. As the position of the decimal point moves, the exponent is rewritten to maintain the correct value. The float type can represent a wider range of numerical values than the integer type even with the same word length. However, the number of significant digits decreases because bits are assigned to the exponent. The decimal point position is automatically adjusted and we are not usually aware of it. When representing a decimal value using an integer type variable, we must adjust the position of the decimal point to ensure the necessary significant digits. The overhead to convert the numerical value increases, and the source code becomes more difficult to read. Therefore, the introduction of programs using integer types is the last.

The sample program is Listing 9.10. At the beginning of the xyz_to_psi function, we obtain the pixel pitch (LCD_DOT_PITCH) divided by the laser wavelength. Both the pitch and the wavelength are $O(10^{-6})$. We can reduce the dynamic range for the coordinate data of an object point by performing this division. Because 10^6

is equivalent to 2^{20}, we can save approximately 20 bits. In the program part that corrects the coordinate data of the object point, we convert the value to an integer while rounding it.

Listing 9.10 xyz_to_psi function using the integer type

```
#define SIMD_ALIGN    (256 / 8)     // avx2: 256 bits
#define FtoI(n)       (((n) < 0.0f) ? (int32_t)((n) - 0.5f) : (int32_t)((n) + 0.5f)) // float to int32_t

#define GDT_FP_POS    (24)          // Fixed-Point Position of Gamma, Delta, and Theta [bit]

int32_t psi[LCD_HEIGHT][LCD_WIDTH] __attribute__((aligned(SIMD_ALIGN)));

void xyz_to_psi(int32_t obj_n, float (*obj_xyz)[3])
{
  const float ppl = LCD_DOT_PITCH / LAMBDA;
  int32_t (*int_xyz)[3] = (int32_t (*)[3])obj_xyz;

#pragma omp parallel for simd
  for (int32_t n = 0; n < obj_n; n++) {
    int_xyz[n][0] = FtoI(obj_xyz[n][0] * XYZ_EXP + X_OFFSET);
    int_xyz[n][1] = FtoI(obj_xyz[n][1] * XYZ_EXP + Y_OFFSET);
    int_xyz[n][2] = FtoI(obj_xyz[n][2] * XYZ_EXP + Z_OFFSET);
  }

  memset(psi, 0, sizeof (psi));

#pragma omp barrier

#pragma omp parallel for // reduction(+:psi)
  for (int32_t ya = 0; ya < LCD_HEIGHT; ya++) {
    int32_t n, xa, xaj, yaj;
    uint32_t rho;
    int32_t gamma, delta, theta;

    for (n = 0; n < obj_n; n++) {
      rho = ppl / int_xyz[n][2] * (0x1UL << (GDT_FP_POS - 1));
      gamma = rho << 1;
      xaj = 0 - int_xyz[n][0];
      yaj = ya - int_xyz[n][1];
      delta = rho * ((xaj << 1) + 1);
      theta = rho * (xaj * xaj + yaj * yaj);

      for (xa = 0; xa < LCD_WIDTH; xa++) {
        psi[ya][xa] += cos_table[(theta >> (GDT_FP_POS - COS_DEPTH)) & COS_MASK];
        theta += delta;
        delta += gamma;
      }
    }
  }
}
```

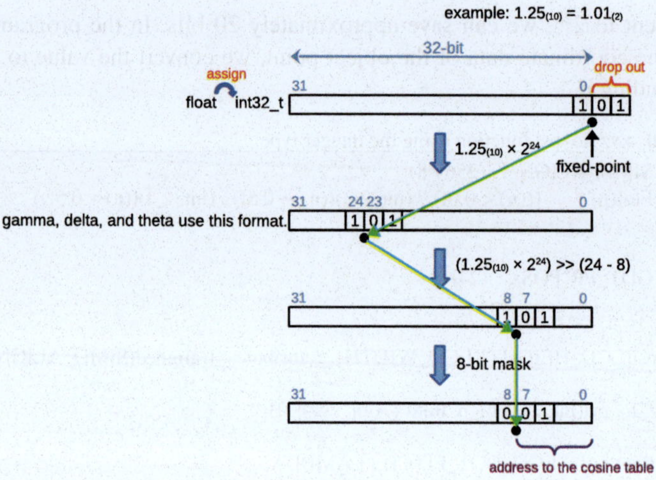

Fig. 9.6 Movement of decimal point

Figure 9.6 shows how the decimal point moves. The figure uses a constant of 1.25 for simplicity. We need to devise ways to retain the significant digits of the variable rho. The result of dividing the float type constant ppl by $2z_j$ (= 2 * int_xyz[n][2]) is also a decimal fraction with the float type. When we assign it to the integer type variable rho, the fractional information is lost. Therefore, the significant digits of the variable rho may be lost. In the sample program, we maintain the significant digits by shifting the decimal point position to the left. Based on the authors' experience, the number of significant digits that can be observed as a good reconstruction image is 24 bits. The specific calculation method is to move the decimal part to the integer part by multiplying by 2^{24}. In the sample program, we multiply by 2^{23}, which is 24 bits minus 1 bit, and then divide by z_j (int_xyz[n][2]). We calculate the integer type variables gamma, delta, and theta using the integer type variable rho.

Listing 9.11 is a function to create the cosine table of integer type data. The cosine table data is an 8-bit integer. This is a word length that can observe an image that is nearly the same as the reconstructed image calculated with the float type. The sampling rate is obtained by dividing one period of the cosine into 256. Therefore, we use the right shift operation to change the decimal point position of theta to the 8th bit. We mask the variable theta with the constant 0xff and remove the integer part of the value. We refer to the cosine table using the normalized variable theta.

Table 9.9 Performance of the xyz_to_psi function using the integer type. The value in parentheses is a magnification that is accelerated from the program shown in previous subsection

N_{obj}	Calculation time [s]	Acceleration rate	Used memory [MiB]	$(12N_{obj} + 4N_{hol})$
710	0.21	137.09 (1.72)	7.92	$(0.01 + 7.91)$
44,647	12.89	141.88 (1.71)	8.42	$(0.51 + 7.91)$
978,416	282.26	144.56 (1.71)	19.11	$(11.20 + 7.91)$

Listing 9.11 make_cos_table function to create LUT of integer type data

```
#define COS_DEPTH     8 // [bit]
#define COS_N         (0x1 << COS_DEPTH)
#define COS_MASK      (COS_N - 1)
typedef int8_t        cos_tbl_t;

cos_tbl_t cos_table[COS_N];

void make_cos_table()
{
  const unsigned long offset = (0x1UL << (sizeof (cos_tbl_t) * 8 - 1)) - 1;
  int i;

  for (i = 0; i < COS_N; i++) cos_table[i] = roundf(cosf(2.0 * M_PI * i / COS_N) * offset);
}
```

Finally, we obtain the computational performance shown in Table 9.9. We succeeded in accelerating the calculation even if the overhead such as the adjustment of the decimal point position, which does not exist in the original hologram calculation increases. This is because we changed from the floating-point arithmetic to the integer arithmetic and reduced the cosine table size. We can calculate a hologram that is practically sufficient. As shown in Chap. 7, the fact that operations can be performed in a fixed-point format means that we can reduce the circuit area compared to arithmetic circuits using floating-point numbers. We obtained excellent results for designing hardware such as special purpose computers.

9.4 Results

Figure 9.7 is a 3D object "chess" image consisting of 44,647 points. The holograms were calculated using the coordinate data of the chess. Figures 9.8, 9.9 and 9.10 show the calculation results for the double-precision type, the single-precision type, and the integer type. Only the calculation result using the integer type is slightly different from the others. Figure 9.11 is a reconstructed image obtained from the hologram in Fig. 9.8 using the numerical Fresnel diffraction in Chap. 8. Similarly, Fig. 9.12 is

Fig. 9.7 Image of the chess constructed of 44,647 points

Fig. 9.8 Hologram image generated by the direct calculation using the double type

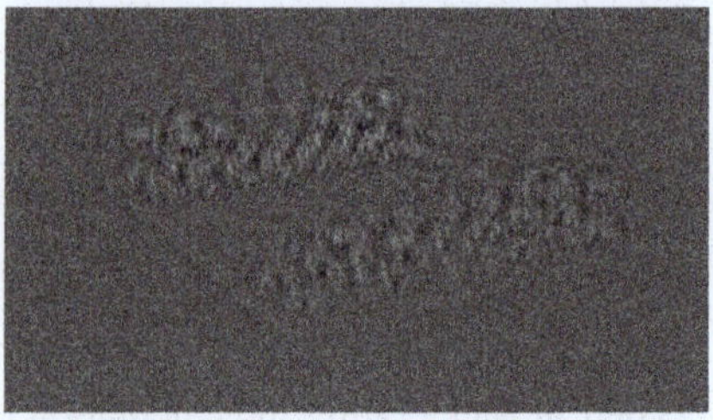

Fig. 9.9 Hologram image generated by the direct calculation using the float type

Fig. 9.10 Hologram image generated by the recurrence algorithm using the integer type

a reconstructed image obtained from Fig. 9.10. We could obtain good reconstructed images from holograms using integer arithmetic.

Table 9.10 summarizes the performance results from the previous section. Furthermore, Table 9.11 also shows the performance of a CPU with eight physical cores. Even if the number of physical cores is doubled, the processing speed is not limited by communication with the system memory. The processing speed nearly doubled as expected.

Fig. 9.11 Reconstructed image from the hologram of Fig. 9.8

Fig. 9.12 Reconstructed image from the hologram of Fig. 9.10

Table 9.10 Performance summary of the xyz_to_psi functions. CPU is i7-7700K (4 cores, 4.5 GHz). "rate" is the rate of speedup relative to the program using the double-precision type

Method	N_{obj} : 710		N_{obj} : 44, 647		N_{obj} : 978, 416	
	Time [s]	Rate	Time [s]	Rate	Time [s]	Rate
Double-precision (Listing 9.2)	28.42	1.00	1,828.25	1.00	40,804.55	1.00
Single-precision (Listing 9.3)	17.40	1.63	1,148.72	1.59	26,663.88	1.53
OpenMP (Listing 9.4)	0.38	74.14	22.04	82.97	1,892.76	21.56
Recurrence Algorithm (Listing 9.5)	0.21	136.96	11.20	163.28	1,908.04	21.39
Look-Up Table (Listing 9.7)	0.11	260.90	6.02	303.81	1,573.55	25.93
Memory Reduction (Listing 9.9)	0.36	79.85	22.08	82.79	481.81	84.69
Integer (Listing 9.10)	0.21	137.09	12.89	141.88	282.26	144.56

Table 9.11 Performance summary of the xyz_to_psi functions. CPU is i7-7820X (8 cores, 4.2 GHz at single thread, 4.0 GHz at multi-threads). "rate" is the rate of speedup relative to the program using the double-precision type.

Method	N_{obj} : 710		N_{obj} : 44, 647		N_{obj} : 978, 416	
	Time [s]	Rate	Time [s]	Rate	Time [s]	Rate
Double-precision (Listing 9.2)	28.73	1.00	1,931.08	1.00	41,101.11	1.00
Single-precision (Listing 9.3)	18.45	1.56	1,215.42	1.59	28,148.78	1.46
OpenMP (Listing 9.4)	0.21	135.20	10.62	181.89	1,114.84	36.87
Recurrence Algorithm (Listing 9.5)	0.16	183.55	6.79	284.44	1,104.35	37.22
Look-Up Table (Listing 9.7)	0.09	303.55	4.21	458.33	804.29	51.10
Memory Reduction (Listing 9.9)	0.22	131.01	12.58	153.48	275.51	149.18
Integer (Listing 9.10)	0.14	204.15	7.36	262.35	160.45	256.16

References

1. The Linux Kernel Archives. https://www.kernel.org/ (Retrieved September 5, 2022).
2. The GNU Compiler Collection. https://gcc.gnu.org/ (Retrieved September 5, 2022).
3. Intel C++ Compiler. https://www.intel.com/content/www/us/en/developer/tools/oneapi/dpc-compiler.html (Retrieved September 5, 2022).
4. Device-Independent Bitmaps. https://docs.microsoft.com/en-us/windows/win32/gdi/device-independent-bitmaps (Retrieved September 5, 2022).
5. Intel. Intel 64 and IA-32 Architectures Optimization Reference Manual. 2:34–35 (February 2022). https://www.intel.com/content/www/us/en/developer/articles/technical/intel-sdm.html (Retrieved September 5, 2022).
6. OpenMP. https://www.openmp.org/ (Retrieved September 5, 2022).
7. OpenMP Reference Guide. https://www.openmp.org/resources/refguides/ (Retrieved September 5, 2022).
8. ISO/IEC 9899:2011. https://www.iso.org/standard/57853.html (Retrieved September 5, 2022).
9. T. Shimobaba and T. Ito. An efficient computational method suitable for hardware ofcomputer-generated hologram with phase computationby addition. *Comp. Phys. Commun.*, **138**, 44–52 (2001).
10. T. Shimobaba, S. Hishinuma and T. Ito. Special-purpose computer for holography HORN-4 with recurrence algorithm. *Comp. Phys. Commun.*, **148**, 160–170 (2002).
11. The GNU C Library. https://www.gnu.org/software/libc/ (Retrieved September 5, 2022).
12. Intel. Intel 64 and IA-32 Architectures Optimization Reference Manual. F:44 (February 2022). https://www.intel.com/content/www/us/en/developer/articles/technical/intel-sdm.html (Retrieved September 5, 2022).
13. The Intel Intrinsics Guide. https://www.intel.com/content/www/us/en/docs/intrinsics-guide/index.html (Retrieved September 5, 2022).

Chapter 10
Computer-Generated Hologram Calculation Employing the Graphics Processing Unit

Takashi Nishitsuji

Abstract As discussed in the previous chapter, a point-cloud is among the simplest 3D model for implementing CGH calculation. Further, since the point-cloud-based CGH calculation is very similar to the ray-tracing process in computer graphics technologies, it can suitably drive graphic processing units (GPUs). In this section, two different approaches, namely, ray-tracing and look-up table (LUT) methods, were introduced to implement the point-cloud-based CGH calculation. Employing these source codes, which were written in CUDA, readers can attempt the point-cloud-based CGH calculation employing GPU on their computers.

10.1 General Instruction for Implementing Point-Cloud-Based CGH on GPU

There are two approaches for accelerating point-cloud-based **CGH** calculation on GPU: (1) the selection of an appropriate algorithm for the calculation and (2) applying effective implementation techniques for the **GPU**. The first approach is common in any device, although it is crucial to consider the affinity between the algorithm and the GPU. Generally, GPUs can suitably execute many homogeneous calculations (it is not suitable for executing complex calculations involving many conditional branches); they exhibit limited memory capacities and bandwidths. Therefore, the selection of a simple algorithm that does not require many memory operations should be among the initial guidelines for designing a GPU program to execute a point-cloud-based CGH calculation. Conversely, regarding the second approach, implementing a memory-based algorithm for calculating point-cloud-based CGH on GPU would be a preferred choice if many calculations of the point-cloud-based CGH can be

---

Supplementary Information The online version contains supplementary material available at https://doi.org/10.1007/978-981-99-1938-3_10.

T. Nishitsuji (✉)
Faculty of Science, Toho University, 2-1-1 Miyama, Funabashi-shi, Chiba, Japan
e-mail: takashi.nishitsuji@is.sci.toho-u.ac.jp

T. Shimobaba and T. Ito (eds.), *Hardware Acceleration of Computational Holography*,
https://doi.org/10.1007/978-981-99-1938-3_10

substituted with precalculated values, which can be stored in the memory of the GPU, and if the transfer speed of the memory of the GPU is sufficiently high.

Thus, two examples are introduced in this section. The first one involves the effective implementation of CGH by modifying the algorithm for its calculation, as well as optimizing the computational operations, e.g., employing the fast-math functions. The second example entails the **LUT** approach in which sequential calculations are substituted with precalculated outputs of the whole or a part of the CGH calculation.

There are three major techniques for effectively implementing point-cloud-based CGH calculations on GPU; they include the following:

- Reduction of access to the global memory.
- Reduction of the computational loads of the operations.
- Reduction of the number of wasted operations.

GPUs comprise a hierarchical memory architecture in which each layer exhibits a different size and access speed. Therefore, a reduction of the frequency to access the global memory, which is the slowest memory space in a GPU, is among the straightforward strategies for accelerating CGH computation. Put differently, the utilization of other faster memory spaces, e.g., the shared memory and register, must be considered to achieve accelerated CGH calculation. In point-cloud-based CGH calculation, three major kinds of buffers are required on the GPU; they include for a point cloud, for a complex amplitude distribution of the wavefront as the intermediate data, and for CGH as the final output. Here, the complex amplitude distribution of the wavefront should be stored in the registers rather than the global memory since it is the intermediate data of each pixel of CGH, which does not require a direct transfer from the GPU. Therefore, the kernels in the following section include the iteration loop for scanning all the point clouds to calculate the wavefront of each one and accumulate them separately in the register of each thread.

Conversely, the utilization of **fast-math operations** is also a straightforward approach for accelerating the CGH calculation since the computational loads of the trigonometric functions, e.g., cos and sin, are generally intensive. Fast math operations proceed via specially designed and implemented circuits on the GPU core; thus, they calculate faster than usual math functions, although with generally low precision. As the CGH calculation is generally robust against noises, the loss of computational precision with the fast-math operations does not significantly degrade the quality of the reconstructed 3D image. Therefore, the fast-math operations, e.g., __cosf() for the cosine function, should be actively employed in point-cloud-based CGH calculations.

Further, the substitution of operations with precalculated values is the easiest method for reducing the number of wasted operations and computational loads. For example, the wave number $k = \frac{2\pi}{\lambda}$ can be substituted with precalculated constants because it is invariant in the video sequence and determined as preliminary. Similar to the example, if the calculated values of apart or the whole of the output in the CGH calculation can be assumed to be in a reasonable range and invariant within a video sequence, the utilization of precalculated values rather than directly calculating the equations will effectively reduce the number of operations and the computational loads. This idea is generally called the LUT whose point-cloud-based CGH calculation on a GPU has been widely studied [1].

10.2 Examples of Point-Cloud-Based CGH Calculation on GPU

10.2.1 Ray-Tracing Method

Ray-tracing method is the simplest method of processing point-cloud-based CGH calculation on GPU since it can separately calculate the wavefront of each point cloud in each pixel. For effective implementation, the tasks or data for point-cloud-based CGH calculation, which is to be allocated to the processing unit in GPU, should be considered. In many cases involving point-cloud-based CGH calculation, the pixel-wise partitioning of the tasks is better because GPU can suitably execute many homogeneous and independent calculations for each pixel. Further, since the resolution of CGH is generally invariable with the same video sequence, the pixel-wise distribution of the tasks favors the point-cloud-based CGH calculation on the GPU, i.e., grid sizes (corresponding to workgroup sizes in OpenCL) should be set as the resolution of CGH so that each thread (corresponding to a work item in OpenCL) can facilitate the calculation of one-pixel value on CGH.

In the following source codes, the Thrust library [2] was employed to simplify the source code, especially in managing the memory. Owing to the permitted page limit, the details of the Thrust APIs are not discussed here. Thus, the official reference on the website can be referenced to understand in detail the Thrust API and other usages [2].

Listing 10.1 is the host code for executing the kernel employing ray-tracing method. According to the list, the point-cloud-based CGH calculation on GPU proceeds, as follows:

- Setting the constant values (Lines 18–21)
- Allocating the host and device buffers as the Thrust vectors (Lines 24–27)
- Reading the point-cloud data to the host-vector (Line 30)
- Transferring the point-cloud data into a device (Line 33)
- Setting the sizes of the grid and threads (Lines 36 and 37)
- Executing the kernel (Lines 40–42)
- Receiving the calculated hologram from the GPU (Line 45).

On this list, the format of the input point-cloud data is sequentially aligned with the coordinates, such as x_j, y_j, z_j, x_{j+1}, and y_{j+1}..., which are read from file employing the original function, *ReadPointCloud()*, whose arguments are the filename of the file and the host-vector for point light sources (PLSs). Here, the coordinates of the

Table 10.1 Arguments of the kernels in ray-tracing method

Type	Name	Description
float3*	dPLS	Raw pointer to the device vector for point light sources
uchar*	dCGH	Raw pointer to the device vector for CGH
const int	numPLS	Number of PLSs stored in dPLS
const float	cgh_width	Width of the CGH

point clouds are assumed to be normalized by the pixel pitch of the display device, p, and distributed in a reasonable range to visualize the 3D object via the numerical simulation.

Listing 10.2 exhibits the kernels of the ray-tracing method in which the arguments are the same (Table 10.1).

Next, the basic structure of the point-cloud-based CGH calculation is described by referencing the *PC_CGH_RT()* function as an example (Lines 5–21 on the Listing 10.2). The kernel calculates all the wavefronts of the point clouds in each pixel and accumulates them via the temporary variables (*cplx tmp*) in the iteration on Lines from 14–18. Notably, the local variables in the kernel without a modifier will be assigned to the registers except if the utilization of the register exceeds the limit of the hardware. *cplx* is the redefined form of *thrust::complex*$<$ *float* $>$ (Line 13), which is a built-in type of variable for a complex number in the float precision of the Thrust library that can be employed along with *thrust/complex.h*. After the iteration, the argument of the complex wavefront is calculated employing the *thrust::arg()* function on Line 20 to obtain the pixel value of the kinoform-type CGH.

To avoid the unnecessary calculation of the invariable constants, the constants, *c2_pi_pp_div_wl* and *c128_div_pi*, were introduced to precalculate $\frac{2\pi p}{\lambda}$ and $\frac{128}{\pi}$, respectively, to map from $-\pi$ to π, representing the output range (*thrust::arg()*), -128 to 127. Line 20 calculates the phase value of the kinoform-type CGH from *tmp*, which utilizes the overflow of the unsigned char to map the output, *thrust::arg()*, i.e., the negative output (-128 to -1) is mapped to $+128$ to $+255$; the positive output (0 to $+127$) is mapped to 0 to $+127$.

To further accelerate the process, several techniques were introduced into the *PC_CGH_RT()* kernel function. Firstly, the fast-mathematical functions were attempted instead of the normal mathematical operations. Considering *PC_CGH_RT()*, it is observed that many basic operations, including the subtraction, multiplication, and cosine and sine functions, were utilized in the code, and all of them can be substituted with the fast functions. *PC_CGH_RT_fastmath()* is the kernel code of ray-tracing method employing the fast-mathematical functions. It employs *__fsub_rd()*, *__fsqrt_rd()*, and *__fmul_rd()* for the subtraction, square root, multiplication, and *__cosf()*, *__sinf()* for the cosine and sine functions, respectively, thus significantly affecting the computational speed. Notably, the prototypes of these functions must be declared in the device code when employing the special mathematical functions, as shown on Listing 10.3.

Another approach for accelerating the process involves the application of an effective algorithm. Similar to a CPU implementation, ray-tracing method can be easily accelerated by applying the Fresnel approximation, which can be applied to calculate the distance between the point light source and CGH pixel, as follows:

$$r = \sqrt{(x_h - x_j)^2 + (y_h - y_j)^2 + z_j^2} - z_j, \approx \frac{(x_h - x_j)^2 + (y_h - y_j)^2}{2z_j}, \quad (10.1)$$

where r is the distance, (x_h, y_h) are the coordinates of CGH, and (x_j, y_j, z_j) are the coordinates of the j-th point light source. Compared with the ray-tracing method without an approximation, the square-root calculation is removed from the equation in the ray-tracing method instead of adding the division operation, and this is generally a computationally intensive operation; thus, the effective implementation of the division operation is the main consideration of the Fresnel approximation. The Fresnel approximation can be effectively implemented in the GPU via two approaches: the precalculation of the reciprocal, z_j, in the host computer before launching the kernel, and the utilization of special mathematical functions. Subsequently, an example involving the special division functions is illustrated. *PC_CGH_RT_Fresnel_fastmath()* is the kernel function of the Fresnel approximation in CUDA employing the special division function. In Line 67, instead of the normal division operation, /, the special division function, *__fdividef()*, is employed. Here, the constant *(pi_pp_div_wl)* in Line 66 corresponds to $\frac{\pi p}{\lambda}$

Conversely, the lowest (among the discussed three methods of implementation) computational load of the Fresnel approximation can be obtained if $1/z_j$ can be precalculated on the host computer (sending the point-cloud data as $(x_j, y_j, 1/z_j)$. Further, if the wavelength, λ, and display pitch, p, are the constant within the video sequence, it will be better to send $\frac{\pi p}{\lambda z_j}$ instead of $\frac{1}{z_j}$.

Listing 10.1 Host code of the point-cloud-based CGH calculation employing the ray-tracing method.

```
1 #include <cuda_runtime.h>
2 #include <device_launch_parameters.h>
3 #include <device_functions.h>
4 #include <cuda.h>
5 #include <thrust/host_vector.h>
6 #include <thrust/device_vector.h>
7 #include <thrust/complex.h>
8
9 //Prototype Declaration
10 __host__ void ReadPointCloud(const char* filename, thrust::host_vector<float3>& obj);
11 __global__ void PC_CGH_RT(float3* dPLS, unsigned char* dCGH, const int numPLS, const
       int cgh_width);
12
13 using cplx = thrust::complex<float>;
14
15 int main() {
16
17   //Constants*****************************************************
```

```
    const float depth = 0.3F; // nearest depth of the PLS [m]
    const int numPLS = 11646; // num of PLSs
    const int cghWidth = 2048; // width of CGH [pixel]
    const int cghHeight = 1024; // height of CGH [pixel]

    //Buffers**************************************************
    thrust::host_vector<unsigned char> hCGH(cghWidth * cghHeight); //quantized CGH
        buffer (host)
    thrust::device_vector<unsigned char> dCGH(cghWidth * cghHeight); //quantized CGH
        buffer (device)
    thrust::host_vector<float3> hPLS(numPLS); //point light sources (host)
    thrust::device_vector<float3> dPLS(numPLS); //point light souces (device)

    //Reading Point-clouds
    ReadPointCloud("tyranno11646.3df", hPLS);

    //Send point-clouds data to GPU
    dPLS = hPLS;

    //Set the grid and block size for a kernel execution
    dim3 threads(256, 1, 1);
    dim3 blocks(cghWidth / threads.x, cghHeight / threads.y, 1);

    //Execute the kernel
    PC_CGH_RT <<< blocks, threads >>> (
      thrust::raw_pointer_cast(dPLS.data()),
      thrust::raw_pointer_cast(dCGH.data()), numPLS, cghWidth);

    //Trasnfer the CGH data from GPU
    hCGH = dCGH;

    return 0;
}
```

Listing 10.2 Kernel code for the point-cloud-based CGH calculation employing the RT method.

```
#define c128_div_pi 40.743665431525205956f
#define c2_pi_pp_div_wl 94.48398958f
#define pi_pp_div_wl 47.24199479f

__global__ void PC_CGH_RT(float3* dPLS, unsigned char* dCGH, const int numPLS, const
    int cgh_width)
{
  int x = blockIdx.x * blockDim.x + threadIdx.x;
  int y = blockIdx.y * blockDim.y + threadIdx.y;
  int addr = x + y * cgh_width;

  cplx tmp(0.0, 0.0);
  float phase;

  for (int i = 0; i < numPLS; i++)
  {
    phase = c2_pi_pp_div_wl * sqrtf(powf(x - dPLS[i].x, 2.0f) + powf(y - dPLS[i].y, 2.0f) +
        powf(dPLS[i].z, 2));
```

```
    tmp += cplx(cosf(phase), sinf(phase));
  }

  dCGH[addr] = (unsigned char)((int)(c128_div_pi * thrust::arg(tmp)));
}


__global__ void PC_CGH_RT_fastmath(float3* dPLS, unsigned char* dCGH, const int
      numPLS, const int cgh_width)
{
  int x = blockIdx.x * blockDim.x + threadIdx.x;
  int y = blockIdx.y * blockDim.y + threadIdx.y;
  int addr = x + y * cgh_width;

  cplx tmp(0.0, 0.0);
  float phase;
  float dx;
  float dy;

  for (int i = 0; i < numPLS; i++)
  {
    dx = __fsub_rd(x, dPLS[i].x);
    dy = __fsub_rd(y, dPLS[i].y);

    phase = __fsqrt_rd(__fmul_rd(dx, dx) + __fmul_rd(dy, dy) + __fmul_rd(dPLS[i].z, dPLS[i].z
        ));
    phase = __fmul_rd(c2_pi_pp_div_wl, phase);
    tmp += cplx(cosf(phase), sinf(phase));
  }

  float arg = __fmul_rd(c128_div_pi, thrust::arg(tmp));
  dCGH[addr] = (unsigned char)((int)arg);
}

__global__ void PC_CGH_Fresnel_fastmath(float3* dPLS, unsigned char* dCGH, const int
      numPLS, const int cgh_width)
{
  int x = blockIdx.x * blockDim.x + threadIdx.x;
  int y = blockIdx.y * blockDim.y + threadIdx.y;

  int addr = x + y * cgh_width;
  cplx tmp(0.0, 0.0);
  float phase;
  float dx;
  float dy;

  for (int i = 0; i < numPLS; i++)
  {
    dx = __fsub_rd(x, dPLS[i].x);
    dy = __fsub_rd(y, dPLS[i].y);

    phase = __fmul_rd(dx, dx) + __fmul_rd(dy, dy);
    phase = __fmul_rd(pi_pp_div_wl, phase);
```

```
    phase = __fdividef(phase, dPLS[i].z);
    tmp += cplx(__cosf(phase), __sinf(phase));
  }

  float arg = __fmul_rd(c128_div_pi, thrust::arg(tmp));
  dCGH[addr] = (unsigned char)((int)arg);
}
```

Listing 10.3 Prototype declarations for the fast mathematical functions in CUDA.

```
__device__ float __cosf(float);
__device__ float __sinf(float);
__device__ float __powf(float, float);
__device__ float __fdividef(float, float);
__device__ float __fsqrt_rd(float);
__device__ float __fmul_rd(float, float);
__device__ float __fsub_rd(float, float);
__device__ float __fadd_rd(float, float);
```

10.2.2 LUT Method

LUT is a well-known method for accelerating various computations employing pre-calculated values that have been stored in the memory. The required memory capacity and access speed for reading and writing the data generally pose practical challenges except if the LUT method theoretically eliminates the computational load.

Many LUT-based methods have been proposed for the CGH calculation [1, 3–6], and almost all of them store the wavefront of each point cloud and process them according to the coordinates of the point cloud. Figure 10.1 shows the fundamental process of LUT [1]. The pattern of the wavefront only depends on z_j; thus, the required memory capacity for LUT is approximately LW^2, where W is the average width of the wavefront of the point cloud, and L is the resolution of the depth direction. Employing LUT, CGH can be calculated by reading the wavefront according to z_j and processing it as the center of the wavefront become (x_j, y_j).

Although the LUT method effectively calculates CGH, the gigabyte order of the memory capacity is required. Unfortunately, this cannot be easily implemented on a GPU owing to its limited memory capacity and bandwidth. Therefore, many compression algorithms for LUT should be studied [1, 4–6]. In this section, the Split-LUT (S-LUT)-based LUT method is introduced [4]. This method (**S-LUT**) can reduce the memory capacity to $2LW$ by dividing the two-dimensional point-cloud-based CGH calculation into two one-dimensional ones, followed by storing the one-dimensional precalculated data in LUT. Since the computation can be divided by the axes, the number of computations can be reduced by grouping the point clouds that exhibit the same x_j or y_j values.

The point-cloud-based CGH calculation employing the Fresnel approximation can also be described, as follows:

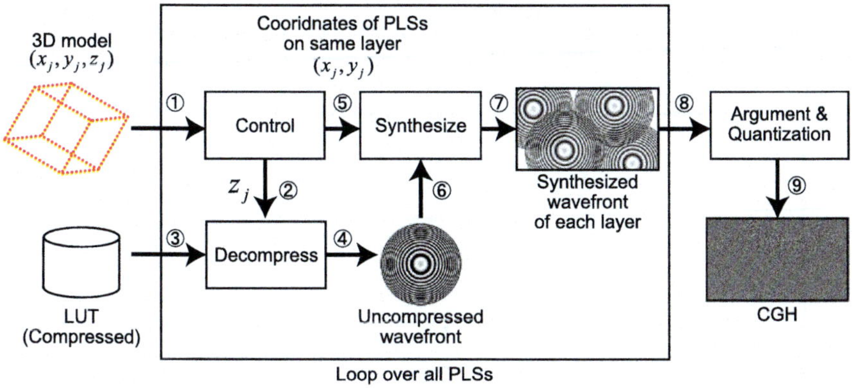

Fig. 10.1 Calculation system for LUT [1]

$$I(x_h, y_h) = \sum_{j=0}^{N-1} a_j \exp\left\{\frac{2\pi i}{\lambda}\left[z_j + \frac{1}{2z_j}\left(\Delta x^2 + \Delta y^2\right)\right]\right\}, \tag{10.2}$$

where $\Delta x = (x_h - x_j)$, $\Delta y = (y_h - y_j)$. the equation can be rewritten as

$$I(x_h, y_h) = \sum_{j=0}^{N-1} a_j \exp\left\{\frac{2\pi i}{\lambda}\left(z_j + \frac{\Delta x^2}{2z_j} + \frac{\Delta y^2}{2z_j}\right)\right\}, \tag{10.3}$$

$$= \sum_{j=0}^{N-1} a_j \exp\left\{\frac{\pi i}{\lambda}\left(z_j + \frac{\Delta x^2}{z_j}\right) + \frac{\pi i}{\lambda}\left(z_j + \frac{\Delta y^2}{z_j}\right)\right\}, \tag{10.4}$$

$$= \sum_{j=0}^{N-1} a_j \exp\left\{\frac{\pi i}{\lambda}\left(z_j + \frac{\Delta x^2}{z_j}\right)\right\} \exp\left\{\frac{\pi i}{\lambda}\left(z_j + \frac{\Delta y^2}{z_j}\right)\right\}, \tag{10.5}$$

$$= \sum_{j=0}^{N-1} a_j H(\Delta x, z_j) \times V(\Delta y, z_j). \tag{10.6}$$

Since Eq. (10.6) becomes the multiples of the functions, $H(\Delta x, z_j)$ and $V(\Delta y, z_j)$, which are independent along the x and y directions, respectively, the data stored in LUT will become two one-dimensonal data. Here, $a_j = 1$ was set for simplification. Further, those functions exhibit the same structure; both LUTs do not require independent preparations.

Listing 10.4 shows the host code of the S-LUT-based method employing CUDA; Listing 10.5 is a kernel code. In this implementation, the LUT data are calculated on the host side, which is described on Lines 51–70 of Listing 10.4 and sent to GPU before the execution of the kernel, which is described on Line 80. The memory capacities of the host and GPU are allocated, as described on Lines 43–48 of Listing 10.4.

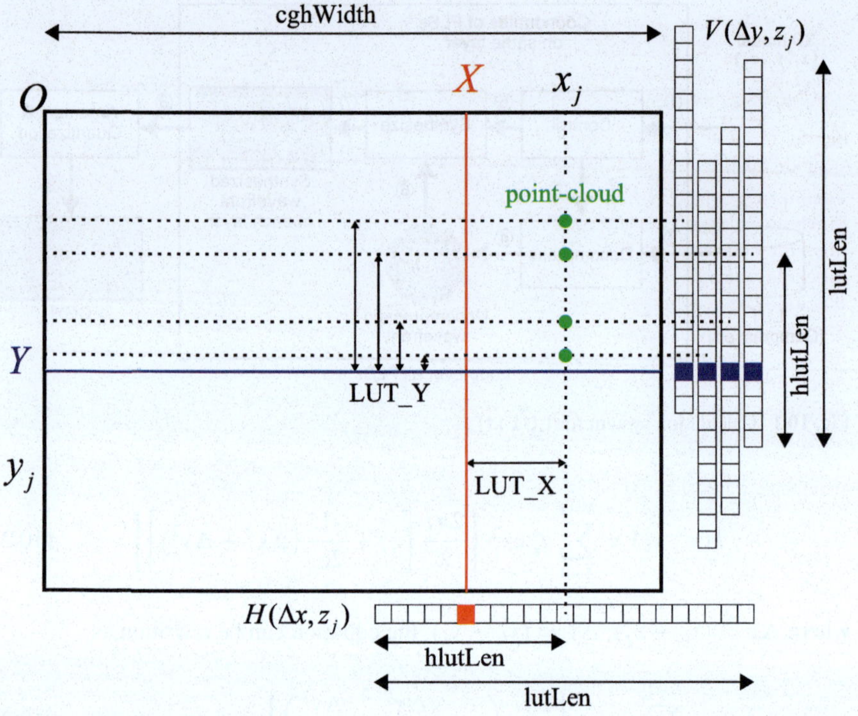

Fig. 10.2 Overview of S-LUT

To determine the required memory capacity, the depth resolution must be defined; it is set as 512 on Line 39 in the host code. Further, to avoid the aliasing noise, the maximum width of the LUT, which is calculated employing the diffraction angle of the spatial light modulator at each depth, is determined on Line 57 of the host code [8]:

$$W_{\max} = 2 \times \text{maxRadius} = 2|z_j| \tan\left(\sin^{-1}\frac{\lambda}{2p}\right). \tag{10.7}$$

Figure 10.2 shows an overview of the process of calculating via the S-LUT method; this overview describes the technique for obtaining the wavefront from four point light sources at (X, Y) and the same z_j, x_j. Dissimilar to the ray-tracing method, S-LUT requires the sorting of the point light sources in the order, z_j and x_j, since Eq. (10.6) shows that the CGH calculation of point light sources with the same z_j, x_j can employ the same $H(\Delta x, z_j)$ at every (X, Y). Therefore, as shown in Fig. 10.2 and Listing 10.5, $V(\Delta y, z_j)$ is first accumulated from four point light sources at $y = Y$, followed by the multiples, $H(\Delta x, z_j)$, at $x = X$. Finally, the complex wavefront is calculated from the four PLSs at (X, Y). To obtain CGH from all the point light sources, the above process should be iterated.

To implement the S-LUT-based method, the structure of point light source is defined on Lines 14–21 of Listing 10.4, which extends the comparison operator as point light sources is sorted in the z_j, x_j order by the *thrust::sort()* function.

Listing 10.4 Host code for the point-cloud-based CGH calculation employing the S-LUT-based method.

```
1 #define _USE_MATH_DEFINES
2 #include <cuda_runtime.h>
3 #include <device_launch_parameters.h>
4 #include <cuda.h>
5 #include <device_functions.h>
6 #include <thrust/host_vector.h>
7 #include <thrust/device_vector.h>
8 #include <thrust/complex.h>
9 #include <thrust/sort.h>
10 #include <cmath>
11
12 using cplx = thrust::complex<float>;
13
14 struct PLS {
15   int x, y, z;
16   bool operator <(const PLS& another) const {
17     if (z != another.z) return z > another.z;
18     if (x != another.x) return x > another.x;
19     if (y != another.y) return y > another.y;
20   }
21 };
22
23 //Prototype Declaration
24 __host__ void ReadPointCloud(const char* filename, thrust::host_vector<PLS>& obj);
25 __global__ void pls_CGH_SLUT(PLS* pls, const int numPLS, unsigned char* dCGH, cplx*
     dSLUT, const int cghWidth, const int lutLen, const int hlutLen);
26
27 int main() {
28
29   //Constants*************************************************
30   const float depth = 0.3F; //reconstruction distance [m]
31   const float pp = 0.000008F; //pixel pitch of display device [m]
32   const float wl = 0.000000532F; //wavelength of incident light [m]
33   const int numPLS = 11646; //num of point cloud
34
35   const int cghWidth = 2048; //width of a CGH [pixel]
36   const int cghHeight = 1024; //height of a CGH [pixel]
37   const int lutLen = 4096; //length of LUT [A.U.]
38   const int hlutLen = 2048; //half length of LUT [A.U.]
39   const int maxDepth = 512; //maximum number of depth layer of Point
       Cloud[A.U.]
40   const float pi_p_div_wl = M_PI*pp/wl;
41
42   //Buffres**************************************************
43   thrust::host_vector<unsigned char> hCGH(cghWidth * cghHeight); //quantized CGH
       buffer (host)
```

```
thrust::device_vector<unsigned char> dCGH(cghWidth * cghHeight); //quantized CGH
     buffer (device)
thrust::host_vector<PLS> hPLS(numPLS);     //point light sources (host)
thrust::device_vector<PLS> dPLS(numPLS);    //point light souces (device)
thrust::host_vector<cplx> hSLUT(lutLen * maxDepth);   //LUT buffer (host)
thrust::device_vector<cplx> dSLUT(lutLen * maxDepth);  //LUT buffer (device)

//Create S-LUT***********************************************
for (int z = 0; z < maxDepth; z++)
{
  //Calculate the depth of layer (normalized by pixelpitch)
  int curz = round((depth + z * pp) / pp);

  //Calculate the maximum length of LUT using diffraction limit
  int maxRadius = curz * tan(asin(wl / pp * 0.5));

  //Calcualte LUT values
  for (int x = 0; x < lutLen; x++)
  {
    cplx tmp(0.0, 0.0);
    if (abs(hlutLen - x) < maxRadius)
    {
      float phase = pi_p_div_wl * (curz + pow(x - hlutLen, 2) / curz);
      tmp = cplx(cos(phase), sin(phase));
    }
    hSLUT[x + z * lutLen] = tmp;
  }
}

//Reading Point-clouds data
ReadPointCloud("tyranno11646.3df", hPLS);

//Sort PLS data
thrust::sort(hPLS.begin(), hPLS.end());

//Send data host to device
dPLS = hPLS;
dSLUT = hSLUT;

//Set the grid and block size for a kernel execution
dim3 threads(1024, 1, 1);
dim3 blocks(cghWidth / threads.x, cghHeight / threads.y, 1);

pls_CGH_SLUT <<< blocks, threads >>> (
  thrust::raw_pointer_cast(dPLS.data()),
  numPLS,
  thrust::raw_pointer_cast(dCGH.data()),
  thrust::raw_pointer_cast(dSLUT.data()),
  cghWidth,
  lutLen,
  hlutLen);

//Obtain CGH data from the device
```

```
  hCGH = dCGH;

  return 0;
}
```

Listing 10.5 Kernel code for the point-cloud-based CGH calculation employing the S-LUT-based method.

```
#define c128_div_pi 40.743665431525205956f
__global__ void pls_CGH_SLUT(PLS* pls, const int numPLS, unsigned char* dCGH, cplx*
    dSLUT, const int cghWidth, const int lutLen, const int hlutLen)
{
  int x = blockIdx.x * blockDim.x + threadIdx.x;
  int y = blockIdx.y * blockDim.y + threadIdx.y;
  int addr = x + y * cghWidth;

  cplx v(0, 0);
  cplx h(0, 0);
  cplx tmp(0, 0);

  int prevX = pls[0].x;
  int prevZ = pls[0].z;

  int LUT_X = hlutLen + x - pls[0].x;
  h = dSLUT[LUT_X + pls[0].z * lutLen];

  for (int n = 0; n < numPLS; n++)
  {
    int xj = pls[n].x;
    int yj = pls[n].y;
    int zj = pls[n].z;

    if (prevX != pls[n].x)
    {
      tmp += h * v;
      v = cplx(0.0, 0.0);
      int LUT_X = hlutLen + x - xj;
      h = dSLUT[LUT_X + zj * lutLen];
    }

    int LUT_Y = hlutLen + y - yj;
    v += dSLUT[LUT_Y + zj * lutLen];

    prevX = pls[n].x;
  }

  dCGH[addr] = (unsigned char)((int)(c128_div_pi * thrust::arg(tmp)));
}
```

10.2.3 Performance Comparison

Table 10.2 compares the calculation times of the point-cloud-based CGH calculations per introduced framework. Here, the computational environment is thus: CPU (host PC): AMD Ryzen9 3950X 3.50 GHz, Memory: DDR4-3200 64 GB, GPU: NVIDIA Geforce RTX-2080 Super. Further, the computational conditions are thus: number of point-cloud: 11,646, CGH resolution: 2048×1024, pixel pitch of the display device (p): 8μm, wavelength of the incident light (λ): 532 nm.

The theortical highest effect of accelerating the computational algorithm was obtained via the S-LUT method. However, the table reveals that the highest practical computational speed was achieved via the Fresnel approximation and fast-mathematical operations in the ray-tracing method. It is proposed that the utilization of the fast-mathematical functions is more effective compared with LUT for accelerating the point-cloud-based CGH calculation on GPUs. Notably, the degradation of the image quality of the reconstructed image was within a reasonable range.

Figure 10.3 shows an example of the point-cloud model that was employed in this experiment; the image was numerically reconstructed via the kinoform-type CGH

Table 10.2 Comparison of the performances of the point-cloud-based CGH calculation methods on GPU

Method	Fastmath	Memory transfer[ms]	Calculation time [ms]
Ray-tracing		0.38	1414
Ray-tracing	Used	0.38	1013
Ray-tracing with the Fresnel approx.		0.38	665.7
Ray-tracing with the Fresnel approx.	Used	0.36	64.99
S-LUT		2.28	111.6

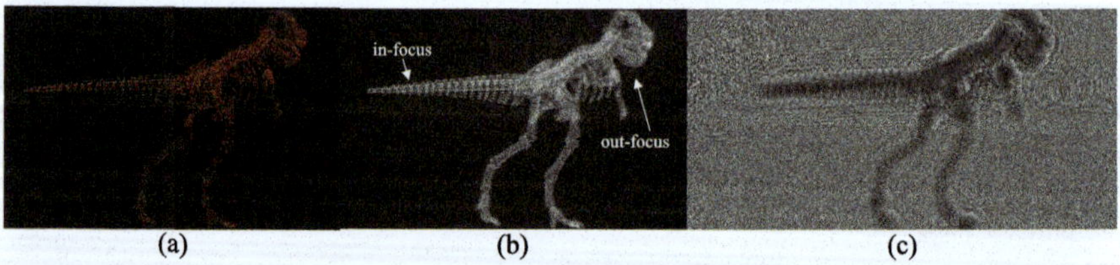

Fig. 10.3 Numerically reconstructed image of CGH that was created via the Fresnel approximation employing the fast-math operations: **a** original point-cloud model, **b** numerically reconstructed image, **c** Kinoform-type CGH

that was created employing the Fresnel approximation and the fast-math operations. The numerically reconstructed image includes in- and out-focus point clouds, indicating that the desired 3D image was replayed.

Fundings This work was supported by JSPS KAKENHI Grant Number 22H03616.

References

1. T. Nishitsuji, T. Shimobaba, T. Kakue, and T. Ito, "Review of fast calculation techniques for computer-generated holograms with the point-light-source-based model," IEEE Trans. Ind. Inf. 13(5), 2447–2454 (2017).
2. Thrust, "https://thrust.github.io/"
3. M. E. Lucente, "Interactive computation of holograms using a look-up table," J. Electron. Imaging 2(1), 28–34 (1993).
4. Y. Pan, X. Xu, S. Solanki, X. Liang, R. B. A. Tanjung, C. Tan, and T. C. Chong, "Fast CGH computation using S-LUT on GPU," Opt. Express 17(21), 18543–18555 (2009).
5. J. Jia, Y. Wang, J. Liu, X. Li, Y. Pan, Z. Sun, B. Zhang, Q. Zhao, and W. Jiang, "Reducing the memory usage for effective computer-generated hologram calculation using compressed look-up table in full-color holographic display," Appl. Opt. 52(7), 1404–1412 (2013).
6. T. Nishitsuji, T. Shimobaba, T. Kakue, and T. Ito, "Fast calculation of computer-generated hologram using run-length encoding based recurrence relation," Opt. Express 23(8), 9852–9857 (2015).
7. S.-C. Kim, X.-B. Dong, M.-W. Kwon, and E.-S. Kim, "Fast generation of video holograms of three-dimensional moving objects using a motion compensation-based novel look-up table," Opt. Express 21(9), 11568–11584 (2013).
8. T. Shimobaba and T. Ito, Computer holography: acceleration algorithms and hardware implementations (CRC Press, 2019).

Chapter 11
Computer-Generated Hologram: Multiview Image Approach

Yasuyuki Ichihashi

Abstract This chapter describes a method for calculating holograms from light-ray information such as multiview images. Light-ray information is three-dimensional (3D) information that includes the intensity and direction of light rays, and there are various formats of light-ray information in accordance with the light field display method that reproduces a 3D image from light-ray information. In addition, by converting light-ray information into wavefront information, calculations of computer-generated holograms can be performed. This chapter describes a method for calculating computer-generated holograms based on a real-time capture and reconstruction system with multiple graphics processing units, which the authors constructed in 2012, for a 3D live scene by a generation from integral-photography images.

11.1 Implementation of CGH Based on Light-Ray Information

In this section, we describe the implementation of a program to generate a hologram based on light-ray information [1]. Examples of **light-ray information** include multi-viewpoint images and combinations of two-dimensional (2D) images and depth images. Moreover, it is easy to imagine that CGHs can be calculated by constructing 3D model information from light-ray information.

For example, when using a time-of-flight camera, 2D information and depth information can be obtained. In general, depth information is often divided into about 256 layers because of the dynamic range of the sensor used, and a three-dimensional (3D) model can be constructed by mapping 2D information corresponding to each depth layer. On the other hand, when the viewpoint is changed, there is a problem that an occlusion hole occurs when there is no 2D information from that viewpoint. Therefore, this problem can be solved by using multiple-depth cameras. A proposal

Y. Ichihashi (✉)
Radio Research Institute, National Institute of Information and Communications Technology (NICT), 4-2-1 Nukuikitamachi, Koganei, Tokyo 184-8795, Japan
e-mail: y-ichihashi@nict.go.jp

T. Shimobaba and T. Ito (eds.), *Hardware Acceleration of Computational Holography*,
https://doi.org/10.1007/978-981-99-1938-3_11

has been made to compress 3D information from multiple parallax images and depth information in Ref. [2]. By adopting this proposal, the holograms can be efficiently produced without considering the problem of occlusion holes of the 3D model.

As another technique, cameras are arranged along the circumference, the subject at the center of the circumference is shot, and 3D information is constructed from the entire circumference image. 3D models are estimated from a homography matrix of each camera by arranging 300 cameras along the circumference [3].

In these methods, 3D information can be obtained in accordance with the characteristics of hardware, such as the performance and number of cameras. On the other hand, there is a problem that various calculation processes are required before the calculation of CGH, such as calibration between cameras and image correction after shooting. There is **integral photography (IP)** as a method for taking a 3D image with relatively low preprocessing costs.

IP is a technology that records 3D information on a photographic plate, and it was invented by Lippmann in 1908 [4]. Figure 11.1 shows an overview of IP. A **lens array** is placed between a 3D object and a capturing medium such as a photographic plate. The lens array is made up of many small lenses, which are called elemental lenses. By placing the photographic plate at the focal points of the elemental lenses, object beams are captured in elemental images on the photographic plate near each elemental lens. The size and location of each elemental image are equal to those of each elemental lens. Since object beams propagating from various angles are recorded

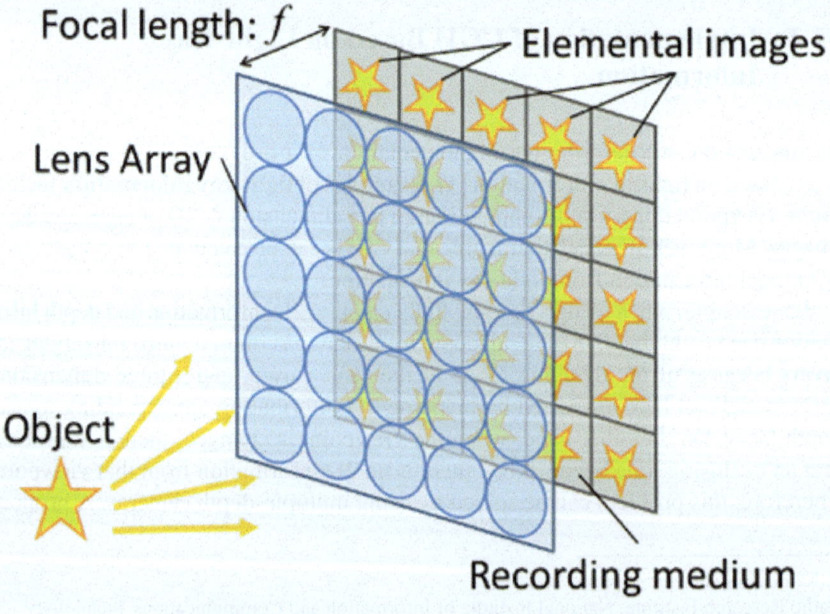

Fig. 11.1 Overview of integral photography. Reprinted with permission from [1] © The Optical Society

in the elemental image, a 3D image can be taken under natural light. Normally, to reproduce a 3D image from this captured IP image, we should rotate each elemental image 180 °C and observe the 3D object with the lens array located at its original position. This is because the image is inverted top to bottom and left to right since the direction used for observing it differs from the direction used when it was captured.

11.2 Converting Light-Ray Information into Wavefront Information

Next, a method for converting light-ray information acquired by IP into wavefront information is described. In Fig. 11.1, the IP image is placed at a location separated from the lens array by a distance equal to the focal length. Similarly, the hologram is placed on the other side of the lens array. f is the focal length of the lens array comprising elemental lenses and d is the distance from the lens array to the hologram. D is the diameter of an elemental image and D_H is the diameter of an elemental hologram. One of the elemental holograms is generated by simulating the propagation of the object light from one of the elemental images. It is evaluated using

$$g_1(x_1, y_1) = \iint_{-\infty}^{+\infty} g_0(x_0, y_0) \cdot e^{jk\left[\frac{(x_1-x_0)^2+(y_1-y_0)^2}{2f}\right]} dx_0 dy_0 \tag{11.1}$$

$$g_2(x_2, y_2) = g_1(x_1, y_1) \cdot e^{-jk\left(\frac{x_1{}^2+y_1{}^2}{2f}\right)} \tag{11.2}$$

$$g_3(x_3, y_3) = \iint_{-\infty}^{+\infty} g_2(x_2, y_2) \cdot e^{jk\left[\frac{(x_3-x_2)^2+(y_3-y_2)^2}{2d}\right]} dx_2 dy_2 \tag{11.3}$$

where $g_0(x_0, y_0)$ is the light intensity distribution of the elemental image, $g_1(x_1, y_1)$ is the light intensity distribution before transmitting object light through the lens array, $g_2(x_2, y_2)$ is the light intensity distribution after transmitting object light through the lens array, and $g_3(x_3, y_3)$ is the light intensity distribution of the elemental hologram. Note that Eqs. (11.1) and (11.3) are the Fresnel diffraction and originally have complex coefficients, but these can be omitted because they do not affect hologram generation. k is the wave number of the object light and λ is the wavelength of the object light. Equations (11.1) and (11.3) are Fresnel diffraction integrals. Equation (11.2) is the phase variation of the object light caused by transmitting the object light through the lens array. If we assume that f is equal to d, the following Fourier transform can be derived from Eqs. (11.1)–(11.3):

$$g_3(x_3, y_3) = \frac{-e^{-2jkf}}{\lambda f} \iint_{-\infty}^{+\infty} g_0(x_0, y_0) \cdot e^{-j2\pi\left[\frac{x_3x_0+y_3y_0}{\lambda f}\right]} dx_0 dy_0 \tag{11.4}$$

Discretizing the variables related to the coordinates, we obtain the following discrete Fourier transform (DFT) corresponding to Eq. (11.4):

$$g_3(X_3\Delta p_m, Y_3\Delta p_n) = \sum_{X_0=1}^{N}\sum_{Y_0=1}^{M} g_0(X_0\Delta p_m, Y_0\Delta p_n)\cdot e^{j2\pi\left\{X_3 Y_0 \frac{\Delta pm^2}{\lambda f} + X_3 Y_0 \frac{\Delta pn^2}{\lambda f}\right\}} \tag{11.5}$$

where M and N are the numbers of pixels in the horizontal and vertical directions, respectively, in the elemental hologram. Δp_m and Δp_n are the horizontal and vertical pixel pitch of the elemental image, respectively. X_0, Y_0, X_3, and Y_3 are discretized variables, and $x_0 = X_0\Delta p_m$, $y_0 = Y_0\Delta p_n$, $x_3 = X_3\Delta p_m$, and $y_3 = Y_3\Delta p_n$, respectively. Moreover, $g_3(x_3, y_3)$ is the complex amplitude distribution of the elemental hologram when the reference light is assumed to be parallel light.

Furthermore, the following equation is derived from Eq. (11.5) when D is equal to D_H and the elemental images and elemental holograms are arranged with no intervening spaces, as shown in Fig. 11.2.

$$M\Delta p_m^2 = N\Delta p_n^2 = \lambda f \tag{11.6}$$

Equation (11.6) shows that the parameters of the IP camera are determined by M and N. Since we can determine the values of M and N arbitrarily, fast Fourier transform (FFT) can be performed efficiently by substituting suitable values for M and N. Moreover, the calculation of each elemental hologram is performed in parallel because elemental images correspond one-to-one with elemental holograms.

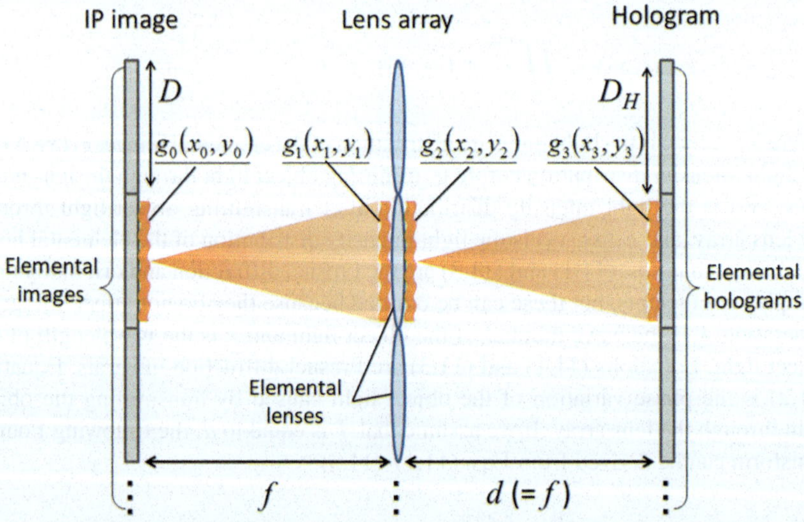

Fig. 11.2 Generation of elemental holograms from elemental images of IP. Reprinted with permission from [1] © The Optical Society

11.3 Implementation of a Program for Generating Holograms from IP Images

Next, the implementation of a program for generating holograms from IP images is described. The size of the entire IP image is defined as "WIDTH", "HEIGHT", and the size of the elemental image is defined as "IZE_OF_EIMAGE" as follows.

```
#define WIDTH (3840) //Width of 4K camera
#define HEIGHT (2160) //Height of 4K camera
#define SIZE_OF_EIMAGE (16) //The size of elemental image
```

The IP image is an 8-bit grayscale bitmap, and the pointer of the array in which the bitmap data is stored is "buf". The pointer of the array for storing the data after converting the IP image into the wavefront is "I". These are declared as global variables for the sake of simplicity.

Listing 11.1 Pointers for an IP image and wavefront.

```
unsigned char *buf;//Pointer of IP image
fftw_complex *I;// Pointer of wavefront plane
```

Here, it is assumed that FFTW is used as a high-speed calculation library for FFT. For the variables declared above, each array can be allocated as follows in the main function.

Listing 11.2 Memory allocation for an IP image and wavefront.

```
buf = (unsigned char *) malloc(sizeof(unsigned char) * WIDTH * HEIGHT);
I = (fftw_complex *) fftw_malloc(sizeof(fftw_complex) * WIDTH * HEIGHT);
```

Listing 11.3 shows the conversion of all elemental images into wavefronts. The function "convIPtoWF(int m, int n)" converts a single elemental image into the corresponding wavefront.

Listing 11.3 Conversion all elemental images into wavefronts.

```
int mmax = WIDTH / SIZE_OF_EIMAGE;
int nmax = HEIGHT / SIZE_OF_EIMAGE;
for(int n = 0; n < nmax; n++){
    for(int m = 0; m < mmax; m++){
        convIPtoWF(m, n);
    }
}
```

Since the size of the IP image is $3,840 \times 2,160$ pixels, when the size of the element image is 16×16 pixels, the number of element images is 240×135. Therefore, m and n are set as variables for the loop, and the (m, n)-th elemental image in the 240×135 element images is sequentially calculated by the function "convIPtoWF (int m, int n)". Since the operation for converting the light-ray information (IP images) into the wavefront is independent for each elemental image, the processing of this function can be sped up by multithreading.

Next, the internal processing of the function “convIPtoWF(int m, int n)” is described in Listing 11.4.

that converts a single elemental image into the corresponding wavefront.,

Listing 11.4 The function “convIPtoWF(int m

```
int size = SIZE_OF_EIMAGE;
int size2 = size * 2;
int size2sq = size2 * size2;
fftwf_plan fp = fftw_plan_dft_2d( size2, size2, in, out, FFTW_FORWARD,
        FFTW_ESTIMATE);
```

To perform FFT on elemental images, an array in and out of an area twice the size of the elemental image is defined. The reason for doubling the area is to eliminate the effects of aliasing based on the sampling theorem. We declare the relevant variables to create the FFTW plan as a forward FFT.

Next, Listing 11.5 shows the preprocessing for wavefront calculation by FFT.

Listing 11.5 Preprocessing for calculating wavefront.

```
int addr;
double theta;
for(int j = 0; j < size; j++){
    for(int i = 0; i < size; i++){
        addr = m * size + i + (n * size + j) * WIDTH; //For sampling plane
        theta = (2. * M_PI * rand()) / RAND_MAX;
        in[(i + size/2) + (j + size/2) * size2][0] = buf[addr] * cos(theta) / size2sq;
        in[(i + size/2) + (j + size/2) * size2][1] = buf[addr] * sin(theta) / size2sq;
    }
}
fftw_execute(fp);
```

Here, elemental image data is input to the FFT array using loop variables “i” and “j”. Since the size of the FFT array is twice as large as the size of the elemental image, it is necessary to input data only to the central portion of the FFT array. The index of the FFT array needs to be devised like the program above. A variable for an address, called “addr”, is declared and used as an index for array “buf” in which IP image data is stored. Thereby, the pixel (i, j) in the (m, n)-th elemental image can be extracted.

Furthermore, by applying a random phase to light-ray information, the reconstructed light is widely diffused and DC light concentration in hologram calculation can be avoided, which greatly improves the quality of the reconstructed image. The variable “theta” in the program is a variable for the phase of random phases. Adding a random phase means including a random initial phase term in a calculation formula for obtaining CGH and is expressed as the following equation:

$$E_i(x_a, y_a) = \frac{A_i}{r_{\alpha i}} e^{j(kr_{\alpha i} + \phi_{i,j})} = \frac{A_i}{r_{\alpha i}} e^{jkr_{\alpha i}} \cdot e^{j\phi_{i,j}} \tag{11.7}$$

The random phase can be added by multiplying $e^{j\phi_{i,j}}$ to the CGH formula from Eq. (11.7). As a result, Eq. (11.7) can be regarded as a spherical wave with each pixel of the elemental image as a light source. Since the array for FFT is a complex number, a value obtained by multiplying the pixel value of the elemental image by a cosine function is input to the real part of the array for FFT, and a value obtained by multiplying the pixel value of the elemental image by a sine function is input to the imaginary part of the array for FFT.

For normalization, it is necessary to divide the input value by the number of FFT elements, that is, the variable "size2sq" After finishing the above processing, FFT is executed by the function "fftw_execute ()".

Finally, the processing after FFT is described.

Listing 11.6 Normalization of the FFT results.

```
for(int j = 0; j < size2; j++){
    for(int i = 0; i < size2; i++){
        in[i + j * size2][0] = out[(i + size) % size2 + ((j + size) % size2) * size2][0];
        in[i + j * size2][1] = out[(i + size) % size2 + ((j + size) % size2) * size2][1];
    }
}
for(int j = 0; j < size; j++){
    for(int i = 0; i < size; i++){
        addr = m * size + i + (n * size + j) * WIDTH; //For sampling plane
        I[addr][0] = in[(i + size/2) + (j + size/2) * size2][0];
        I[addr][0] = in[(i + size/2) + (j + size/2) * size2][1];
    }
}
```

After the FFT, the quadrants must be replaced because the low-frequency region is at the periphery and the high-frequency region is at the center. This is often called "FFT Shift". In this case, the quadrant of "out" after the FFT is changed and input to "in" again. Then, the result converted to the wavefront is input to array "I". At this time, it is necessary to input data with the same care as when reading from the array "buf".

The basic calculation process is as described above. The light-ray information can be created optically using lens arrays and computationally by computer graphics models. However, when input obtained from an actual camera is used, a technique such as the extraction of an elemental image is required. In addition, in the case of the IP image, there is a problem in that the resolution decreases when the object moves away from the light-ray sampling plane. Therefore, it is possible to calculate CGH from a photorealistic model with a large depth by regarding the acquisition of light-ray information by IP as a light-ray sampling plane and propagating the wavefront converted from light-ray information to the hologram plane. This technique is referred to as the ray-sampling method [5]. Specifically, wavefront information $g_3(X_3\Delta p_m, Y_3\Delta p_n)$ of the light-ray sampling plane in Eq. (11.5) is regarded as $o(x_i, y_i, z_i)$, and wavefront propagation is calculated using the following formula:

$$O(x_\alpha, y_\alpha) = o(x_\alpha, y_\alpha, z_\alpha) * g(x_\alpha - x_i, y_\alpha - y_i), \tag{11.8}$$

$$g(x_\alpha - x_i, y_\alpha - y_i) = \frac{e^{jk|z_i|}}{jkz_i} \exp\left[jk \frac{(x_\alpha - x_i)^2 + (y_\alpha - y_i)^2}{2|z_i|} \right] \tag{11.9}$$

For implementation details of propagation calculation, refer to Chap. 8.

References

1. Y. Ichihashi, R. Oi, T. Senoh, K. Yamamoto, and T. Kurita, "Real-time capture and reconstruction system with multiple GPUs for a 3D live scene by a generation from 4K IP images to 8K holograms," Opt. Express **20**, 21645–21655 (2012).
2. T. Senoh, K. Wakunami, Y. Ichihashi, H. Sasaki, R. Oi, and K. Yamamoto, "Multiview image and depth map coding for holographic TV system," Opt. Eng. **53**, 112302 (2014).
3. K. Yamamoto, Y. Ichihashi, T. Senoh, R. Oi, and T. Kurita, "Interactive electronic holography and 300-camera array in dense arrangement," 2012 IEEE International Conference on Acoustics, Speech and Signal Processing (ICASSP), 5453–5456 (2012).
4. G. Lippmann, "Epreuves reversibles donnant la sensation du relief," J. Phys. Theor. Appl. **7**, 821–825 (1908).
5. K. Wakunami and M. Yamaguchi, "Calculation for computer generated hologram using ray-sampling plane," Opt. Express **19**, 9086–9101 (2011).

Chapter 12
Hologram Calculation Using Layer Methods

Harutaka Shiomi

Abstract Recently, many augmented and virtual reality (AR/VR) devices with RGB cameras and depth cameras have been developed. The combination of these cameras enables us to acquire three-dimensional information. This chapter explains the fundamental method for calculating holograms from layer images represented by RGB-D images, and finally, we provide an overview of related methods.

12.1 Introduction

RGB-D images express the three-dimensional (3D) scene with a pair of color and monochrome images, as shown in Fig. 12.1a, b. RGB and monochrome images express the color and depth of a 3D scene, respectively. The RGB-D images were acquired using a **depth camera**, 3D graphics libraries, and a dataset published on the Internet [1, 2].

Because RGB-D images represent the color and depth of each pixel in a 3D scene, we can also treat each pixel as a point light source and calculate the hologram using a point light source-based method. When we treat RGB-D images with Full-HD (1920 × 1080 pixels) resolution as a set of point light sources, there are approximately two million object points and results in a time-consuming calculation. HORN-8 [3, 4], state-of-the-art dedicated processor for hologram calculation, can calculate a Full-HD hologram at 60 frames per second from 60 thousand object points. Because the computational complexity of hologram calculation is proportional to the number of object points, the calculation time for a hologram with two

Supplementary Information The online version contains supplementary material available at https://doi.org/10.1007/978-981-99-1938-3_12.

H. Shiomi (✉)
Graduate School of Engineering, Chiba University, 1-33 Yayoi-cho, Inage-ku, Chiba-shi, Chiba, Japan
e-mail: h-shiomi@chiba-u.jp

T. Shimobaba and T. Ito (eds.), *Hardware Acceleration of Computational Holography*,
https://doi.org/10.1007/978-981-99-1938-3_12

(a) (b)

Fig. 12.1 RGB (**a**) and Depth (**b**) images

million object points is approximately 30 times longer than that with 60 thousand object points. Even if we use the HORN-8 processor, we only achieve hologram calculation from two million object points with two frames per second. However, this method was unrealistic.

This is a faster hologram calculation method for treating RGB-D images as **layer images** parallel to the hologram and calculating the diffraction of each layer image [5–7]. This method is known as the **layer method**. Diffraction calculation methods, such as the angular spectrum method and Fresnel diffraction introduced in Chap. 1, can be used to calculate the object wave on the hologram plane from each layer image. We can calculate the object wave on the hologram plane from the 3D scene represented by RGB-D images by summing all the diffractions of the layer images.

This method can calculate holograms faster because the diffraction calculation can be accelerated by fast Fourier transforms (FFTs). However, because **zero padding** is required for **linear convolution** using FFT, high memory usage is a problem, particularly when calculating high-resolution holograms. When we observe the reconstructed image from a hologram calculated by the layer method from another viewpoint, incorrect occlusion is also a problem because the RGB-D images are captured from a certain viewpoint. The layer method can be regarded as a specialized method for near-eye and head-mounted holographic displays [8–10].

In this section, we first explain the method used to calculate the hologram from RGB-D images and analyze the computational complexity. Second, we show a programming source code written in C++ and the reconstructed image from a hologram calculated using the code. Finally, we introduce the problems of hologram calculation using the FFT-based layer method and discuss recent researches to resolve these problems.

12.2 Method

We explain the method used to calculate the hologram from the RGB-D images using diffraction calculations. The method consists of three steps:

1. Decomposing RGB-D images into layer images
2. Calculating the diffraction calculation of each layer image
3. Converting to the amplitude hologram or phase-only hologram.

We explain each step in the following subsections. In this chapter, we treated the resolution and pixel pitch of the RGB image, depth image, and hologram as the same for simplification. The method described in this section was used to calculate the color hologram.

12.2.1 Decomposing RGB-D Images into Layer Images

As mentioned above, RGB-D images represent the color and distance of each pixel from the hologram, respectively. For example, the depth image shown in Fig. 12.1 shows that the black pixels are close to, and the white pixels are far from the hologram. The layer images consist of the pixels of the RGB images, which are at the same distance from the hologram. When the color values of the RGB and depth images are denoted by Red, $Green$, $Blue$, and $Depth$, respectively, and the ith layer image of the RGB color is denoted by $LayerRed_i$, $LayerGreen_i$, and $LayerBlue_i$, respectively, it is expressed as

$$LayerRed_i(x, y) = \begin{cases} Red(x, y) & Depth(x, y) = i \\ 0 & Depth(x, y) \neq i \end{cases} \tag{12.1}$$

$$LayerGreen_i(x, y) = \begin{cases} Green(x, y) & Depth(x, y) = i \\ 0 & Depth(x, y) \neq i \end{cases} \tag{12.2}$$

$$LayerBlue_i(x, y) = \begin{cases} Blue(x, y) & Depth(x, y) = i \\ 0 & Depth(x, y) \neq i. \end{cases} \tag{12.3}$$

We extracted the layer image of a certain depth from the RGB-D images using the code in Listing 12.1.

Listing 12.1 The function extracts a layer image at a certain depth from RGB-D images.

```
// The below function extracts the specified layer image from
    RGB-D images.
// Red, Green, Blue, and Depth are the image whose pixel is
    represented by 8-bit.
// LayerRed, LayerGreen, and LayerBlue are the list of std::
    complex<float> for later calculations.
// i is the layer number.
// nx and ny are the horizontal and vertical resolutions,
    respectively.
void extract_layer(const uint8_t* Red, const uint8_t* Green, const uint8_t* Blue,
    const uint8_t* Depth, std::complex<float>* LayerRed,
    std::complex<float>* LayerGreen, std::complex<float>* LayerBlue,
    const uint32_t i, const uint32_t ny, const uint32_t nx) {

    for (uint32_t y = 0; y < ny; ++y) {
        for (uint32_t x = 0; x < nx; ++x) {
            // Calculation the position in the one dimensional
                memory.
            uint32_t pos = y * nx + x;

            if (Depth[pos] == i) {
                LayerRed[pos] = std::complex<float>(Red[pos], 0);
                LayerGreen[pos] = std::complex<float>(Green[pos], 0);
                LayerBlue[pos] = std::complex<float>(Blue[pos], 0);
            }
            else {
                LayerRed[pos] = std::complex<float>(0, 0);
                LayerGreen[pos] = std::complex<float>(0, 0);
                LayerBlue[pos] = std::complex<float>(0, 0);
            }
        }
    }
}
```

12.2.2 Calculating the Diffraction Calculation of Each Layer Image

We treated each layer image as the **sectional images** (layers) of a 3D scene and summed the diffracted results from each layer image to obtain the hologram. In this calculation, it was necessary to set the initial phase of the optical waves. **Random phases** are often used. Another method is called the **compensate phase** [11], which sets the phase corresponding to the distance of each layer from the hologram. The compensating phase optimizes the phase information to improve the image reconstruction quality [12]. This section shows the code using the random and compensate phases in Listing 12.2.

Listing 12.2 The function for the random and compensate phases.

```
void random_phase(std::complex<float>* LayerRed, std::complex<float>* LayerGreen,
    std::complex<float>* LayerBlue, const uint32_t ny, const uint32_t nx) {

    // Initial setting for uniformly distributed random number
    std::random_device seed_gen;
    std::mt19937 engine(seed_gen());
    std::uniform_real_distribution<float> dist(-0.5 * M_PI, 0.5 * M_PI);
    std::complex<float> ImaginaryUnit(0, 1);
    for (uint32_t y = 0; y < ny; ++y) {
        for (uint32_t x = 0; x < nx; ++x) {

            uint32_t pos = y * nx + x;
            LayerRed[pos] *= std::exp(ImaginaryUnit * dist(engine));
            LayerGreen[pos] *= std::exp(ImaginaryUnit * dist(engine));
            LayerBlue[pos] *= std::exp(ImaginaryUnit * dist(engine));
        }
    }
}

void compensate_phase(std::complex<float>* LayerRed,
    std::complex<float>* LayerGreen, std::complex<float>* LayerBlue,
    const float distance, const float lambda_red, const float lambda_green,
    const float lambda_blue, const uint32_t ny, const uint32_t nx) {

    // Calculating the wave number.
    const float wavenumber_red = 2 * M_PI / lambda_red;
    const float wavenumber_green = 2 * M_PI / lambda_green;
    const float wavenumber_blue = 2 * M_PI / lambda_blue;
    std::complex<float> ImaginaryUnit(0, 1);
    for (uint32_t y = 0; y < ny; ++y) {
        for (uint32_t x = 0; x < nx; ++x) {

            uint32_t pos = y * nx + x;
            LayerRed[pos] *= std::exp(-ImaginaryUnit
                                * wavenumber_red * distance);
            LayerGreen[pos] *= std::exp(-ImaginaryUnit
                                * wavenumber_green * distance);
            LayerBlue[pos] *= std::exp(-ImaginaryUnit
                                * wavenumber_blue * distance);
        }
    }
}
```

As mentioned above, diffraction calculations can be performed using the angular spectrum method and Fresnel diffraction. In this study, we used the angular spectrum method. $\mathcal{F}[\cdot]$ and $\mathcal{F}^{-1}[\cdot]$ denote the Fourier and its inverse transforms, respectively. The transfer function for each color of the ith layer image is denoted by H_i^{red}, H_i^{green}, and H_i^{blue}. The hologram calculation is expressed as

$$HologramRed(x,y)=\sum_{i=0}^{255}\left(\mathcal{F}^{-1}\left[\mathcal{F}[LayerRed_i]\odot H_i^{red}\right]\right) \tag{12.4}$$

$$HologramGreen(x,y)=\sum_{i=0}^{255}\left(\mathcal{F}^{-1}\left[\mathcal{F}[LayerGreen_i]\odot H_i^{green}\right]\right) \tag{12.5}$$

$$HologramBlue(x,y)=\sum_{i=0}^{255}\left(\mathcal{F}^{-1}\left[\mathcal{F}[LayerBlue_i]\odot H_i^{blue}\right]\right), \tag{12.6}$$

where $\odot$ denotes the Hadamard product. In the computational process, $\mathcal{F}$ and $\mathcal{F}^{-1}$ are performed using the FFT. FFT is the most time-consuming process in this calculation. We can focus on the linearity of the FFT and change the order of the summation and IFFT to reduce the number of FFTs for a faster calculation.

$$HologramRed(x,y)=\mathcal{F}^{-1}\left[\sum_{i=0}^{255}(\mathcal{F}[LayerRed_i]\odot H_i^{red})\right] \tag{12.7}$$

$$HologramGreen(x,y)=\mathcal{F}^{-1}\left[\sum_{i=0}^{255}\mathcal{F}[LayerGreen_i]\odot H_i^{green})\right] \tag{12.8}$$

$$HologramBlue(x,y)=\mathcal{F}^{-1}\left[\sum_{i=0}^{255}\mathcal{F}[LayerBlue_i]\odot H_i^{blue})\right]. \tag{12.9}$$

We show the implemented code in Listing 12.3. In this code, we use the functions such as zeropadding(), fft(), AsmTransferF(), fftshift(), mult(), multscalar(), ifft(), and crop() introduced in Chap. 8.

Listing 12.3 The function to calculate the hologram from RGB-D images.

```
ï»¿void add_complex(std::complex<float>* in1, std::complex<float>* in2,
    std::complex<float>* out, int32_t ny, int32_t nx) {
    for (int32_t n = 0; n < ny; n++) {
        for (int32_t m = 0; m < nx; m++) {
            out[m + n * nx] = in1[m + n * nx] + in2[m + n * nx];
        }
    }
}

void calculate_hologram(std::complex<float>* HologramRed,
    std::complex<float>* HologramGreen, std::complex<float>* HologramBlue,
    const uint8_t* Red, const uint8_t* Green, const uint8_t* Blue,
    const uint8_t* Depth, const float lambda_red, const float lambda_green,
    const float lambda_blue, const float pitch_y, const float pitch_x,
    const float zmin, const float dz, const uint32_t ny, const uint32_t nx) {

    // The memory for the layer image.
    std::complex<float>* LayerRed = new std::complex<float>[ny * nx];
    std::complex<float>* LayerGreen = new std::complex<float>[ny * nx];
```

```
std::complex<float>* LayerBlue = new std::complex<float>[ny * nx];

// The memory for the zeropadding data.
std::complex<float>* LayerRed_pad = new std::complex<float>[2 * ny * 2 * nx];
std::complex<float>* LayerGreen_pad = new std::complex<float>[2 * ny * 2 * nx];
std::complex<float>* LayerBlue_pad = new std::complex<float>[2 * ny * 2 * nx];

// The memory for the transfer function.
std::complex<float>* H_red = new std::complex<float>[2 * ny * 2 * nx];
std::complex<float>* H_green = new std::complex<float>[2 * ny * 2 * nx];
std::complex<float>* H_blue = new std::complex<float>[2 * ny * 2 * nx];

// The memory for the summation before IFFT.
std::complex<float>* HologramRed_pad
   = new std::complex<float>[2 * ny * 2 * nx];
std::complex<float>* HologramGreen_pad
   = new std::complex<float>[2 * ny * 2 * nx];
std::complex<float>* HologramBlue_pad
   = new std::complex<float>[2 * ny * 2 * nx];

for (uint32_t i = 0; i < 256; ++i) {
   std::cout << i << " / " << 255 << "\r";

   // The distance from the i-th layer image and the
       hologram.
   const float z = zmin + i * dz;

   // Extracting the layer image from RGB-D images.
   extract_layer(Red, Green, Blue, Depth,
      LayerRed, LayerGreen, LayerBlue, i, ny, nx);

   // Setting the phase information.
   // In the case of random phase.
   random_phase(LayerRed, LayerGreen, LayerBlue, ny, nx);
   // In the case of compensate phase.
   /* compensate_phase(LayerRed, LayerGreen, LayerBlue, z,
      lambda_red, lambda_green, lambda_blue, ny, nx); */

   // Zeropadding.
   zeropadding(LayerRed, LayerRed_pad, ny, nx);
   zeropadding(LayerGreen, LayerGreen_pad, ny, nx);
   zeropadding(LayerBlue, LayerBlue_pad, ny, nx);

   // The Fourier transform.
   fft(LayerRed_pad, LayerRed_pad, 2 * ny, 2 * nx);
   fft(LayerGreen_pad, LayerGreen_pad, 2 * ny, 2 * nx);
   fft(LayerBlue_pad, LayerBlue_pad, 2 * ny, 2 * nx);

   // Calculation of the transfer function.
   const float dv = 1 / (pitch_y * 2 * ny), du = 1 / (pitch_x * 2 * nx);
   AsmTransferF(H_red, 2 * ny, 2 * nx, dv, du, lambda_red, z);
   AsmTransferF(H_green, 2 * ny, 2 * nx, dv, du, lambda_green, z);
   AsmTransferF(H_blue, 2 * ny, 2 * nx, dv, du, lambda_blue, z);
```

```
        fftshift(H_red, 2 * ny, 2 * nx);
        fftshift(H_green, 2 * ny, 2 * nx);
        fftshift(H_blue, 2 * ny, 2 * nx);

        // Hadamard products.
        mult(LayerRed_pad, H_red, LayerRed_pad, 2 * ny, 2 * nx);
        mult(LayerGreen_pad, H_green, LayerGreen_pad, 2 * ny, 2 * nx);
        mult(LayerBlue_pad, H_blue, LayerBlue_pad, 2 * ny, 2 * nx);

        // Normalization
        multscalar(LayerRed_pad, 1.0 / (2 * ny * 2 * nx), 2 * ny, 2 * nx);
        multscalar(LayerGreen_pad, 1.0 / (2 * ny * 2 * nx), 2 * ny, 2 * nx);
        multscalar(LayerBlue_pad, 1.0 / (2 * ny * 2 * nx), 2 * ny, 2 * nx);

        // Summation
        add_complex(HologramRed_pad, LayerRed_pad,
            HologramRed_pad, 2 * ny, 2 * nx);
        add_complex(HologramGreen_pad, LayerGreen_pad,
            HologramGreen_pad, 2 * ny, 2 * nx);
        add_complex(HologramBlue_pad, LayerBlue_pad,
            HologramBlue_pad, 2 * ny, 2 * nx);
    }

    // The inverse Fourier transform
    ifft(HologramRed_pad, HologramRed_pad, 2 * ny, 2 * nx);
    ifft(HologramGreen_pad, HologramGreen_pad, 2 * ny, 2 * nx);
    ifft(HologramBlue_pad, HologramBlue_pad, 2 * ny, 2 * nx);

    // Cropping
    crop(HologramRed_pad, HologramRed, 2 * ny, 2 * nx);
    crop(HologramGreen_pad, HologramGreen, 2 * ny, 2 * nx);
    crop(HologramBlue_pad, HologramBlue, 2 * ny, 2 * nx);

    // Deallocation of the used memory.
    delete[] LayerRed;
    delete[] LayerGreen;
    delete[] LayerBlue;
    delete[] LayerRed_pad;
    delete[] LayerGreen_pad;
    delete[] LayerBlue_pad;
    delete[] H_red;
    delete[] H_green;
    delete[] H_blue;
    delete[] HologramRed_pad;
    delete[] HologramGreen_pad;
    delete[] HologramBlue_pad;

}
```

12.2.3 Converting to an Amplitude Hologram or Phase-Only Hologram

The object light O calculated above has complex values. Spatial light modulators (SLM) that can display complex amplitudes are not generally used. The generally available SLM is amplitude-modulated or phase-modulated SLM. Here, we describe a method to convert the complex amplitude of the object light into data that is suitable for these SLMs.

First, we explain the conversion to an **amplitude hologram**. When the reference light is denoted by R, the interference pattern between the object and the reference light is expressed as

$$I(x, y) = |O(x, y) + R(x, y)|^2. \quad (12.10)$$

In an inline hologram, the reference light is a plane wave. When its amplitude and phase are regarded as units, the reference light can be $R(x, y) = 1$, and the interference pattern I is

$$I(x, y) = |O(x, y) + 1|^2 = |O(x, y)|^2 + 1 + O(x, y) + O^*(x, y). \quad (12.11)$$

The first term $|O(x, y)|^2 + 1$ is negligible because it is constant. The remaining term can be $O(x, y) + O^*(x, y) = 2Re[O(x, y)]$. The constant coefficient 2 was also negligible. Therefore, an amplitude hologram can be obtained from the real part of the complex amplitude.

However, in the case of conversion to the **phase-only hologram**, all the amplitudes on the hologram are assumed to be equal, and only phase information is used. Therefore, the phase-only hologram is calculated as $\tan^{-1}(\mathrm{Im}\{O\}/\mathrm{Re}\{O\})$ where $\mathrm{Re}\{\}$ and $\mathrm{Im}\{\}$ denote the operators to extract the real and imaginary parts from a complex value, respectively.

In these three steps, the hologram is calculated from the RGB-D images. Faster calculation of diffraction with FFT is a good point of the layer method. We compared the computational complexity between the **point-cloud method** and the layer method in the case of a single layer. In the point cloud-based method, each pixel is regarded as an isolated point-light source. The hologram calculation is the summation of all spherical waves from the point light sources. When the resolution of the layer image and hologram was $H \times W$, the computational complexity was $O(H^2W^2)$. In the layer method, diffraction calculations, such as the angular spectrum method and Fresnel diffraction, are performed using FFT. In this case, a two-dimensional FFT of resolution $H \times W$ is performed. The computational complexity is $O(HW \log(HW))$. Thus, the hologram calculation is much faster with the FFT.

12.3 Results

We show the holograms and reconstructed images calculated using the aforementioned method. We calculated the amplitude hologram from the RGB-D images, as shown in Fig. 12.1. The calculation conditions are as follows: The red, green, and blue wavelengths are 650, 532, and 450 nm, respectively. The pixel pitch is 3.74 mm. The distance between the hologram and its nearest layer image was 10 cm, and the distance between each layer image was 0.1 mm. The resolution of the hologram and RGB-D images was 2048 × 2048 pixels. We set the phase information of all layer images to the random phase. Figure 12.2 shows the amplitude hologram with the random phase for each color. We calculated the inverse diffraction from the holograms of each color and showed the reconstructed images at each distance from the hologram. The reconstructed images in Figs. 12.3 contain the speckle noise. This was caused by the random phase. Figure 12.4 shows the reconstructed images from the hologram calculated using the compensate phase, which sets the constant phase corresponding to the distance from the hologram.

We present the reconstructed images from the phase-only hologram under the same calculation conditions. Figures 12.5 and 12.6 show the reconstructed images from the hologram with random and compensate phases, respectively

Finally, we show the reconstructed images from a **complex hologram**, which contains amplitude and phase information, in Figs. 12.7 and 12.8.

Fig. 12.2 The amplitude holograms of each color calculated with the random phase

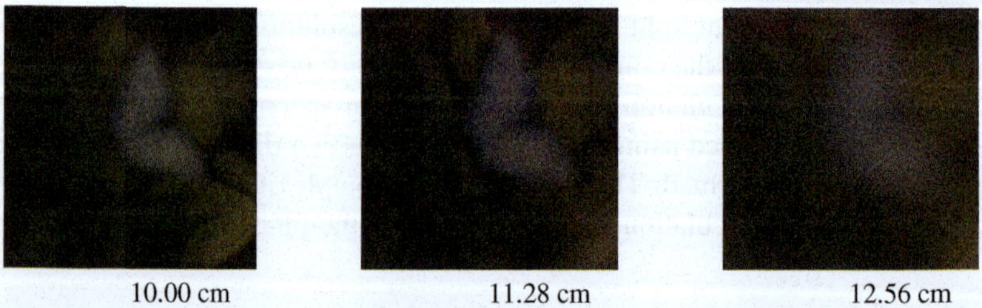

Fig. 12.3 The reconstructed images at each distance from the amplitude hologram calculated with the random phase

10.00 cm 11.28 cm 12.56 cm

Fig. 12.4 The reconstructed images at each distance from the amplitude hologram calculated with the compensate phase

10 cm 11.28 cm 12.56 cm

Fig. 12.5 The reconstructed images at each distance from the phase-only hologram calculated with the random phase

10.00 cm 11.28 cm 12.56 cm

Fig. 12.6 The reconstructed images at each distance from the phase-only hologram calculated with the compensate phase

10 cm 11.28 cm 12.56 cm

Fig. 12.7 The reconstructed images at each distance from the complex hologram calculated with the random phase

Fig. 12.8 The reconstructed images at each distance from the complex hologram calculated with the compensate phase

Using the computational environment of an Intel Core i7-6500U CPU, Windows 10 Home operating system, and Microsoft Visual C++2019 compiler, the calculation time was 365 s.

12.4 Discussion

The layer method can calculate a hologram from RGB-D images faster because of the FFT. However, it is not a silver bullet. FFTs are the most time-consuming with this method. The number of FFTs significantly affects the calculation time. The number of FFTs depends on the number of layer images. Therefore, the calculation time was proportional to the number of layer images. When we calculate from three-dimensional scenes with many layer images, for example, depth images with a deep bit-depth, the calculation time increases.

FFTs can be accelerated by parallel computation using graphics processing units (GPUs). In this method, the memory usage should be four times larger than the original resolution because of zero padding for the linear convolution with the FFT. When calculating a high-resolution hologram, it may be difficult to use the GPU because of the limitation of GPU memory. To address this memory usage issue, a method called implicit convolution has been proposed [13]. This method enables the calculation of linear convolutions without zero padding.

Finally, the efficiency of the layer method is discussed. When the depth of the three-dimensional scenes is greater, the layer images tend to be sparse. The calculation time of FFT does not rely on the sparsity of the data. In the case of sparse data, this method also calculates the area without light waves, resulting in a decrease in efficiency. The look-up table (LUT) method [14], which uses LUT, and the wavelet shrinkage-based superposition (WASABI) [15], which calculates the diffraction in the wavelet space with the wavelet transform, has been proposed as a fast calculation method that focuses on the sparsity of the data and decreases memory usage. Both methods can accelerate the hologram calculation compared to FFT-based layer methods when 3D scenes are close to the hologram.

Other methods for calculating a hologram from RGB-D images have been suggested using machine learning [16]. These methods allow for much faster calculations; however, they must be learned in advance and are limited to near-eye holographic display systems.

References

1. Honauer, Katrin and Johannsen, Ole and Kondermann, Daniel and Goldluecke, Bastian, "A dataset and evaluation methodology for depth estimation on 4D light fields," Asian conference on computer vision, 19–34 (2016).
2. Johannsen, Ole and Honauer, Katrin and Goldluecke, Bastian and Alperovich, Anna and Battisti, Federica and Bok, Yunsu and Brizzi, Michele and Carli, Marco and Choe, Gyeongmin and Diebold, Maximilian and others, "A taxonomy and evaluation of dense light field depth estimation algorithms," Proceedings of the IEEE Conference on Computer Vision and Pattern Recognition Workshops, 82–99 (2017).
3. Yamamoto, Yota and Nakayama, Hirotaka and Takada, Naoki and Nishitsuji, Takashi and Sugie, Takashige and Kakue, Takashi and Shimobaba, Tomoyoshi and Ito, Tomoyoshi, "Large-scale electroholography by HORN-8 from a point-cloud model with 400,000 points," Opt. Express **26**, 34259–34265 (2018).
4. Nishitsuji, Takashi and Yamamoto, Yota and Sugie, Takashige and Akamatsu, Takanori and Hirayama, Ryuji and Nakayama, Hirotaka and Kakue, Takashi and Shimobaba, Tomoyoshi and Ito, Tomoyoshi, "Special-purpose computer HORN-8 for phase-type electro-holography," Opt. Express **26**, 26722–26733 (2018).
5. Okada, Naohisa and Shimobaba, Tomoyoshi and Ichihashi, Yasuyuki and Oi, Ryutaro and Yamamoto, Kenji and Oikawa, Minoru and Kakue, Takashi and Masuda, Nobuyuki and Ito, Tomoyoshi, "Band-limited double-step Fresnel diffraction and its application to computer-generated holograms," Opt. Express **21**, 9192–9197 (2013).
6. "Rapid hologram generation utilizing layer-based approach and graphic rendering for realistic three-dimensional image reconstruction by angular tiling," Journal of Electronic Imaging **23**, 023016 (2014).
7. Zhao, Yan and Cao, Liangcai and Zhang, Hao and Kong, Dezhao and Jin, Guofan, "Accurate calculation of computer-generated holograms using angular-spectrum layer-oriented method," Opt. Express **23**, 25440–25449 (2015).
8. Yoneyama, Takuo and Yang, Chanyoung and Sakamoto, Yuji and Okuyama, Fumio, "Eyepiece-type full-color electro-holographic display for binocular vision," Practical Holography XXVII: Materials and Applications **8644**, 251–258 (2013).
9. Maimone, Andrew and Georgiou, Andreas and Kollin, Joel S, "Holographic near-eye displays for virtual and augmented reality," ACM Transactions on Graphics (Tog) **36**, 1–16 (2017).
10. Chang, Chenliang and Bang, Kiseung and Wetzstein, Gordon and Lee, Byoungho and Gao, Liang, "Toward the next-generation VR/AR optics: a review of holographic near-eye displays from a human-centric perspective," Optica **7**, 1563–1578 (2020).
11. Chun Chen and Byounghyo Lee and Nan-Nan Li and Minseok Chae and Di Wang and Qiong-Hua Wang and Byoungho Lee, "Multi-depth hologram generation using stochastic gradient descent algorithm with complex loss function," Opt. Express **29**, 15089–15103 (2021).
12. Gerchberg, Ralph W, "A practical algorithm for the determination of plane from image and diffraction pictures," Optik **35**, 237–246 (1972).
13. Tomoyoshi Shimobaba and Takayuki Takahashi and Yota Yamamoto and Takashi Nishitsuji and Atsushi Shiraki and Naoto Hoshikawa and Takashi Kakue and Tomoyoshi Ito, "Efficient diffraction calculations using implicit convolution," OSA Continuum **2**, 642–650 (2018).

14. Kim, Seung-Cheol and Kim, Eun-Soo, "Effective generation of digital holograms of three-dimensional objects using a novel look-up table method," Appl. Opt. **47**, D55–D62 (2008).
15. Shimobaba, Tomoyoshi and Ito, Tomoyoshi, "Fast generation of computer-generated holograms using wavelet shrinkage," Opt. Express **25**, 77–87 (2017).
16. Shi, Liang and Li, Beichen and Kim, Changil and Kellnhofer, Petr and Matusik, Wojciech, "Towards real-time photorealistic 3D holography with deep neural networks," Nature **591**, 234–239 (2021).

Chapter 13
Polygon-Based Hologram Calculation Methods

Fan Wang

Abstract In computer-graphics (CG) technologies, three-dimensional (3D) objects are generally considered to comprise a set of micro-polygons. Therefore, the polygon-based method is significant in generating holograms. This chapter introduces six methods of polygon-based holograms implemented in a MATLAB environment. All these can be classified using numerical-and analytical-based methods. We theoretically analyze the differences in each approach comprehensively and compare their performances on the same calculation platform. This chapter addresses only computational issues based on triangular meshes, and not rendering issues.

13.1 General Instruction of Polygon-Based Holograms

Unlike the point-based hologram calculation method, a **polygon-based hologram** cannot be generated by ray tracing. This is because each light-emitting mesh comprises three vertices of an oblique triangle, and the pixels inside the triangle are continuous rather than discrete. Hence, the critical technique of the polygon-based method considers the diffraction field distribution calculation of an arbitrarily tilted triangle in the hologram plane according to three vertices.

In general, the polygon-based method simulates the diffraction process in the frequency domain to obtain spectra in the hologram plane. For a 3D object that includes N triangles in the global system (x, y, z), as illustrated in Fig. 13.1a, we first calculated the spectra of the ith triangle to the target plane, defined as $FH_i(f_x, f_y)$, where (f_x, f_y) represent the frequency coordinates corresponding to the global system. Subsequently, the spectra of all triangles were summed to obtain the spectral and diffraction field distributions of the entire 3D object, while $E(x, y)$ is solved using

Supplementary Information The online version contains supplementary material available at https://doi.org/10.1007/978-981-99-1938-3_13.

F. Wang (✉)
Graduate School of Engineering, Chiba University, 1-33 Yayoi-cho, Inage-ku, Chiba, Japan
e-mail: wangfan@chiba-u.jp

T. Shimobaba and T. Ito (eds.), *Hardware Acceleration of Computational Holography*,
https://doi.org/10.1007/978-981-99-1938-3_13

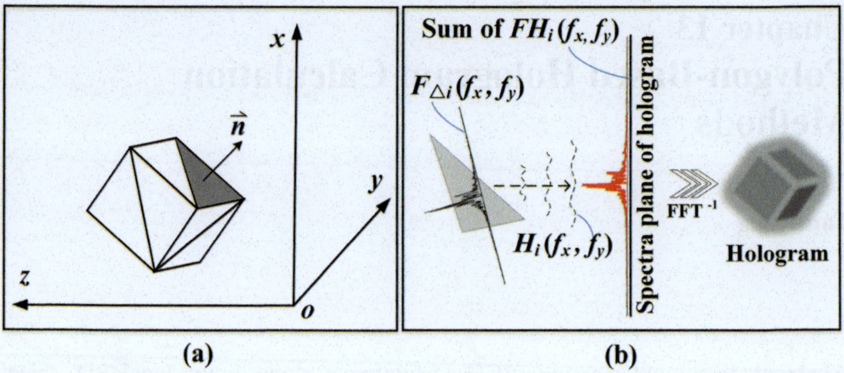

Fig. 13.1 Overview of polygon-based methods

an inverse fast Fourier transform (FFT), as illustrated in Fig. 13.1b. The following equation expresses the physical profile of this process:

$$E(x, y) = \mathcal{F}^{-1}\left[\sum_{i=1}^{N} FH_i(f_x, f_y)\right], \tag{13.1}$$

where operation $\mathcal{F}^{-1}$ denotes the inverse FFT of its argument. Based on the angular spectrum theory, the spectra on the hologram plane, $FH_i(f_x, f_y)$, can be calculated by

$$FH_i(f_x, f_y) = F_{\Delta_i}(f_x, f_y) \cdot H_i(f_x, f_y), \tag{13.2}$$

where $F_{\Delta_i}(f_x, f_y)$ denotes the frequency spectrum of the ith triangle in the global system. $H_i(f_x, f_y) = e^{j2\pi f_z z}$ is the transfer function, where $j = \sqrt{-1}$ and $f_z(f_x, f_y) = \sqrt{1/\lambda^2 - f_x^2 - f_y^2}$, and λ is the wavelength.

From Fig. 13.1b, the $F_{\Delta_i}(f_x, f_y)$ does not exactly match the spectra plane of the hologram because the triangle is not parallel to the hologram plane. Therefore, obtaining the spectra of oblique triangles $F_{\Delta_i}(f_x, f_y)$ corresponding to the frequency coordinates of the hologram plane, is the main issue of polygon-based holograms. In this section, the generation methods of polygon-based holograms are classified into six categories according to different methods of obtaining the tilted spectra, as illustrated in Fig. 13.2. In the next section, we comprehensively describe each method's theory and implementation process, as illustrated in Fig. 13.2.

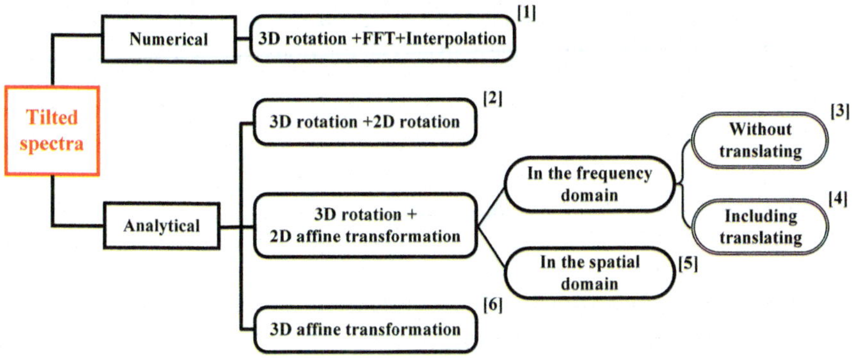

Fig. 13.2 A framework for six polygon-based methods

13.2 Overview of Six Methods for Polygon-Based

As illustrated in Fig. 13.3a, an arbitrary △ABC in the global system (x, y, z) is tilted on the hologram plane. It was assumed that the triangular mesh is a self-luminous object that emits a uniform plane light wave E_0 into the hologram plane. We define

$$E_0 = a_0 \cdot e^{j2\pi(f_{x_0}x+f_{y_0}y+f_{z_0}z)}, \tag{13.3}$$

where a_0 is the light amplitude, generally regarded as constant. $f_{x_0} = \cos\alpha/\lambda$, $f_{y_0} = \cos\beta/\lambda$ and $f_{z_0} = f_z(f_{x_0}, f_{y_0})$ are the angular spectrum components of the light source along the x-, y-, and z-axes, respectively, where α and β denote the propagation direction angles of E_0 along the x- and y- axes, respectively. By building the local coordinate system with **n** as the z' axis, △ABC in the global system can be rotated to the local system as △A′B′C′, as illustrated in Fig. 13.3, where θ

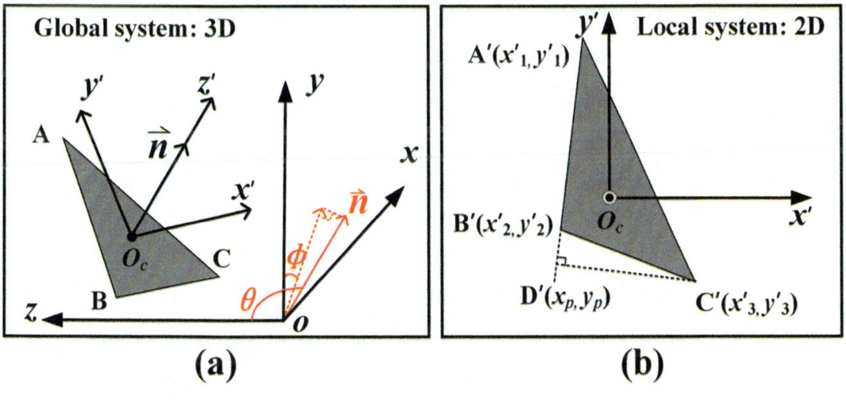

Fig. 13.3 Arbitrary triangle in the global (**a**) and in the local (**b**) coordinates system

denotes the angle of the normal **n** to the z-axis, and ϕ is the angle of the projection vector of **n** in the xOy plane to the y-axis. The rotational relationship from $\triangle$ABC to $\triangle$A′B′C′ can be expressed as

$$[x', y', 0]^\mathsf{T} = R[x - x_c, y - y_c, z - z_c]^\mathsf{T}, \tag{13.4}$$

where $R = \begin{bmatrix} \cos\phi\cos\theta & \sin\phi\cos\theta & -\sin\theta \\ -\sin\phi & \cos\phi & 0 \\ \cos\phi\sin\theta & \sin\phi\sin\theta & \cos\theta \end{bmatrix}$ denotes the 3D **rotation matrix**. We note that R is an orthogonal matrix. (x_c, y_c, z_c) represents the coordinate of the center of gravity O_c of the $\triangle$ABC, as illustrated in Fig. 13.3.

In general, the spectrum of $\triangle$ABC in a global system can be calculated as

$$F_{\triangle \mathrm{ABC}}(f_x, f_y) = \iint_{\triangle\,\mathrm{ABC}} E_0 e^{-j2\pi(f_x x + f_y y)}\, dxdy. \tag{13.5}$$

Suppose that the light emitted from the triangular mesh propagates along the negative direction of z axis, as illustrated in Fig. 13.3a, $\alpha = \beta = 0$, and the light amplitude is defined as $a_0 = 1$, while Eq. (13.3) is $E_0 = e^{j2\pi z/\lambda}$. Therefore, the spectral distribution of the hologram plane in Eq. (13.2) is

$$\begin{aligned} FH_i(f_x, f_y) &= \iint_{\triangle\,\mathrm{ABC}} e^{j2\pi z/\lambda} \cdot e^{-j2\pi(f_x x + f_y y)}\, dxdy \cdot H_i(f_x, f_y) \\ &= \iint_{\triangle\,\mathrm{ABC}} e^{-j2\pi(f_x x + f_y y + f_z z - z/\lambda)}\, dxdy\,, \end{aligned} \tag{13.6}$$

where $H_i(f_x, f_y) = e^{-j2\pi f_z z}$, and the negative sign in the index indicates a negative propagation. Based on the rotational relationship given in Eq. (13.4), the equation above can be rewritten as

$$FH_i(f_x, f_y) = J_r E_1 \iint_{\triangle \mathrm{A'B'C'}} e^{-j2\pi(\hat{f}_x x' + \hat{f}_y y')} dx'dy', \tag{13.7}$$

where

$$J_r = (\cos\phi\cos\theta\cos\phi + \sin\phi\cos\theta\sin\phi)\,, \tag{13.8}$$

is the **Jacobian** determinant resulting from the coordinate transformation, and

$$E_1 = e^{-j2\pi(f_x x_c + f_y y_c + f_z z_c - z_c/\lambda)}\,, \tag{13.9}$$

is a constant factor for the triangular spatial position. $\hat{f}_x$ and $\hat{f}_y$ in Eq. (13.7) are the rotated frequency coordinates, which are determined by

$$\begin{cases} \hat{f}_x = R_{11} f_x + R_{12} f_y + R_{13} f_z - R_{13}/\lambda \\ \hat{f}_y = R_{21} f_x + R_{22} f_y + R_{23} f_z - R_{23}/\lambda \,. \end{cases} \tag{13.10}$$

In Eq. (13.7), the integral term $\iint_{\Delta A'B'C'} e^{-j2\pi(\hat{f}_x x' + \hat{f}_y y')} dx'dy'$ can be considered the frequency spectrum of ΔA′B′C′ in the local system, which is defined as $F_{\Delta A'B'C'}(\hat{f}_x, \hat{f}_y)$. Therefore, Eq. (13.7) can be simplified to

$$FH_i(f_x, f_y) = J_r \cdot E_1 \cdot F_{\Delta A'B'C'}(\hat{f}_x, \hat{f}_y). \tag{13.11}$$

This equation indicates that the spectrum of ΔABC on the hologram plane depends primarily on the spectra of ΔA′B′C′ in the local system.

Hence, the key task is to solve the tilted spectrum $F_{\Delta A'B'C'}(\hat{f}_x, \hat{f}_y)$. All of the above 3D rotations can be considered as preparation for obtaining the value of $F_{\Delta A'B'C'}(\hat{f}_x, \hat{f}_y)$. Matsushima et al. [1], Kim et al. [2], Ahrenberg et al. [3], Zhang et al. [4], Liu et al. [5] and Pan et al. [6] proposed various methods for solving this problem. All these are introduced in the next subsection.

Listing 13.1 provides a **MATLAB** program for implementing from Eqs. (13.1) to (13.10). A graphic processing unit was used to accelerate the calculations.

Listing 13.1 Common codes used in the 3D rotation.

```
%% Basic parameters definition: /All the physical units are
    millimeter
dp=0.008; % hologram pixel size
lambda=532e-6; k=2*pi/lambda; % wavelength and wave number
Nx=1920; Ny=1080; % hologram sampling numbers
Lx=Nx*dp; Ly=Ny*dp; % hologram plane size in physics
%% frequency coordinates in the global system
fx=linspace(-1/dp/2,1/dp/2-1/Lx,Nx);
fy=linspace(1/dp/2,-1/dp/2+1/Ly,Ny);
[fx,fy]=meshgrid(fx,fy');
fx=gpuArray(fx);
fy=gpuArray(fy); % import data into GPU
fz=@(fx,fy)1/lambda-lambda/2.*(fx.^2+fy.^2);
%% light source from the triangle mesh
fx0=0; fy0=0;
a0=1;
%% triangle ABC in the global system
A=[5,2,286]'; B=[3,-1,308]'; C=[-4,-3,334]';
V=[A B C];
center=mean(V,2); % center of gravity
xc=center(1); yc=center(2); zc=center(3);
AB=[B(1)-A(1) B(2)-A(2) B(3)-A(3)];
BC=[C(1)-B(1) C(2)-B(2) C(3)-B(3)];
n=cross(AB,BC)'; % normal vector: n(nx,ny,nz)
%% 3D rotation transformtion
if abs(n(3))>=cosd(89.9) % Remove extremely tilted triangles
  R=eye(3) ;
  phi=0; theta=0; % predefine rotation parameters
  if n(1)~=0 || n(2)~=0
    ez=[0,0,1];
    if n(2)<0
      n=-n; % Counterclockwise rotation
    end
    theta=acos(ez*n/norm(n));
    ex=[1,0];
    phi=acos(ex*[n(1);n(2)]/norm([n(1);n(2)]));
```

```
  R=[cos(phi)*cos(theta) cos(theta)*sin(phi) -sin(theta);
      -sin(phi) cos(phi) 0 ;
    cos(phi)*sin(theta) sin(phi)*sin(theta) cos(theta)];
  end
end
Vr=R*(V-center); % A'B'C' in the local system
Jr=R(1,1)*R(2,2)-R(1,2)*R(2,1); % Rotation Jacobbian factor
E1=exp(-1j*2*pi*(xc*fx+yc*fy-zc*fz(fx,fy)+zc/lambda));
fx_hat=R(1,1)*fx+R(1,2)*fy+R(1,3)*fz(fx,fy)-R(1,3)/lambda;% fx'
fy_hat=R(2,1)*fx+R(2,2)*fy+R(2,3)*fz(fx,fy)-R(2,3)/lambda;% fy'
```

13.2.1 *Numerical-Based Method*

In 2003, Matsushima [1] proposed an numerical-based method for calculating the tilted spectrum, which comprised three steps: drawing a rasterized triangle, FFT operation, and interpolating the spectrum. Listing 13.2 presents Matsushima's method.

Step 1: drawing. Based on the introduction in Sect. 13.2, the tilted $\triangle$ABC shown in Fig. 13.4a is rotated into the $\triangle$A′B′C′, where the vertexes coordinates of the $\triangle$A′B′C′ are obtained by the matrix R expressed by Eq. (13.4). $\triangle$A′B′C′ is rasterized and drawn on a two-dimensional (2D) canvas sampled at the same interval as the hologram. For simplicity, we assigned each pixel inside the triangle to "1" and outside to "0", as presented in Fig. 13.4b. Figure 13.4c presents an extended sampling window via **zero padding** to avoid convolution errors (see chap. 8) triggered by the angular spectrum propagation. The rasterized triangle $\triangle$A′B′C′ is generated by calling a subfunction called `@plot_tri(vertex,pitch)` in the program, shown in line 1 of Listing 13.2.

Step 2: performing FFT. The spectrum of $\triangle$A′B′C′ can be calculated easily by performing an FFT operation on the image of $\triangle$A′B′C′. However, note that the origin O' of the local system (x', y') in Fig. 13.4c does not generally coincide with the center of gravity O_c of $\triangle$A′B′C′, and we define this offset as the vector $\overrightarrow{O_c O'}$, as shown in Fig. 13.4c. Therefore, the spectrum of $\triangle$A′B′C′ in the local system can be represented as:

$$F_{\triangle \mathrm{A'B'C'}}(f'_x, f'_y) = \mathcal{F}(\triangle \mathrm{A'B'C'}) \exp\left(j2\pi \overrightarrow{O_c O'} \cdot \mathbf{f}'\right), \tag{13.12}$$

where $\mathcal{F}$ denotes the Fourier transformation, f'_x and f'_y represent the regular sampling grids of the 2D canvas in the frequency domain, as shown in the black area in Fig. 13.5a, and the vector $\mathbf{f}' = [f'_x, f'_y]$. Line 7 of Listing 13.2 implements this operation.

Step 3: interpolation. Equation (13.12) gives the spectrum of $\triangle$A′B′C′, $F_{\triangle \mathrm{A'B'C'}}(f'_x, f'_y)$, which is sampled with regular grids (f'_x, f'_y), and we expect to obtain the spectrum $F_{\triangle \mathrm{A'B'C'}}(\hat{f}_x, \hat{f}_y)$ shown in Eq. (13.11), which is sampled with irregular grids

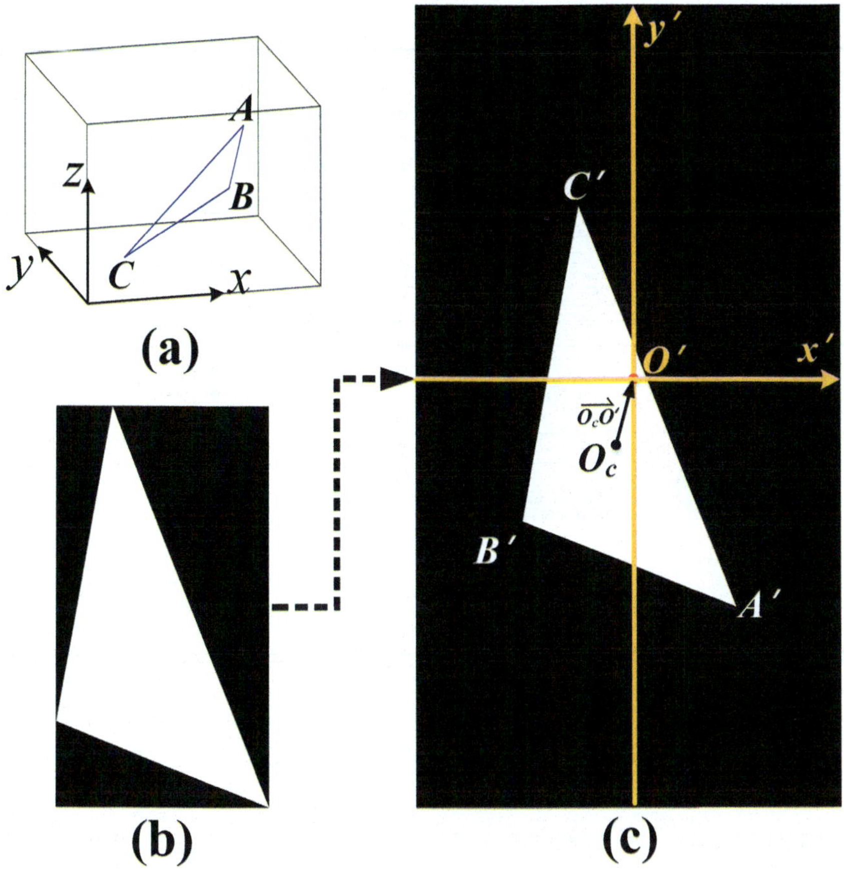

Fig. 13.4 **a** Schematic diagram of the tilted △ABC in the global system. **b** Rasterized △A′B′C′ drawn on the 2D canvas of the local system. **c** Extended sampling window by zero-padding, O' is the origin of the local system and O_c is the center of gravity of the △A′B′C′

$(\hat{f}_x, \hat{f}_y)$ given in Eq. (13.10). Therefore, the **interpolation** method is used to obtain $F_{\triangle A'B'C'}(\hat{f}_x, \hat{f}_y)$ based on $F_{\triangle A'B'C'}(f'_x, f'_y)$, which is represented as

$$F_{\triangle A'B'C'}(\hat{f}_x, \hat{f}_y) = \text{Interpolate}\left(F_{\triangle A'B'C'}(f'_x, f'_y)\right), \tag{13.13}$$

where Interpolate(·) denotes an interpolation operation. As shown in Fig. 13.5a, the black area (f'_x, f'_y) is a regular rectangle with equal sampling intervals, whereas the red area $(\hat{f}_x, \hat{f}_y)$ is an irregular quadrilateral with variable sampling intervals. Figure 13.5b shows an enlarged view of the specific portion, indicating that the red circles are addressed by interpolation based on the black dots. Commonly used interpolation methods include linear interpolation, spline interpolation, and cubic

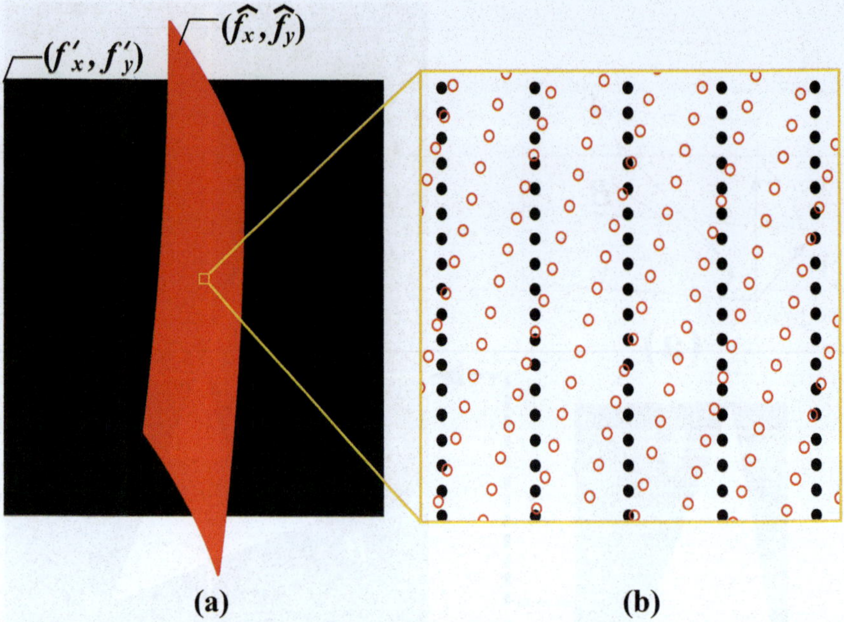

Fig. 13.5 Schematic of the spectrum interpolation. **a** The black area represents the spectral region of (f'_x, f'_y) and the red area represents that of $(\hat{f}_x, \hat{f}_y)$. **b** The enlarged view of the specific portion of (**a**), where "black dots" are the regular samples with equal intervals and "red circle" are the target samples with irregular intervals obtained by interpolation

interpolation, etc. These methods cause differences in accuracy and computational effort.

With the three steps above, we solved the tilted spectrum $F_{\triangle A'B'C'}(\hat{f}_x, \hat{f}_y)$, so that the hologram can be obtained using Eqs. (13.11) and (13.1), as shown in lines 9–10 of Listing 13.2. This method allows the flexible rendering of objects, such as applying random phases and textures, because the surface information of each triangle can be customized in the *step 1 (drawing)*.

Listing 13.2 Matsushima's method: FFT solution.

```
1  [tri0,L_p,N_p,sft]=plot_tri(Vr,dp); % draw a rasterized triangle
2  tri=padarray(tri0,[N_p(2)/2 N_p(1)/2],0,'both'); % extended the sampling
       window
3  fx_p=linspace(-1/2/dp,1/2/dp-1/L_p(1)/2,N_p(1)*2);
4  fy_p=linspace(1/2/dp,-1/2/dp+1/L_p(2)/2,N_p(2)*2);
5  [fx_p,fy_p]=meshgrid(fx_p,fy_p); % the regular sampling grid
6  F_p=fftshift(fft2(fftshift(tri)));
7  F_p=F_p.*exp(1j*2*pi*(sft(1)*fx_p+sft(2)*fy_p)); % FFT: Eq. (1.12)
8  F_hat=interp2(fx_p,fy_p,F_p,fx_hat,fy_hat,'spline',0); % interpolate:
       Eq. (1.13)
9  FH=Jr.*E1.*F_hat; % spectrum: eq.(1.11)
10 E=fftshift(ifft2(FH)); % hologram: eq.(1.1)
```

13.2.2 2D Rotation-Based Method

In 2008, Kim et al. [2] proposed an **analytical method** based on a 2D rotation in a local system to solve the tilted spectra, which was implemented in Listing 13.3.

As illustrated in Fig. 13.3b, the point $\mathrm{D}'(x_p, y_p)$ is the perpendicular foot of the side A′B′. The subfunction `@pft(vertex)` in line 2 of Listing 13.3 is used to solve the perpendicular foot D′. We translated $\mathrm{D}'(x_p, y_p)$ to the origin of the local system, thus becoming $\mathrm{D}''(0, 0)$, as shown in Fig. 13.6a. Then, we rotated the angle ψ counterclockwise around the point D″, so that the point C′ falls on the y-axis and becomes the point C″, defining the new $\Delta\mathrm{A}''\mathrm{B}''\mathrm{C}''$, as shown in Fig. 13.6b. The vertices of the $\Delta\mathrm{A}''\mathrm{B}''\mathrm{C}''$ are $\mathrm{A}''(a, 0)$, $\mathrm{B}''(b, 0)$, and $\mathrm{C}''(0, c)$, respectively. The program in Listing 13.3 provides an approach to obtaining the values of a, b and c (see lines 15–17).

Furthermore, the frequency spectrum of the $\Delta\mathrm{A}''\mathrm{B}''\mathrm{C}''$ can be derived relatively easily as an analytical expression. A sub-function named `@F_kim` was built in line 22 of Listing 13.3 to calculate the analytical spectrum, defined as $F_{\Delta\mathrm{A}''\mathrm{B}''\mathrm{C}''}(f_x'', f_y'')$, where (f_x'', f_y'') are the frequency coordinates used to sample $\Delta\mathrm{A}''\mathrm{B}''\mathrm{C}''$. From the 2D rotated relationship, the frequency sampling coordinates (f_x'', f_y'') can be expressed as follows:

$$\begin{cases} f_x'' = \cos\psi \hat{f}_x - \sin\psi \hat{f}_y\,, \\ f_y'' = \sin\psi \hat{f}_x + \cos\psi \hat{f}_y \end{cases} \tag{13.14}$$

where $(\hat{f}_x, \hat{f}_x)$ is expressed by Eq. (13.10). Therefore, the tilted spectra of $\Delta\mathrm{A}'\mathrm{B}'\mathrm{C}'$ can be calculated as:

$$F_{\Delta\mathrm{A}'\mathrm{B}'\mathrm{C}'}(\hat{f}_x, \hat{f}_y) = E_2 \cdot F_{\Delta\mathrm{A}''\mathrm{B}''\mathrm{C}''}(f_x'', f_y'')\,, \tag{13.15}$$

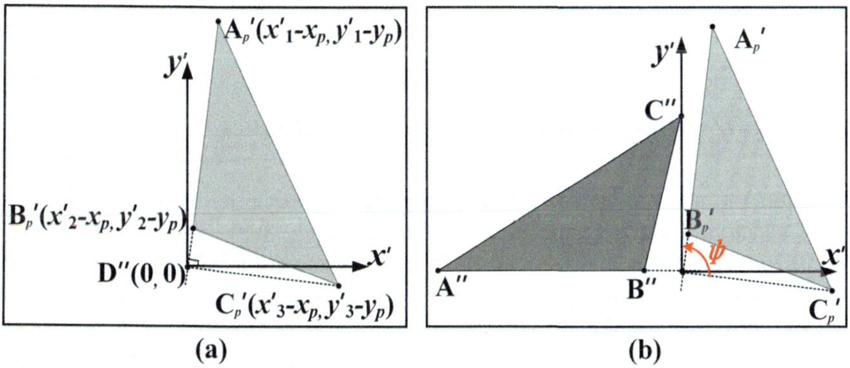

Fig. 13.6 **a** Translating the perpendicular foot D′ to the origin of the local system, defined as D″. **b** Rotating the $\Delta\mathrm{A}'\mathrm{B}'\mathrm{C}'$ around the point D″, so that A″B″ coincides with x'-axis, defined as the new $\Delta\mathrm{A}''\mathrm{B}''\mathrm{C}''$

where $E_2 = e^{-j2\pi(\hat{f}_x x_p + \hat{f}_y y_p)}$ is a compensation factor resulting from the translation and analytical solution of $F_{\Delta A''B''C''}(f_x'', f_y'')$ can be obtained using Eq. (A4) of Ref. [2].

It is also worth noting that $F_{\Delta A''B''C''}(f_x'', f_y'')$ also depends on the values of a, b, and c, because each △ABC corresponds to a different △A″B″C″. This differs from the method of obtaining the tilted spectra using a 2D affine, which will be introduced in the next subsection.

Listing 13.3 Kim's method: 2D rotation solution analytically.

```
%% solve perpendicular foot
pft= pft(Vr); % sub-function
xp=pft(1); yp=pft(2); % D ' (x ' ,y')
%% 2D rotation: angle-->\psi
DC=[Vr(1,3)−xp,Vr(2,3)−yp];
if DC(1)<0
   DC=−DC; % counterclockwise rotation
end
ey=[0,1];
psi=acos(ey∗DC'/norm(DC)); % rotation angle
Tr2=[cos(psi) −sin(psi) 0;
     sin(psi) cos(psi) 0;
        0 0 1] ;
Vr2=Tr2∗(Vr−[xp,yp,−1]');
a=Vr2(1,1);
b=Vr2(1,2);
c=Vr2(2,3);
fx_2p=fx_p∗cos(psi)−fy_p∗sin(psi);%fx''
fy_2p=fx_p∗sin(psi)+fy_p∗cos(psi);%fy''
%% spectrum calculation
E2=exp(−1j∗2∗pi∗(fx_p∗xp+fy_p∗yp));
F_hat=E2.∗F_kim(fx_2p,fy_2p,a,b,c); % Spectra of A'B'C'
FH=Jr.∗E1.∗F_hat; % Spectra: eq.(1.11)
E=fftshift(ifft2(FH)); % Hologram: eq.(1.1)
```

13.2.3 2D Affine Transformation-Based Method

In 2008, Ahrenberg et al. [3] proposed another analytical method based on a 2D affine transformation [7], which was implemented in Listing 13.4.

Affine theory states that there must be a mapping relationship between two triangles containing translation, rotation, and scaling information. As shown in Fig. 13.7, the fixed primitive triangle △uvw is located in the local coordinate system with the vertex coordinates $(0, 0)$, $(1, 0)$, and $(1, 1)$. Then, △A′B′C′ can be mapped as △uvw using an **affine matrix** $[T_a]_{2\times3}$, which is

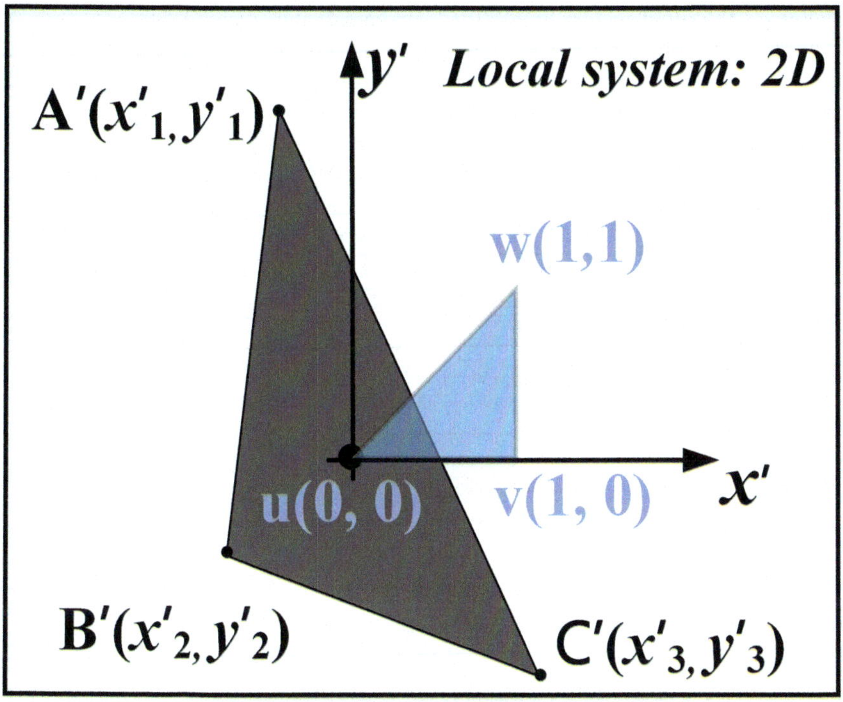

Fig. 13.7 2D affine transformation in the local system from △A′B′C′ to the primitive triangle △uvw

$$\begin{bmatrix} x_1' & x_2' & x_3' \\ y_1' & y_2' & y_3' \end{bmatrix} = \begin{bmatrix} T_{a_{11}} & T_{a_{12}} & T_{a_{13}} \\ T_{a_{21}} & T_{a_{22}} & T_{a_{23}} \end{bmatrix} \begin{bmatrix} 0 & 1 & 1 \\ 0 & 0 & 1 \\ 1 & 1 & 1 \end{bmatrix} . \tag{13.16}$$

From the above equation, the matrix T_a is determined by the vertex values of △A′B′C′, as shown in lines 3–4 of Listing 13.4. In contrast to Kim's method [2] in Sect. 13.2.2, the vertex coordinates of △uvw are constant, and its frequency spectrum can be solved analytically, defined as $F_{\triangle \mathrm{uvw}}(f_u, f_v)$, where (f_u, f_v) are the frequency coordinates used to sample △uvw. The program in Listing 13.4 invokes a sub-function named `@F_pri` to calculate $F_{\triangle \mathrm{uvw}}(f_u, f_v)$, whose analytical expression is given in Eq. (14) of Ref. [3].

From the 2D affine relationship, the frequency sampling coordinates (f_u, f_v) can be expressed as

$$\begin{cases} f_u = T_{a_{11}} \hat{f}_x + T_{a_{21}} \hat{f}_y \\ f_v = T_{a_{12}} \hat{f}_x + T_{a_{22}} \hat{f}_y . \end{cases} \tag{13.17}$$

Therefore, the tilted spectra of $\Delta A'B'C'$ can be calculated as:

$$F_{\Delta A'B'C'}(\hat{f}_x, \hat{f}_y) = J_a \cdot E_2 \cdot F_{\Delta uvw}(f_u, f_v)\,, \tag{13.18}$$

where $J_a = Ta_{11} \cdot Ta_{22} - Ta_{12} \cdot Ta_{21}$ is the Jacobian factor and $E_2 = e^{-j2\pi(T_{a_{13}}\hat{f}_x + T_{a_{23}}\hat{f}_y)}$.

Listing 13.4 Ahrenberg's method: 2D affine transformation solution analytically.

```
%% 2D affine transformation
xr=Vr(1,:); yr=Vr(2,:);
Ta=[xr(2)−xr(1) xr(3)−xr(2) xr(1);
   yr(2)−yr(1) yr(3)−yr(2) yr(1)];
Ja=Ta(1,1)*Ta(2,2)−Ta(1,2)*Ta(2,1);
%% affine frequency coordinates
fu=Ta(1,1)*fx_hat+Ta(2,1)*fy_hat;
fv=Ta(1,2)*fx_hat+Ta(2,2)*fy_hat;
E2=exp(−1j*2*pi*(fx_hat*Ta(1,3)+fy_hat*Ta(2,3)));
F_hat=Ja*E2.*F_pri(fu,fv);
FH=Jr.*E1.*F_hat;
E=fftshift(ifft2(FH)); % Hologram
```

13.2.4 2D Affine Transformation-Based Method with Translation

In 2018, Zhang et al. [4] proposed an affine analytical method similar to Ahrenberg's method [3]; however, it was implemented in the same global system and required translation. Listing 13.5 provides a program for this method.

As illustrated in Fig. 13.8, the original triangle ΔABC is rotated around the center of gravity to be $\Delta A'B'C'$ that is in the gravity plane ($z = z_c$). Therefore, the rotational relationship in Eq. (13.4) can be rewritten as

$$[x', y', z']^\mathsf{T} = R[x - x_c, y - y_c, z - z_c]^\mathsf{T} + [x_c, y_c, z_c]^\mathsf{T}\,. \tag{13.19}$$

Substituting the above equation into Eq. (13.6), the hologram spectra $FH_i(f_x, f_y)$ of Eq. (13.11) are expressed as follows:

$$FH_i(f_x, f_y) = J_r \cdot E_1 \cdot E_1' \cdot F_{\Delta A'B'C'}(\hat{f}_x, \hat{f}_y)\,. \tag{13.20}$$

where $E_1' = e^{j2\pi(\hat{f}_x x_c + \hat{f}_y y_c)}$ which is the translational factor resulting from the 3D rotation, as shown in line 1 of Listing 13.5.

Furthermore, primitive Δuvw with constant vertices $(0, 0, z_c)$, $(1, 0, z_c)$ and $(1, 1, z_c)$ in Fig. 13.8 is located in the gravity plane. There is a 2D affine transformation relationship between $\Delta A'B'C'$ and Δuvw given by Eq. (13.16). The difference

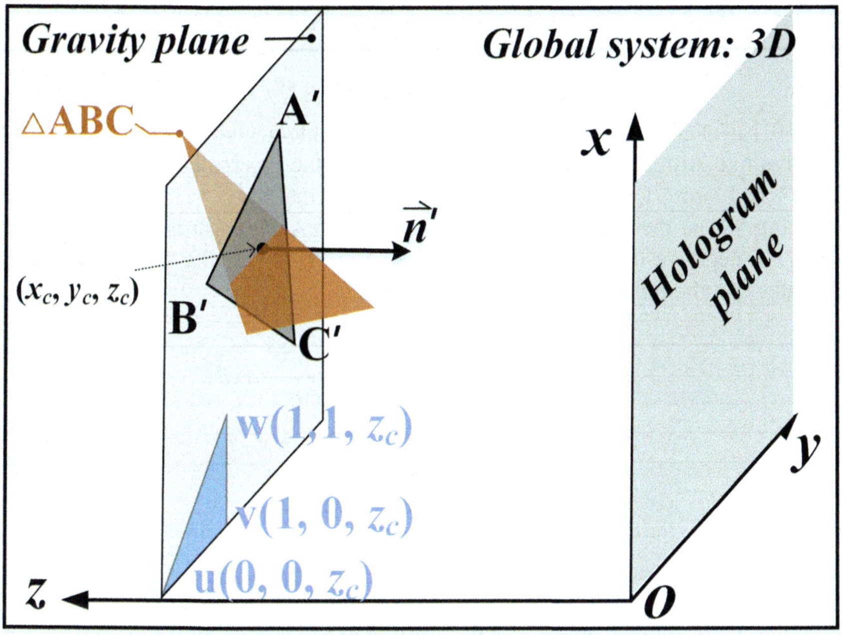

Fig. 13.8 2D affine transformation in the global system including translation

is that △A′B′C′ is not in the local system but the $z = z_c$ plane of the global system. The tilted spectra of △A′B′C′ can still be calculated using Eqs. (13.17), and (13.18).

Listing 13.5 Zhang's method: 2D affine transformation solution analytically with a translation.

```
%% 2D affine transformation
E1_p=exp(1j*2*pi*(fx_hat.*xc+fy_hat.*yc)); % E1'
Vr=Vr+center; % translate to the gravity plane
xr=Vr(1,:); yr=Vr(2,:);
Ta=[xr(2)-xr(1) xr(3)-xr(2) xr(1);
   yr(2)-yr(1) yr(3)-yr(2) yr(1)];
Ja=Ta(1,1)*Ta(2,2)-Ta(1,2)*Ta(2,1);
%% affine frequency coordinates
fu=Ta(1,1)*fx_hat+Ta(2,1)*fy_hat;
fv=Ta(1,2)*fx_hat+Ta(2,2)*fy_hat;
E2=exp(-1j*2*pi*(fx_hat*Ta(1,3)+fy_hat*Ta(2,3)));
F_hat=Ja*E2.*F_pri(fu,fv);
FH=Jr.*E1.*E1_p.*F_hat;
E=fftshift(ifft2(FH)); % Hologram
```

13.2.5 2D Affine Transformation-Based Method in the Spatial Domain

In 2010, Liu et al. proposed an analytical method implemented in the spatial domain, which is like Ahrenberg's method, but eliminates the inverse Fourier transform in Eq. (13.1). Listing 13.6 provides a program for this method.

In Fig. 13.3a, the light wave emitted from the $\triangle$ABC mesh is set to propagate along the negative direction of z-axis, then $E_0 = e^{j\frac{2\pi}{\lambda}z}$. Scalar diffraction theory indicates that the light field in a hologram is

$$\begin{aligned} E(x_h, y_h) &= \frac{1}{j\lambda}\iint_{\triangle ABC} a_0 E_0 \frac{e^{-j\frac{2\pi}{\lambda}r}}{r} dxdy \\ &= \frac{a_0}{j\lambda r}\iint_{\triangle ABC} e^{j\frac{2\pi}{\lambda}(z-r)} dxdy\,, \end{aligned} \tag{13.21}$$

where (x_h, y_h) is the coordinate of the hologram pixel and the minus sign in the wavefront $\frac{e^{-j\frac{2\pi}{\lambda}r}}{r}$ indicates negative propagation, where $r=\sqrt{(x_h-x)^2+(y_h-y)^2+z^2}$ is the distance between points (x, y, z) within $\triangle$ABC and pixels $(x_h, y_h, 0)$. The 3D rotational relationship between $\triangle$ABC and $\triangle$A′B′C′ is given by Eq. (13.4); thus, by replacing (x, y, z) in Eq. (13.21) with (x', y'), $(z-r)$ can be expanded as

$$z - r \approx (z_c - r_0) - \frac{x'^2 + y'^2}{2r_0} - \frac{\hat{x_h}x' + \hat{y_h}y'}{r_0}\,, \tag{13.22}$$

where $r_0=\sqrt{(x_h-x_c)^2+(y_h-y_c)^2+z_c^2}$ is the distance between the center of gravity of $\triangle$ABC and hologram pixel. The new coordinates resulting from the 3D rotation are

$$\begin{cases} \hat{x_h} = R_{11}(x_c-x_h)+R_{12}(y_c-y_h)+R_{13}z_c - R_{13}r_0 \\ \hat{y_h} = R_{21}(x_c-x_h)+R_{22}(y_c-y_h)+R_{23}z_c - R_{23}r_0\,, \end{cases} \tag{13.23}$$

where R_{ij} denotes an element of the rotational matrix given in Eq. (13.4).

Therefore, $E(x_h, y_h)$ in Eqs. (13.21) is

$$E(x_h, y_h) = J_r \frac{a_0 e^{j\frac{2\pi}{\lambda}(z_c-r_0)}}{j\lambda r_0}\iint_{\triangle A'B'C'} e^{-j2\pi(\frac{\hat{x_h}}{\lambda r_0}x'+\frac{\hat{y_h}}{\lambda r_0}y')} dx'dy'\,, \tag{13.24}$$

where J_r denotes the rotational Jacobian factor given by Eq. (13.8). Note that the equation above omits the quadratic phase factor $e^{-j\frac{2\pi}{\lambda}\frac{x'^2+y'^2}{2r_0}}$ because it does not contribute to hologram reconstruction. The integral term above can be considered as the frequency of $\triangle$A′B′C′:

$$F_{\triangle A'B'C'}(\hat{f_x}, \hat{f_y}) = \iint_{\triangle A'B'C'} e^{-j2\pi(\frac{\hat{x_h}}{\lambda r_0}x'+\frac{\hat{y_h}}{\lambda r_0}y')} dx'dy'\,, \tag{13.25}$$

where $\hat{f}_x = \frac{\hat{x}_h}{\lambda r_0}$ and $\hat{f}_y = \frac{\hat{y}_h}{\lambda r_0}$. Furthermore, similar to Ahrenberg's method [3], $F_{\Delta A'B'C'}(\hat{f}_x, \hat{f}_y)$ can be calculated by solving $F_{\Delta uvw}(f_u, f_v)$ which is the spectrum of the primitive Δuvw, as stated from Eqs. (13.16) to (13.18) according to the 2D affine transformation. Line 17 of Listing 13.6 uses the same sub-function `@F_pri` to calculate Eq. (13.25).

Unlike other methods, this method calculates the hologram directly from Eq. (13.24) without inverse FFT, as shown in line 19 of Listing 13.6.

Listing 13.6 Liu's method: 2D affine transformation solution analytically in the spatial domain.

```
1  %% 3D rotation coordinates transformation
2  xh=-dp/2-(Nx/2-1)*dp:dp:dp/2+(Nx/2-1)*dp;
3  yh=dp/2+(Ny/2-1)*dp:-dp:-dp/2-(Ny/2-1)*dp;
4  [xh,yh]=meshgrid(xh,yh);
5  r0=sqrt((xh-xc).^2+(yh-yc).^2+zc^2);
6  xh_hat=R(1,1)*(xc-xh)+R(1,2)*(yc-yh)+R(1,3)*zc-R(1,3)*r0; %xh'
7  yh_hat=R(2,1)*(xc-xh)+R(2,2)*(yc-yh)+R(2,3)*zc-R(2,3)*r0; %yh'
8  %% 2D affine
9  xr=Vr(1,:); yr=Vr(2,:);
10 Ta=[xr(2)-xr(1),xr(3)-xr(2),xr(1);
11    yr(2)-yr(1),yr(3)-yr(2),yr(1)];
12 Ja=Ta(2,2)*Ta(1,1)-Ta(1,2)*Ta(2,1);
13 %% affine frequency coordinates
14 fu=(Ta(1,1)*xh_hat+Ta(2,1)*yh_hat)./lambda./r0;
15 fv=(Ta(1,2)*xh_hat+Ta(2,2)*yh_hat)./lambda./r0;
16 E2=exp(-1j*2*pi*(Ta(1,3)*xh_hat+Ta(2,3)*yh_hat)/lambda./r0);
17 F_hat=Ja*E2.*F_pri(fu,fv);
18 E1=exp(1j*k*(-r0+zc))./(1j*lambda*r0);
19 E=Jr.*E1.*F_hat; % hologram, Eq. (24)
```

13.2.6 3D Affine Transformation-Based Method

In 2014, Pan et al. proposed an analytical method based on **3D affine transformation**, which successfully avoided all processes, such as 3D rotation, 2D affine, or translation. Listing 13.7 provides a program for this method. Here, we adopted a new approach to derive the 3D affine method.

As illustrated in Fig. 13.9, an arbitrary ΔABC is located in the global coordinates system (x, y, z). We establish a local Cartesian system: (x', y', z') using the normal ΔABC as the z'-axis, and a primitive triangle with vertices $u(0, 0, 0)$, $v(1, 0, 0)$, and $w(1, 1, 0)$ is defined in the local system. Then, there must be an affine relationship between ΔABC and Δuvw, which is expressed as:

$$[x, y, z]^{\mathsf{T}} = T[x', y', z', 1]^{\mathsf{T}}, \tag{13.26}$$

where T is a matrix with 3×4 representing the 3D affine transformation, and the superscript ^{T} denotes the transposition of the matrix. T can be solved by

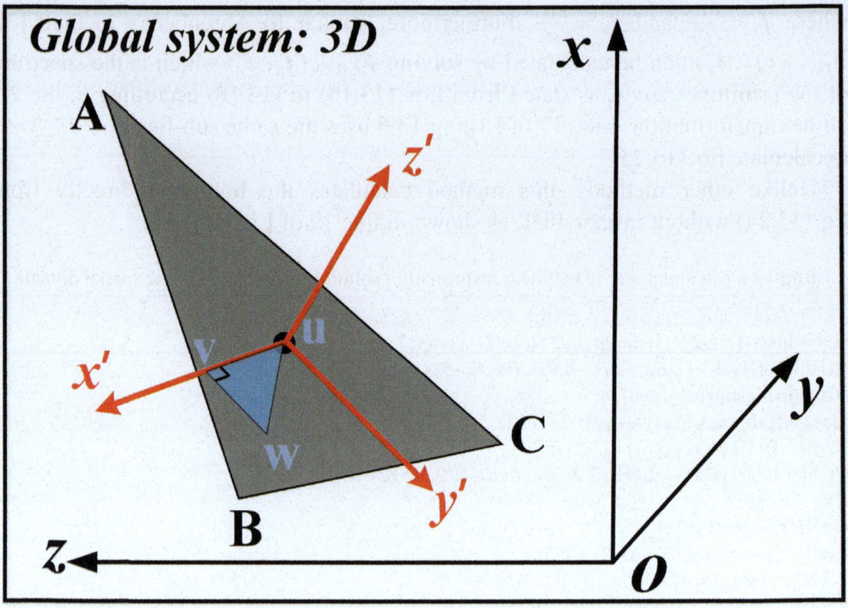

Fig. 13.9 3D affine transformation. △ABC is the arbitrary triangle in the global system; △uvw is the primitive triangle in the local system

$$T = [x, y, z]^{\mathsf{T}}[x', y', z', 1]^{\mathsf{T}^{\dagger}}, \tag{13.27}$$

where $[\cdot]^{\dagger}$ denotes the pseudo-inverse matrix of the argument. $[x', y', z', 1]^{\mathsf{T}}$ is a singular matrix, owing to $z' \equiv 0$.

According to the mapping relationship between (x, y, z) and (x', y', z') in Eq. (13.27), the frequency spectrum of △ABC mentioned in Eq. (13.6) can be expressed as follows:

$$\begin{aligned} FH_i(f_x, f_y) &= J \cdot E_1 \iint_{\triangle \mathrm{uvw}} e^{-j2\pi(\hat{f}_x x' + \hat{f}_y y')} dx'dy' \\ &= J \cdot E_1 \cdot F_{\triangle \mathrm{uvw}}(\hat{f}_x, \hat{f}_y), \end{aligned} \tag{13.28}$$

where

$$\begin{cases} J = T_{11}T_{22} - T_{12}T_{21} \\ E_1 = e^{-j2\pi(f_x T_{14} + f_y T_{24} - f_z T_{34} + T_{34}/\lambda)} \\ \hat{f}_x = T_{11} f_x + T_{21} f_y - T_{31} f_z + T_{31}/\lambda \\ \hat{f}_y = T_{12} f_x + T_{22} f_y - T_{32} f_z + T_{32}/\lambda \end{cases}. \tag{13.29}$$

$F_{\triangle \mathrm{uvw}}(\hat{f}_x, \hat{f}_y)$ in Eq. (13.28) is the spectrum of primitive △uvw, which can be solved analytically as $F_{\triangle \mathrm{uvw}}(f_u, f_v)$ in Eq. (13.18).

The 3D affine method does not adopt the standard codes given in Listing 13.1 because no 3D rotation step is described in the initial part of Sect. 1.2. Instead, the 3D affine method program is the most concise, as shown in Listing 13.7.

Listing 13.7 Pan's method: 3D affine transformation solution analytically.

```
V=[A B C];
P=[0 1 1;0 0 1;0 0 0;1,1,1]; %primitive triangle
T=V*pinv(P); % 3D affine matrix
%% affine frequency coordinates
fx_hat=T(1,1)*fx+T(2,1)*fy-T(3,1)*fz(fx,fy)+T(3,1)/lambda;%fx'
fy_hat=T(1,2)*fx+T(2,2)*fy-T(3,2)*fz(fx,fy)+T(3,2)/lambda;%fy'
F_hat=F_pri(fx_hat,fy_hat); % Spectra of triangle uvw
J=T(1,1).*T(2,2)-T(1,2).*T(2,1);
E1=exp(-1j*2*pi*(T(1,4)*fx+T(2,4)*fy-T(3,4)*fz(fx,fy)+T(3,4)/lambda));
FH=J.*E1.*F_hat;
E=fftshift(ifft2(FH)); % Hologram
```

13.3 Results for all Methods

In summary, the main issue in calculating the hologram of a 3D object composed of triangles is obtaining the frequency spectrum of the tilted triangle on the hologram plane. The six methods introduced in this chapter were used to solve this issue using various approaches. We analyzed the theory and implemented steps of every method, pointed out the similarities and differences between them, and listed the main program to implement them in MATLAB.

The holograms of a single triangle based on each method were reconstructed at vertex A, as shown in Fig. 13.10. They all show the same reconstructed images, indi-

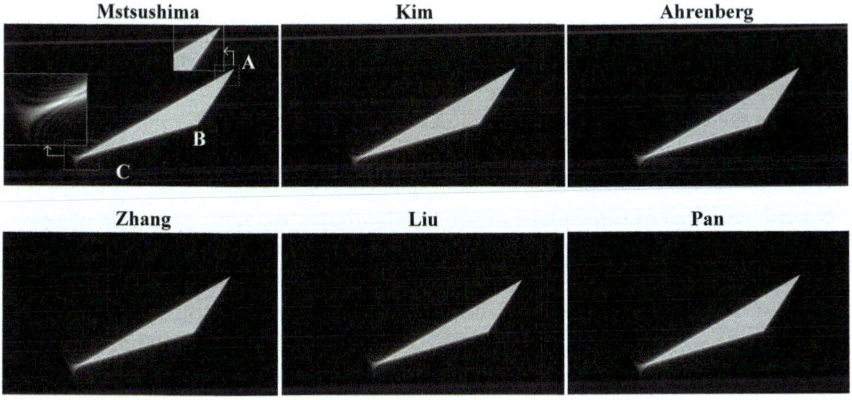

Fig. 13.10 Reconstructed images of the single triangle hologram based on six polygon methods introduced above

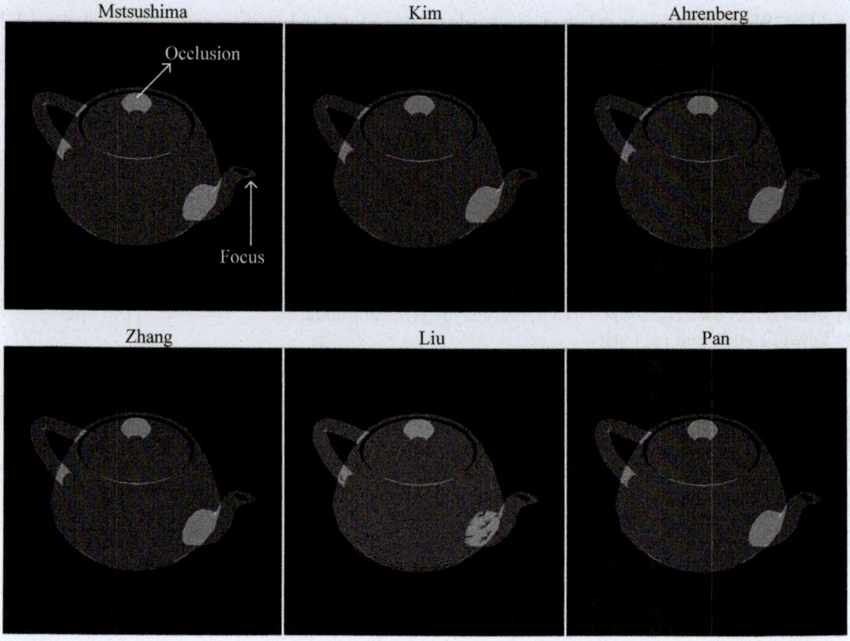

Fig. 13.11 Reconstructed images of the teapot consisting of 1902 triangles. All holograms with 1000×1000 pixels are sampled at 8μm intervals and reconstructed at the plane of the spout of teapot

cating that all methods can be considered computationally identical. However, there must be some differences in the computational performance owing to different ways of solving the spectra of the tilted triangles. Pan's method [6] presents a more concise program than the former five methods because it directly calculates the hologram spectrum from the analytical spectrum expression of the primitive triangle without 3D rotation. The former five methods first need to rotate the triangle from the global system to the local system and then obtain the hologram spectrum by rotating the frequency spectrum in the local system. Among them, Matsushima's method [1] addresses the frequency spectrum in the local system using FFT; however, it faces limitations in computational efficiency owing to the requirements for interpolation and an additional FFT in solving the local spectrum of a triangle. Kim's method [2] is more efficient than Matsushima's in solving the frequency spectrum by an analytical expression rather than FFT, but it still needs to perform a 2D rotation once again. Ahrenberg's [3], Zhang's [4] and Liu's [5] methods show similar calculational performance but are more efficient than Kim's method because a 2D affine transformation instead of 2D rotation is used to solve the analytical spectrum expression.

Using the six methods, a 3D object of a teapot consisting of 1902 triangles was used to generate holograms. Figure 13.11 shows the reconstructions of each method. Backface culling [8] was performed on the object by determining the triangular

Table 13.1 The calculation time required for the teapot using different methods

Methods	Matsushima's	Kim's	Ahrenberg's	Zhang's	Liu's	Pan's
Time (s)	11.56	14.31	7.62	7.53	4.09	5.34

Table 13.2 Calculation parameters used

Total triangles	Backface culling	Holograms	Pixel size	Distance
1908	1028	1000×1000	8 μm	200 ± 3 mm

normal before calculating the hologram. However, occlusion culling was not considered here; therefore, the reconstruction shown in Fig. 13.11 shows some particularly bright parts due to overlap, such as the lid and spout of the teapot. The teapot's spout was focused and clear, whereas the handle was blurred. Kim's, Ahrenberg's, Zhang's, and Pan's methods perform consistently because they all compute holograms in the frequency domain and use analytical spectral equations. Because Liu's method is computed in the spatial domain, the pixel size varies with the reconstruction distance. The gaps between adjacent triangles are shown in Fig. 13.11, which are caused by the different scaling of each triangle when it is reconstructed on a certain plane. This problem can be solved using smaller triangles until they are less noticeable. Matsushima's method first requires drawing a rasterized triangle, which causes the edges of adjacent triangles not to match exactly in the pixels. The non-uniform amplitudes shown in the reconstruction of Matsushima's method in Fig. 13.11 reflect these non-matching edges. This problem can be mitigated by reducing the hologram pixel size.

Table 13.1 shows the computational time of the teapot hologram with each method, and the calculation environments are as follows: CPU: AMD Ryzen 5-3600, GPU: NVIDIA GeForce RTX 3070, and MATLAB (2021a). The hologram parameters are presented in Table 13.2. The number of triangles used for the calculation after backface culling was 1028, and the holograms were 1000×1000 pixels. To avoid ringing errors caused by circular convolution in the frequency domain, in the actual computation, all methods except the Liu's method are calculated for the 2000×2000 pixel domain. In contrast, the Liu's method calculates the same pixel domain as the hologram because it performs the diffraction process in the spatial domain. This is why Liu's method is relatively faster.

In conclusion, all methods for generating holograms have rigorous theoretical derivations and similar grounds; however, they differ in their performance in terms of computational efficiency and reconstruction results. Pan's method generally generates the best holograms most efficiently and concisely. In contrast, Matsushima's method compensates for the former's shortcomings in rendering (e.g., textures and shading) at the expense of efficiency.

References

1. Kyoji Matsushima, Hagen Schimmel, and Frank Wyrowski, "Fast calculation method for optical diffraction on tilted planes by use of the angular spectrum of plane waves," J. Opt. Soc. Am. A **20**, 1755-1762 (2003).
2. Hwi Kim, Joonku Hahn, and Byoungho Lee, "Mathematical modeling of triangle-mesh-modeled three-dimensional surface objects for digital holography," Appl. Opt. **47**, D117-D127 (2008).
3. Lukas Ahrenberg, Philip Benzie, Marcus Magnor, and John Watson, "Computer generated holograms from three dimensional meshes using an analytic light transport model," Appl. Opt. **47**, 1567-1574 (2008).
4. Ya-Ping Zhang, Fan Wang, Ting-Chung Poon, Shuang Fan, and Wei Xu, "Fast generation of full analytical polygon-based computer-generated holograms," Opt. Express **26**, 19206-19224 (2018).
5. Yuan-Zhi Liu, Jian-Wen Dong, Yi-Ying Pu, Bing-Chu Chen, He-Xiang He, and He-Zhou Wang, 'High-speed full analytical holographic computations for true-life scenes," Opt. Express **18**, 3345-3351 (2010).
6. Yijie Pan, Yongtian Wang, Juan Liu, Xin Li, and Jia Jia, "Improved full analytical polygon-based method using Fourier analysis of the three-dimensional affine transformation," Appl. Opt. **53**, 1354-1362 (2014).
7. Bracewell, R. N., Chang, K. Y., Jha, A. K., and Wang, Y. H., Affine theorem for two-dimensional Fourier transform. Electronics Letters, **29**, 304-304 (1993).
8. Underkoffler J S. Occlusion processing and smooth surface shading for fully computed synthetic holography[C]//Practical Holography XI and Holographic Materials III. SPIE, **3011**, 19-30 (1997).

Chapter 14
Real-Time Electroholography Based on Multi-GPU Cluster

Naoki Takada

Abstract The calculation of a computer-generated hologram (CGH) has become computationally prohibitive. A high-performance computational power is indispensable for realizing a three-dimensional (3D) television using electroholography. A graphic processing unit provides a high-performance computational power for floating-point calculation at a low cost. The parallel calculations of large-pixel-count CGHs are suitable for a multiple-graphics processing unit (multi-GPU) cluster system. However, a multi-GPU cluster system cannot easily accomplish fast CGH calculations when CGH transfers among personal computers (PCs) are required. Consequently, the CGH transfer among PCs becomes a bottleneck. This problem usually occurs in multi-GPU cluster systems with a single spatial light modulator. To overcome this problem, we propose herein a simple method using the InfiniBand network. The computational speed of the proposed method using 13 GPUs (NVIDIA GeForce GTX TITAN X) is more than 3000 times faster than that of a central processing unit (Intel Core i7 4770) when the number of 3D object points exceeds 20,480. In practice, the effective performance of the proposed system is approximately 45 TFLOPS when the number of 3D object points exceeds 40,960. The proposed method can reconstruct a real-time video of a 3D object comprising approximately 100,000 points.

14.1 Introduction

The calculation of a **computer-generated hologram** (**CGH**) has become computationally prohibitive. Real-time electroholography must calculate and display at least 30 CGHs on a spatial light modulator (SLM) within a second. Thus, a high-performance computational power is indispensable for realizing a three-dimensional (3D) television based on real-time electroholography. A **graphic processing unit** (**GPU**) provides a high-performance computational power for floating-point calculation at a low cost.

N. Takada (✉)
Kochi University, 2-5-1 akebono-cho, kochi-shi, Kochi, Japan
e-mail: ntakada@kochi-u.ac.jp

T. Shimobaba and T. Ito (eds.), *Hardware Acceleration of Computational Holography*,
https://doi.org/10.1007/978-981-99-1938-3_14

A personal computer (PC) equipped with several GPUs, which is referred to as a multiple-GPU (multi-GPU) PC, is treated in the CGH calculation [1–3]. A PC cluster is a set of multiple PCs connected to a network, which performs parallel and distributed processing. Each PC constituting a **PC cluster** is referred to as a **node**. The PC cluster comprising many multi-GPU PCs is specially referred to as a multi-**GPU cluster**. A fast computation of a 20-megapixel CGH using a multi-GPU cluster system with 12 GPUs and 12 SLMs has been reported [4]. The results of Ref. [4] showed that a multi-GPU cluster could achieve high scalability in large-pixel count CGH calculations. Thus, multi-GPU clusters have been adopted in various approaches for the accelerated calculation of large-pixel count CGHs [5–8].

A multi-GPU cluster system with multiple SLMs is very expensive and large to be practical. Calculated CGH transfers among the nodes of the multi-GPU cluster system are required when a system with a single SLM is used. The calculated CGH transfers are a bottleneck in the parallel computation using this system [8].

To overcome this problem, this study proposes a multi-GPU cluster system with a single SLM and an InfiniBand network [9]. This chapter introduces real-time electroholography using the proposed system. The final section of this chapter introduces the latest study on high-speed CGH calculation based on the proposed method using a multi-GPU cluster system.

14.2 Computer-Generated Hologram

The following formula acquired by **Fresnel approximation** is used in the CGH calculation of a 3D object expressed by a point cloud:

$$I(x_h, y_h, 0) = \sum_{j=1}^{N_p} A_j \cos\left\{\frac{\pi}{\lambda z_j}\left[(x_h - x_j)^2 + (y_h - y_j)^2\right]\right\}, \tag{14.1}$$

where $I(x_h, y_h, 0)$ denotes a CGH pixel $(x_h, y_h, 0)$; (x_j, y_j, z_j) and A_j are the coordinate and the amplitude of the jth object point on a 3D object comprising N_p points, respectively; and λ is the reconstructing light wavelength.

The value calculated from Eq. (14.1) for each point on the CGH is binarized by a threshold value of 0. The binary CGH is generated by the binarized value for each point on the CGH. The resolution of the **binary CGH** displayed on SLM is $H \times W$, where H and W are the CGH height and width, respectively. The computational complexity of Eq. (14.1) is $O(N_p HW)$. Thus, the CGH calculation becomes prohibitively large.

14.3 Multiple-GPU Cluster System with a Single Spatial Light Modulator

Figure 14.1 shows multi-GPU cluster system with a single SLM. The multi-GPU cluster system is composed of a CGH display node (PC 0) connected to a single SLM and N CGH calculation nodes (PCs 1-N). Each node of multi-GPU cluster system has a CPU. The CGH display node and each N CGH calculation node have a GPU and three GPUs, respectively. As shown in Fig. 14.1, the multi-GPU cluster system has $3N + 1$ GPUs.

The CGH calculation nodes calculate the CGHs for all frames in a 3D video using pipeline processing. The calculated CGHs are then sent to the CGH display node via a network of the multi-GPU cluster system. The CGH display node receives the calculated CGHs from the CGH calculation nodes and displays the CGHs on a single SLM. The 3D object points for all frames in the 3D video are stored in the CGH display node, which also plays the role of the network file system (NFS) server. Section 14.4 describes in detail the pipeline processing for real-time electroholography using the multi-GPU cluster system.

In Fig. 14.1, each CGH calculation node has three GPUs. However, the proposed method can be applied to any number of GPUs on the respective CGH calculation nodes.

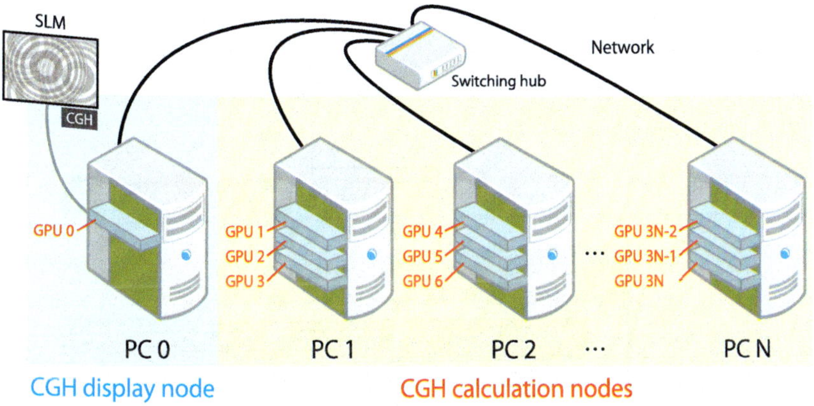

Fig. 14.1 Multi-GPU cluster system with a single SLM. H. Niwase, N. Takada, H. Araki, Y. Maeda, M. Fujiwara, H. Nakayama, T. Kakue, T. Shimobaba, T. Ito, "Real-time electroholography using a multiple-graphics processing unit cluster system with a single spatial light modulator and the InfiniBand network," Optical Engineering, Vol. 55, Issue 9, 093108 (2016)

14.4 Pipeline Processing for Real-Time Electroholography Using Multiple-GPU Cluster System with a Single SLM

Figure 14.2 shows the **pipeline** processing for real-time electroholography using the multi-GPU cluster system with a single SLM. The pipeline processing proceeds as follows:

Step 1 GPUs 1–3 on PC 1 calculate the CGHs for frames 1–3, respectively, in the original 3D video. The GPUs (GPUs 4-3N) on the other CGH calculation nodes calculate the CGHs for frames 4 to 3N in the same manner as the CGH calculations using the GPUs (GPUs 1–3) on PC 1.

Step 2 After the CGH for Frame 1 is calculated using GPU 1 on PC 1, PC 1 sends the calculated CGH for Frame 1 to the CGH display node (PC 0). Similarly, the CGH calculation nodes send the calculated CGHs for the frames from frames 2 to 3N to PC 0 one by one.

Step 3 PC 0 receives the calculated CGH for Frame 1 from PC 1 within the constant time interval T (i.e., display time interval). GPU 0 displays the CGH on the SLM for a constant time T. Similarly, PC 0 receives the calculated CGHs from the CGH calculation nodes. GPU 0 on PC 0 displays the received CGHs for the frames from frames 2 to 3N on the SLM for a constant time T.

After Step 3, the CGHs for frames 3N + 1 to 6N are calculated using the GPUs from GPUs 1 to 3N and are sent to PC 0. PC 0 receives the calculated CGHs from

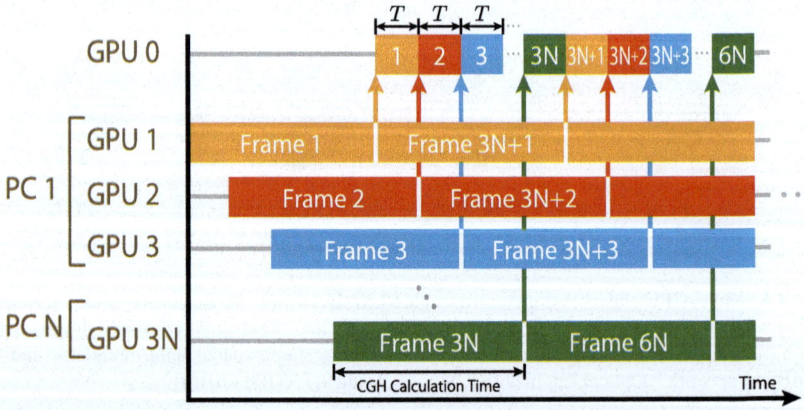

Fig. 14.2 Pipeline processing of real-time electroholography using the multi-GPU cluster system with a single SLM. H. Niwase, N. Takada, H. Araki, Y. Maeda, M. Fujiwara, H. Nakayama, T. Kakue, T. Shimobaba, T. Ito, "Real-time electroholography using a multiple-graphics processing unit cluster system with a single spatial light modulator and the InfiniBand network," Optical Engineering, Vol. 55, Issue 9, 093108 (2016)

the CGH calculation nodes. GPU 0 then displays the received CGHs for frames 3N + 1 to 6N on the SLM for a constant time T. The pipeline processing is repeated until the last frame of the original 3D video is reached. The computation time for a single GPU is $3N \times T$, but the CGH updates can be reduced to T by pipeline processing using multiple GPUs.

14.5 Implementation

Figure 14.3 shows the block diagram of the CGH computation using the proposed method. We used the **message passing interface** (**MPI**)[10] to implement the pipeline processing shown in Fig. 14.2 in a multi-GPU cluster system. The MPI is a well-known communication protocol for parallel and distributed programming on a PC cluster. MPI processes are units of processes executed on the respective nodes of a PC cluster by the MPI program that independently use separate CPU resources and memory space. The identification numbers of MPI processes are called **ranks**. Ranks are expressed as integer 0–1 less than the number of MPI processes.

In Fig. 14.3, "RANK i" depicts the rank with the identification number i. CPU 0 and GPU 0 show a CPU and a GPU on the CGH display node, respectively. Rank 0 is executed on the CGH display node and concentrates on displaying the CGH calculated by the CGH calculation nodes. Each CGH calculation node (PCs 0-N) has a CPU. In each CGH calculation node, a CPU executes three processes allocated to three GPUs. Thus, the total number of ranks is equal to the total number of GPUs.

Ranks 1–3 are executed on the CGH calculation node PC 1. Rank 1 calculates the CGH data as follows using GPU 1.

Step 1 Initialize for the GPU computation.
Step 2 The 3D object data are loaded from the NFS server through the network. Here, the 3D object data are binary data stored in the hard disk of the NFS server in advance.
Step 3 The 3D object data are sent to the global memory on GPU 1.
Step 4 The kernel for the CGH calculation based on Eq. (14.1) is invoked. In GPU 1, the 3D object data stored in the **global memory** are moved to the **shared memory** to realize high-speed memory access. GPU 1 calculates Eq. (14.1) using the 3D object data stored in the shared memory. The calculated CGH data are then stored in the global memory on GPU 1.
Step 5 Rank 1 copies the calculated CGH data stored in the global memory to the main memory on PC 1 and sends the calculated CGH data to Rank 0.

Here, "MPI_Send" is used to send the CGH data to Rank 0. "MPI_Send" performs the blocking send operation in the point-to-point communication (Table 14.1) and waits until the message is received by Rank 0. After the CGH data transfer is completed, Rank 1 begins to generate the next frame CGH data.

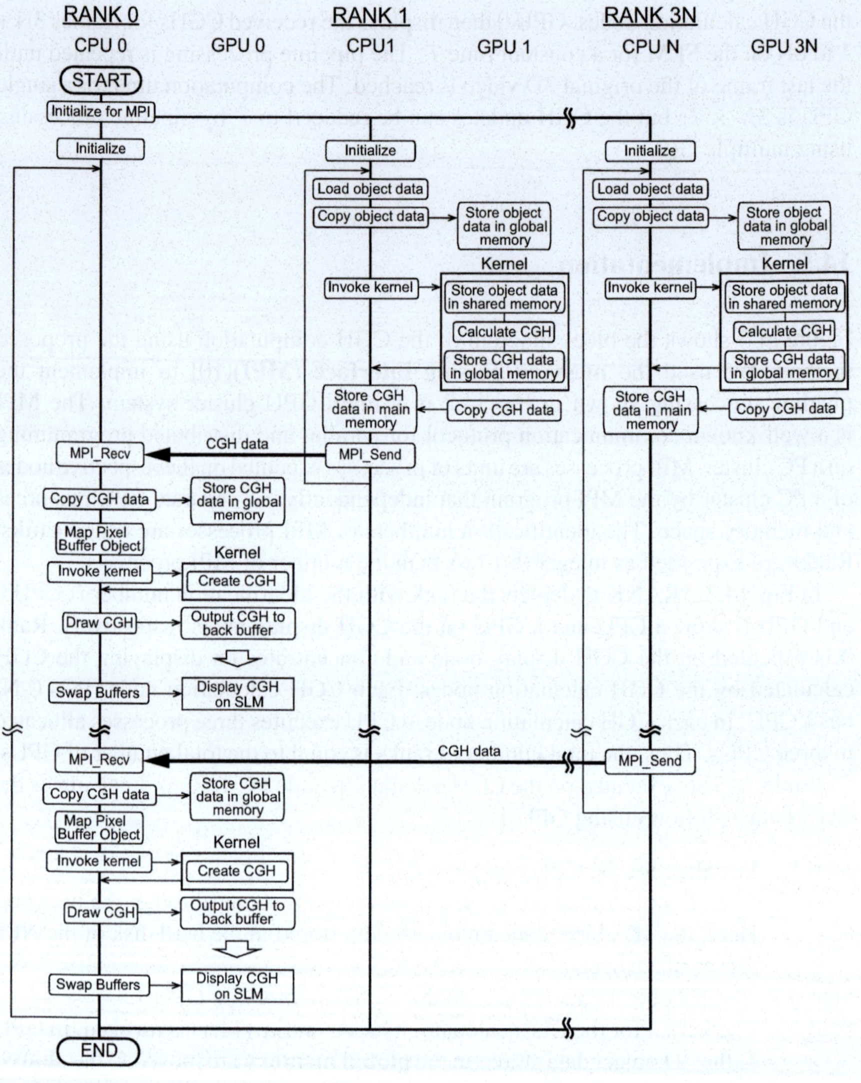

Fig. 14.3 Block diagram of the CGH computation using the proposed method. H. Niwase, N. Takada, H. Araki, Y. Maeda, M. Fujiwara, H. Nakayama, T. Kakue, T. Shimobaba, T. Ito, "Real-time electroholography using a multiple-graphics processing unit cluster system with a single spatial light modulator and the InfiniBand network," Optical Engineering, Vol. 55, Issue 9, 093108 (2016)

Table 14.1 MPI_Send function—performs a standard-mode blocking send

Name:	MPI_Send
Synopsis: int MPI_Send(const void *buf, int count, MPI_Datatype datatype, int dest, int tag, MPI_Comm comm)	
Input parameters:	
buf	Initial address of send buffer
count	Number of elements in send buffer
datatype	Datatype of each send buffer element
dest	Rank of destination
tag	Message tag
comm	Communicator

Table 14.2 MPI_Recv function—performs a standard-mode blocking receive

Name:	MPI_Recv
Synopsis: int MPI_Recv(void *buf, int count, MPI_Datatype datatype, int source, int tag, MPI_Comm comm, MPI_Status *status)	
Input parameters:	
buf	Initial address of receive buffer
count	Maximum number of elements in receive buffer
datatype	Datatype of each receive buffer element
source	Rank of source
tag	Message tag
comm	Communicator

Rank 0 receives the CGH data from Rank 1 and copies the CGH data to the global memory on GPU 0. Here, "MPI_Recv" is used to receive the CGH data from Rank 1. "MPI_Recv" performs the blocking receive operation in the point-to-point communication (Table 14.2) and waits until the message is sent from Rank 1. Rank 0 invokes the kernel to create a CGH image from the received CGH data. GPU 0 creates the CGH image and displays the CGH image on the SLM connected to GPU 0 for the constant time interval T.

Ranks 2 to 3N are performed similarly. In displaying the calculated CGH, double buffering is used to reduce the graphic flicker. Figure 14.4 shows the **double buffering** outline. It requires two buffers: front and back buffers. At Frame N, buffers 1 and 2 play the roles of the front and back buffers, respectively. The CGH of Frame N is stored in the front buffer (Buffer 1) and displayed on an SLM. The GPU draws the CGH of Frame N + 1 on the back buffer (Buffer 2). Buffers 1 and 2 are swapped after the GPU finishes drawing the CGH of Frame N + 1. At Frame N + 1, buffers 2 and 1

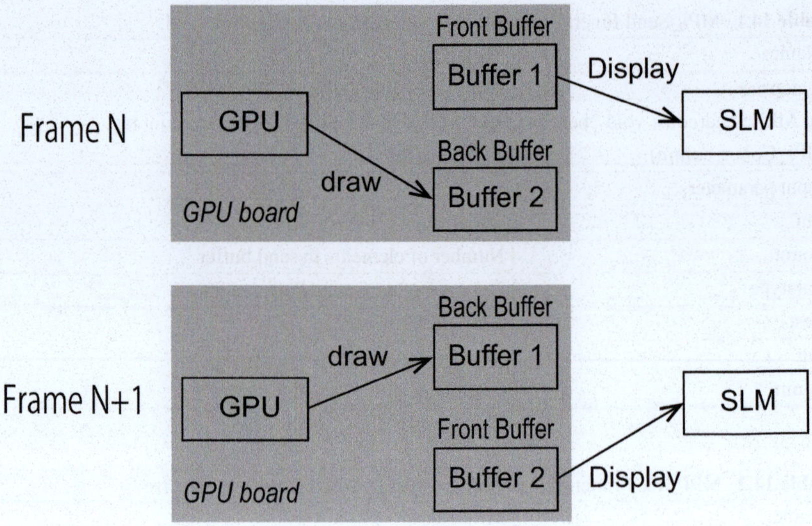

Fig. 14.4 Double buffering

play the roles of the front and back buffers, respectively. The CGH of Frame N + 1 is stored in the front buffer (Buffer 2) and displayed on an SLM. The GPU draws the CGH of Frame N + 2 on the back buffer (Buffer 1). Buffers 1 and 2 are swapped after the GPU finishes drawing the CGH of Frame N + 2. The buffer swap is repeated until the last frame of the original video is reached. The constant time interval T is equal to the periodic time interval of the vertical synchronizing signal when the swap buffer shown in Fig. 14.3 is synchronized with the vertical blanking interval [11]. The synchronization can be performed by the MPI functions "MPI_Send" and "MPI_Recv" and the setting tool of the GPU (e.g., NVIDIA X Server Settings) [12].

To achieve real-time electroholography, at least 30 CGHs must be displayed on the SLM within a second. Both CGH calculation and display must be performed within 33 ms. A CGH image is expressed with 32 bits per pixel, such that the proposed method can be easily applied to phase-only, color, and binary CGHs. Thus, the transferred data of a CGH image are $32(bits/px) \times 1920(px) \times 1024(px) \approx 62.9(Mbits)$ when the CGH image resolution is $1920(px) \times 1024(px)$. The CGH transfer time between the CGH display node and each CGH calculation node is 62.9 ms if a gigabit Ethernet is used as a network in the multi-GPU cluster shown in Fig. 14.1. The CGH transfer time is over 33 ms and becomes a bottleneck. A simple method of overcoming this bottleneck is using a high-speed network instead of a gigabit Ethernet. In this section, the InfiniBand quad data rate (QDR) (40 Gbps) is used as the high-speed network.

14.6 Results and Discussion

A five-node multi-GPU cluster system comprising a CGH display node and four CGH calculation nodes was used. The CGH display node had a GPU. Each CGH calculation node had three GPUs. Therefore, the multi-GPU cluster system had 13 GPUs. In each node of the multi-GPU cluster system, the PC was equipped with Intel Core i7 4770 (clock speed: 3.4 GHz, quad-core) and Linux (Cent OS 7.1) operating system. In the system, NVIDIA GeForce GTX TITAN X and InfiniBand QDR (40 Gbps) were used as the GPU and the network, respectively. The program was written in C language using the CUDA 7.0 software development kit and the Open GL 4.5.0 and Open MPI v1.8.7 libraries.

A green semiconductor laser with 535 nm wavelength was used as the reconstructing light. In the original 3D video, the 3D object was located 1.5 m from the CGH. The liquid crystal display (LCD) panel extracted from a projector (EMP-TW1000, L3C07U series, Epson Inc.) was used as the SLM. The LCD panel specifications were a pixel interval of 8.5 μm, a resolution of 1920 × 1080, and a size of 16 mm × 9 mm. In this study, 1920 × 1024 pixel CGH was used to apply the optimized CGH calculation algorithm to the multi-GPU cluster system.

Table 14.3 shows the display time interval T of the calculated CGH in the reconstructed 3D video using the multi-GPU cluster system against the number of 3D object points. The number of the GPUs shows the total number of the GPUs of the CGH calculation nodes. The synchronization between the swapping buffer on a GPU board and the vertical blanking interval of the SLM was not used herein because the display time interval T was not equal to the periodic time interval of the vertical synchronizing signal. In the program for executing pipeline processing on the multi-GPU cluster system, the sleep function was used to adjust the CGH display timing instead of the synchronization between the swapping buffer and the vertical blanking

Table 14.3 Display time interval T of electroholography using the multi-GPU cluster system. H. Niwase, N. Takada, H. Araki, Y. Maeda, M. Fujiwara, H. Nakayama, T. Kakue, T. Shimobaba, T. Ito, "Real-time electroholography using a multiple-graphics processing unit cluster system with a single spatial light modulator and the InfiniBand network," Optical Engineering, Vol. 55, Issue 9, 093108 (2016)

Object	Display time interval T [ms]				
points	1 GPU	3 GPUs	6 GPUs	9 GPUs	12 GPUs
10,240	39.4	14.6	7.6	6.4	4.3
20,480	78.5	28.6	14.6	10.1	7.3
40,960	153.8	56.7	29.1	19.3	14.4
61,440	232.4	84.6	43.7	28.7	21.6
81,920	309.0	111.5	58.1	38.9	28.6
102,400	385.0	138.2	72.3	47.8	35.5

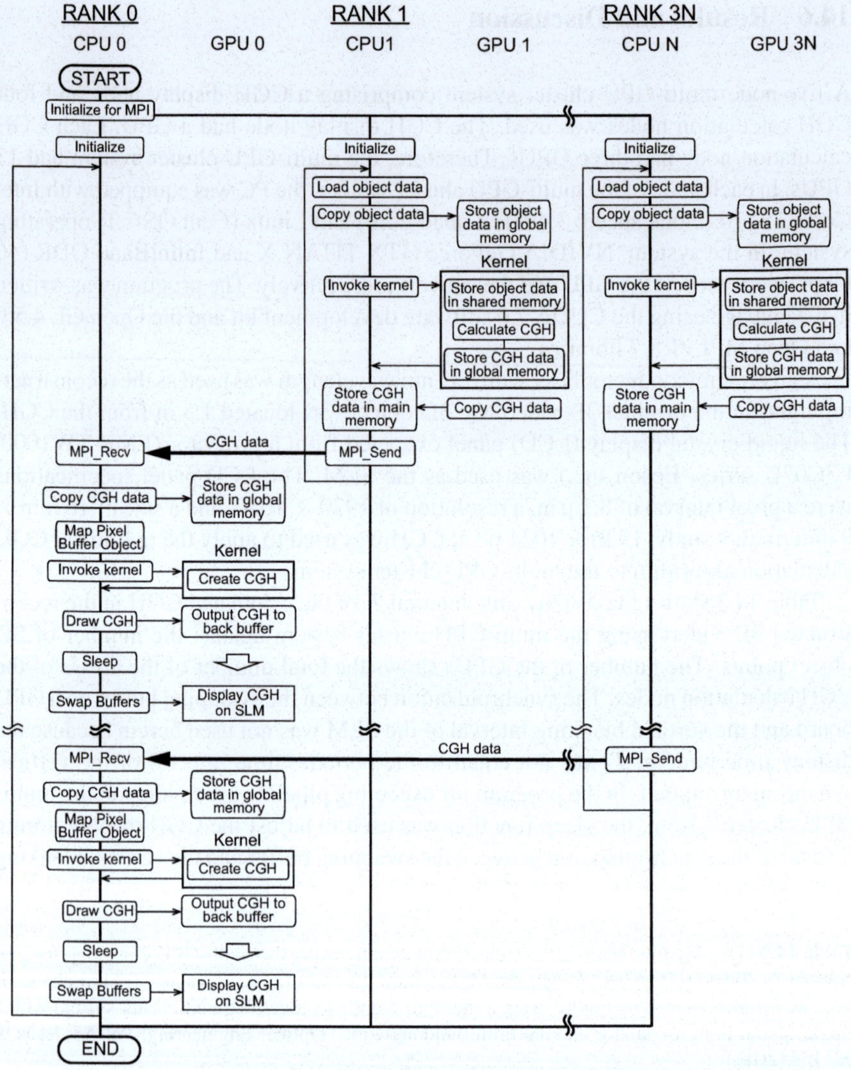

Fig. 14.5 Block diagram of the CGH computation using the proposed method when the sleep function is used to adjust the CGH display timing instead of the synchronization between the swapping buffer and the vertical blanking interval. H. Niwase, N. Takada, H. Araki, Y. Maeda, M. Fujiwara, H. Nakayama, T. Kakue, T. Shimobaba, T. Ito, "Real-time electroholography using a multiple-graphics processing unit cluster system with a single spatial light modulator and the InfiniBand network," Optical Engineering, Vol. 55, Issue 9, 093108 (2016)

interval (Fig. 14.5). The suspension time of the sleep function was obtained by the experimental rule based on the CGH calculation time using a single GPU.

Table 14.4 Frame rate of electroholography using the multi-GPU cluster system. H. Niwase, N. Takada, H. Araki, Y. Maeda, M. Fujiwara, H. Nakayama, T. Kakue, T. Shimobaba, T. Ito, "Real-time electroholography using a multiple-graphics processing unit cluster system with a single spatial light modulator and the InfiniBand network," Optical Engineering, Vol. 55, Issue 9, 093108 (2016)

Object	Frame rate [fps]				
points	1 GPU	3 GPUs	6 GPUs	9 GPUs	12 GPUs
10,240	25.4	68.5	131.6	156.3	232.6
20,480	12.7	35.0	68.5	99.0	137.0
40,960	6.5	17.6	34.4	51.8	69.4
61,440	4.3	11.8	22.9	34.8	46.3
81,920	3.2	9.0	17.2	25.7	35.0
102,400	2.6	7.2	13.8	20.9	28.2

Table 14.5 Effective performance of electroholography using the multi-GPU cluster system

Object	Effective performance [TFLOPS]				
points	1 GPU	3 GPUs	6 GPUs	9 GPUs	12 GPUs
10,240	4.1	11.0	21.1	25.0	37.2
20,480	4.1	11.2	22.0	32.0	44.0
40,960	4.2	11.4	22.2	33.3	44.7
61,440	4.2	11.4	22.1	33.6	44.7
81,920	4.2	11.6	22.2	33.1	45.1
102,400	4.2	11.7	22.3	33.7	45.4

Table 14.4 shows the frame rate of the CGH calculated using the multi-GPU cluster system. The frame rates shown in Table 14.4 were derived from the display time intervals T shown in Table 14.3. The multi-GPU cluster system, in which the CGH calculation nodes had 12 GPUs, achieved approximately 30 fps when the number of object points was 102,400.

Table 14.5 shows the effective performance of the CGH calculation using multi-GPU cluster system. In the CGH calculation of Eq. (14.1), all coefficients $\pi/\lambda z_j$ for the 3D object data are precalculated. The intensity of the object point A_j is also set to 1.0. Equation (14.1) consists of one addition, two subtractions, three multiplications, one cosine function, and one summation. In calculating Eq. (14.1) using the GPU, addition, subtraction, multiplication, and cosine functions are counted as one floating-point operation [12]. The number of the floating-point operations of Eq. (14.1) is $7 \times N_p + (N_p - 1)$. Here, $N_p - 1$ is the number of floating operations derived from the summation in Eq. (14.1). In the $H \times W$ pixel CGH, the total number of the floating-point operations becomes $(7 \times N_p + (N_p - 1)) \times H \times W$. Therefore, the number of floating-point operations per second (FLOPS) is estimated as $(7 \times N_p + (N_p - 1)) \times H \times W/T$, where T is the display time interval. The effective performance of

Table 14.6 Comparison of the performances of the multi-GPU cluster system and a CPU (i.e., Intel Core i7 4770). H. Niwase, N. Takada, H. Araki, Y. Maeda, M. Fujiwara, H. Nakayama, T. Kakue, T. Shimobaba, T. Ito, "Real-time electroholography using a multiple-graphics processing unit cluster system with a single spatial light modulator and the InfiniBand network," Optical Engineering, Vol. 55, Issue 9, 093108 (2016)

Object	Calculation time [ms]	Speed-up (1 CPU / *N* GPUs)				
points	1 CPU	1 GPU	3 GPUs	6 GPUs	9 GPUs	12 GPUs
10,240	12,483	317	855	1,643	1,950	2,903
20,480	23,228	296	812	1,591	2,300	3,182
40,960	44,569	290	786	1,532	2,309	3,095
61,440	66,537	286	786	1,523	2,318	3,080
81,920	88,602	287	795	1,525	2,278	3,098
102,400	107,707	280	779	1,490	2,253	3,034

the proposed system was approximately 45 TFLOPS when the number of 3D object points exceeded 40,960.

Table 14.6 shows the performance of the multi-GPU cluster system compared with that of a CPU (Intel Core i7 4770). The speed-up was estimated to be the CGH calculation time using the CPU divided by the display time interval T shown in Table 14.3. The CPU code for the CGH computation was written in C language and OpenMP and compiled using Intel C compiler version 15.0.3 with the -openmp -O3 options. Eight threads were used in the CGH computation. The computational speed of the multi-GPU cluster system was more than 3000 times faster than that of the CPU when the number of the 3D object points exceeded 20,480.

14.7 Related Work

This section briefly introduces the latest study of real-time electroholography using multi-GPU cluster system. In 2017, our research group proposed fast time-division color electroholography using a multi-GPU cluster system based on the system shown in this chapter, with an SLM and a controller to switch the color of the reconstructing light [13]. The controller comprised a universal serial bus module to drive the liquid crystal optical shutters. The Infiniband QDR (40 Gbps) and NVIDIA GeForce GTX TITAN X were used as the high-speed network and the GPU, respectively. Using the controller, the CGH display node of the multi-GPU cluster system synchronized the display of the CGH with the color switching of the reconstructing light. Fast time-division color electroholography at 20 fps was realized for a 3D object comprising 21,000 points per color when 13 GPUs are used in a multi-GPU cluster system. In

2019, real-time color electroholography was realized for a 3D color object comprising approximately 21,000 points per color using this multi-GPU cluster and three SLMs corresponding to the respective red-, green-, and blue-colored reconstructing lights [14]. Also in 2019, real-time electroholography was realized for a 3D video presenting a point-cloud 3D object composed of approximately 200,000 points using a multi-GPU cluster system with 13 GPUs (NVIDIA GeForce 1080 Ti) and a cost-effective gigabit Ethernet network [15].

In 2014, we proposed spatiotemporal division multiplexing electroholography utilizing the persistence of vision to accelerate the CGH calculation using a single GPU [16]. The method is very simple and easy to handle. In 2019, we implemented the spatiotemporal division multiplexing method on a cluster system with 13 GPUs (NVIDIA GeForce 1080 Ti) connected by a gigabit Ethernet network. In summary, we realized herein a real-time holographic video of a 3D object comprising $\approx$ 1,200,000 object points using the system [17].

References

1. Ahrenberg, L., Benzie, P., Magnor, M., and Watson, J., "Computer generated holography using parallel commodity graphics hardware," Opt. Express **14**, 7636–7641 (2006).
2. Kang, H., Yaraş, F., and Onural, L., "Graphics processing unit accelerated computation of digital holograms," Appl. Opt. **48**, H137–H143 (2009).
3. Zhang, Y., Liu, J., Li, X., and Wang, Y., "Fast processing method to generate gigabyte computer generated holography for three-dimensional dynamic holographic display," Chin. Opt. Lett. **14**, 030901 (2016).
4. Takada, N., Shimobaba, T., Nakayama, H., Shiraki, A., Okada, N., Oikawa, M., Masuda, N., and Ito, T., "Fast high-resolution computer-generated hologram computation using multiple graphics processing unit cluster system," Appl. Opt. **51**, 7303–7307 (2012).
5. Pan, Y., Xu, X., and Liang, X., "Fast distributed large-pixel-count hologram computation using a GPU cluster," Appl. Opt. **52**, 6562–6571 (2013).
6. Jackin, B., J., Miyata, H., Ohkawa, T., Ootsu, K., Yokota, T., Hayasaki, Y., Yatagai, T., and Baba, T., "Distributed calculation method for large-pixel-number holograms by decomposition of object and hologram planes," Opt. Lett. **39**, 6867–6870 (2014).
7. Jackin, B., J., Watanabe, S., Ootsu, K., Ohkawa, T., Yokota, T., Hayasaki, Y., Yatagai, T., and Baba, T., "Decomposition method for fast computation of gigapixel-sized Fresnel holograms on a graphics processing unit cluster," Appl. Opt. **57**, 3134–3145 (2018).
8. Baba, T., Watanabe, S., Jackin, B., J., Ootsu, K., Ohkawa, T., Yokota, T., Hayasaki, Y., and Yatagai, T., "Fast Computation with Efficient Object Data Distribution for Large-Scale Hologram Generation on a Multi-GPU Cluster," IEICE Trans. Inf. & Sys., E102-D, 1310–1320 (2019).
9. Niwase, H., Takada, N., Araki, H., Maeda, Y., Fujiwara, M., Nakayama, H., Kakue, T., Shimobaba, T., and Ito, T., "Real-time electroholography using a multiple-graphics processing unit cluster system with a single spatial light modulator and the InfiniBand network," Opt. Eng. **55**, 093108 (2016).
10. Open MPI Documentation. Available via DIALOG. https://www.open-mpi.org/doc/ .
11. Araki, H., Takada, N., Niwase, H., Ikawa, S., Fujiwara, M., Nakayama, H., Kakue, T., Shimobaba, T., and Ito, T., "Real-time time-division color electroholography using a single GPU and a USB module for synchronizing reference light," Appl. Opt. **54**, 10029–10034 (2015).
12. NVIDIA, CUDA Toolkit Documentation. Available via DIALOG. https://docs.nvidia.com/cuda/. Cited 28 November 2019.

13. Araki, H., Takada, N., Ikawa, S., Niwase, H., Maeda, Y., Fujiwara, M., Nakayama, H., Oikawa, M., Kakue, T., Shimobaba, T., and Ito, T., "Fast time-division color electroholography using a multiple-graphics processing unit cluster system with a single spatial light modulator," Chin. Opt. Lett. **15**, 120902 (2017).
14. Ikawa, S., Takada, N., Araki, H., Niwase, H., Sannomiya, H., Nakayama, H., Oikawa, M., Mori, Y., Kakue, T., Shimobaba, T., and Ito, T., "Real-time color holographic video reconstruction using multiple-graphics processing unit cluster acceleration and three spatial light modulators," Chin. Opt. Lett. **18**, 010901 (2020).
15. Sannomiya, H., Takada, N., Sakaguchi, T., Nakayama, H., Oikawa, M., Mori, Y., Kakue, T., Shimobaba, T., and Ito, T., "Real-time electroholography using a single spatial light modulator and a cluster of graphics-processing units connected by a gigabit Ethernet network," Chin. Opt. Lett. **18**, 020902 (2020).
16. Niwase, H., Takada, N., Araki, H., Nakayama, H., Sugiyama, A., Kakue, T., Shimobaba, T., and Ito, T., "Real-time spatiotemporal division multiplexing electroholography with a single graphics processing unit utilizing movie features," Opt. Express. **22**, 28052–28057 (2014).
17. Sannomiya, H., Takada, N., Suzuki, K., Sakaguchi, T., Nakayama, H., Oikawa, M., Mori, Y., Kakue, T., Shimobaba, T., and Ito, T., "Real-time spatiotemporal division multiplexing electroholography for 1,200,000 object points using multiple-graphics processing unit cluster," Chin. Opt. Lett. **18**, 070901 (2020).

Chapter 15
GPU Acceleration of Compressive Holography

Yutaka Endo

Abstract Compressive holography is an application of compressed sensing to digital holography to reconstruct three-dimensional scattering density from a single two-dimensional hologram. Although the method can effectively remove unwanted out-of-focus objects, twin images, and autocorrelation terms in the reconstructed images, its computational cost is high. This section describes the acceleration of signal reconstruction for compressive holography using a graphics processing unit. We outlined compressed sensing and compressive holography and describe its implementation based on the fast iterative shrinkage-thresholding method with ℓ_1 norm and total variation.

15.1 Introduction

Compressed sensing (CS) is a signal acquisition framework that enables the recovery of **sparse** signals using far fewer samples than conventional methods based on the **Nyquist–Shannon sampling theorem** [6, 10, 13]. Since its establishment, CS has been applied in many fields such as magnetic resonance imaging [20, 21], radar imaging [11] and optical imaging [22]. CS has been used in many applications in holography [4, 19, 24, 26, 27].

An application is compressive holography, which is used to reconstruct three-dimensional (3D) scattering density (i.e., image slices) from a single 2D hologram [4]. In holographic 3D imaging, image slices obtained from the conventional back-propagation technique suffer from out-of-focus objects, twin images, and autocorrelation terms, while compressive holography can remove these unwanted terms.

Supplementary Information The online version contains supplementary material available at https://doi.org/10.1007/978-981-99-1938-3_15.

Y. Endo (✉)
Institute of Science and Engineering, Kanazawa University, Kakuma-machi, Kanazawa, Ishikawa 920-1192, Japan
e-mail: endo@se.kanazawa-u.ac.jp

T. Shimobaba and T. Ito (eds.), *Hardware Acceleration of Computational Holography*, https://doi.org/10.1007/978-981-99-1938-3_15

The high computational cost of signal reconstruction is a concern in **compressive holography**. The data size to be reconstructed in compressive holography is considerable. For example, assuming that we have a hologram with 1000×1000 pixels and attempt to reconstruct a 3D scattering density with 10 depths, the reconstructed signal has 10 million variables. CS reconstruction is performed using an optimization problem (e.g., ℓ_1-norm minimization), and a large-scale optimization problem with 10 million variables is solved, which incurs a high computational cost.

In this section, we describe fast signal reconstruction for compressive holography by using a **graphics processing unit** (**GPU**) [15]. GPUs are parallel computing devices that are suitable for data-parallel computing; the same program is executed on many data elements in parallel. GPU computing is suitable for CS reconstruction because it can be efficiently computed using a data-parallel model. We describe compressive holography, signal reconstruction, and its implementation on GPUs. We implemented the fast iterative shrinkage-thresholding algorithm [2, 3] with ℓ_1 norm and total variation [25] regularization. Our evaluation revealed that GPU-based implementation is more than ten times faster than CPU-based implementation.

15.2 Compressed Sensing

CS is a sampling paradigm that enables the acquisition and recovery of sparse signals from far fewer samples than conventional methods based on the Nyquist–Shannon sampling theorem do [6, 10, 13]. A sparse signal is a signal whose most components are zero, and many natural signals (e.g., audio, images, and videos) are sparse or have a sparse representation after appropriate transformation. Therefore, CS can be effectively applied to numerous natural signals. In the following section, we outline the CS framework. For a detailed description of CS, please refer to [1, 7, 12, 16].

Consider a signal of interest $\mathbf{x} \in \mathbb{C}^N$ that has N discrete values acquired by a linear measurement, and the measured data $\mathbf{y} \in \mathbb{C}^M$ with M discrete values are obtained. The linear measurement process can be expressed as follows:

$$\mathbf{y} = \Phi\mathbf{x}, \tag{15.1}$$

where $\Phi \in \mathbb{C}^{M \times N}$ is the sensing matrix, which is a linear measurement model. The reconstruction of signal $\mathbf{x}$ is an inverse problem. For $M \geq N$, a least-square solution is easily obtained, but, for $M < N$, the inverse problem is ill-posed and does not achieve a unique solution. CS theory shows the conditions under which the signal $\mathbf{x}$ is perfectly reconstructed, even if $M < N$. Such ill-posed problems can be solved through signal sparsity and incoherent sensing.

The assumption of the target signals $\mathbf{x}$ for CS reconstruction is that it is sparse. Formally, the signal is S-sparse if it has at most S nonzero components. Many natural signals are sparse or **compressible** such that they are well approximated by a sparse representation $\mathbf{z} \in \mathbb{C}^N$ that has a few nonzero components in the proper basis $\Psi \in \mathbb{C}^{N\times N}$ such as $\mathbf{x} = \Psi\mathbf{z}$. Using the representation basis, Eq. (15.1) can be expressed as follows:

$$\mathbf{y} = \Phi\Psi\mathbf{z} = \mathbf{A}\mathbf{z},$$

where $\mathbf{A} \in \mathbb{C}^{M\times N}$.

A popular method for sparse-signal reconstruction in the CS framework is ℓ_1-minimization.

$$\min_{\mathbf{z}} \|\mathbf{z}\|_1 \quad \text{subject to} \quad \mathbf{A}\mathbf{z} = \mathbf{y}, \tag{15.2}$$

where $\|\cdot\|_p$ denotes ℓ_p norm. ℓ_1 norm promotes the sparsity of the solution and can be easily computed by efficient algorithms. For many image processing applications, instead of solving the problem Eq. (15.2), ℓ_1 regularization, which is the Lagrangian relaxation of ℓ_1-minimization, is typically used

$$\min_{\mathbf{z}} \frac{1}{2}\|\mathbf{A}\mathbf{z} - \mathbf{y}\|_2^2 + \lambda\|\mathbf{z}\|_1, \tag{15.3}$$

where λ is a regularization parameter that balances the first and second terms (i.e., data fidelity and sparsity).

The sensing matrix Φ affects the success probability of signal reconstruction. **Incoherence** is one of the requirements for the sensing matrix to recover a sparse vector, which is measured by **mutual coherence** [14]. The mutual coherence μ between two orthonormal bases Φ and Ψ for $\mathbb{C}^{N\times N}$ is defined by

$$\mu(\Phi, \Psi) := \max_{1\le i,j\le N} |\langle \boldsymbol{\phi}_i, \boldsymbol{\psi}_j\rangle|,$$

where $\langle\cdot,\cdot\rangle$ denotes the inner product, and $\boldsymbol{\phi}_i$ and $\boldsymbol{\psi}_i$ are the i-th column vectors of Φ and Ψ, respectively. Mutual coherence is bounded by $1/\sqrt{N} \le \mu(\Phi, \Psi) \le 1$. When mutual coherence is low (i.e., incoherent), the number of measurements M required to reconstruct a sparse vector is also low [5, 7]. Another condition is the **restricted isometry property (RIP)** [6]. Matrix $\mathbf{A} \in \mathbb{C}^{M\times N}$ holds the RIP if the condition

$$(1 - \delta_S)\|\mathbf{x}\|_2^2 \le \|\mathbf{A}\mathbf{x}\|_2^2 \le (1 + \delta_S)\|\mathbf{x}\|_2^2$$

is satisfied for all S-sparse vectors $\mathbf{x}$ with small δ_S. Thus, matrix $\mathbf{A}$ approximately preserves the Euclidean length of any S-sparse vectors. If $\mathbf{A}$ holds RIP, ℓ_1-minimization can accurately recover any S-sparse vectors [8]. Random matrices hold the RIP. For example, an $M \times N$ i.i.d. **Gaussian random matrix** can be shown to have a high probability of RIP if $M \ge C \cdot S\ln(N/S)$, where C is a small constant. This result allows us to recover S-sparse vectors from $M \ge C \cdot S\ln(N/S)$ random Gaussian measurements.

15.3 Compressive Holography

The first application of CS to holography is *compressive holography*, which enables the reconstruction of image slices from a single **Gabor hologram** [4]. In holographic imaging, reconstructed image slices suffer from out-of-focus objects, twin images, and autocorrelation terms, whereas compressive holography can remove these unwanted terms through sparse optimization. In this section, we describe compressive holography. For details, please refer to [4].

Figure 15.1 shows a schematic of compressive holography. The 3D object is illuminated by a plane wave, and the Gabor hologram is recorded by an image sensor. The hologram is an interference pattern between the light scattered by an object and the unscattered light that serves as the reference beam [18]. Let (x, y, z) be the coordinates and z be the distance from the image sensor (i.e., $z = 0$ is at the image sensor plane). The scattered field on the image sensor is expressed as follows:

$$O(x, y) = \int h * s(x, y; z) dz, \tag{15.4}$$

where $s(x, y, z)$ is the scattering density of the 3D object, $*$ denotes convolution, and $h(x, y; z)$ is the convolution kernel of diffraction for distance z [17]. The Gabor hologram is the intensity of the sum of the scattered field $O(x, y)$ and uniform reference wave R on the image:

$$g(x, y) = |R + O(x, y)|^2 = 2\mathrm{Re}\{R^* O(x, y)\} + |R|^2 + |O(x, y)|^2,$$

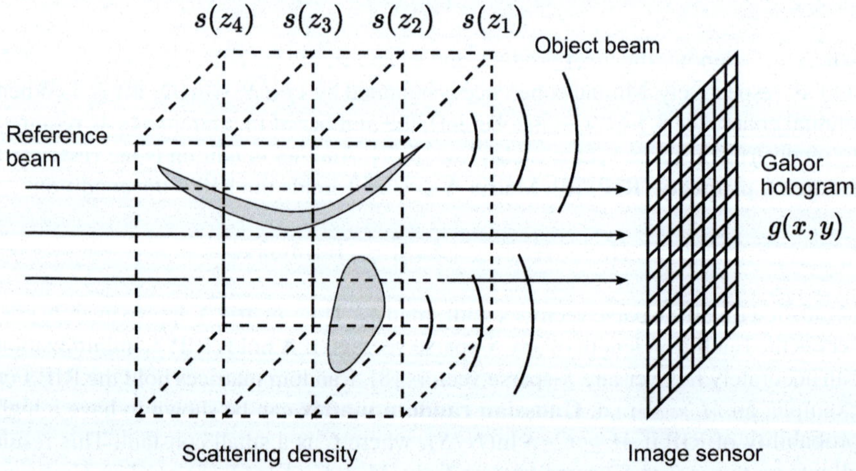

Fig. 15.1 Schematic of Gabor holography

where Re$\{\cdot\}$ is an operator that takes the real part. This measurement model has a constant term $|R|^2$ and nonlinear term $|O(x, y)|^2$. To rewrite this model as a linear model, we make two assumptions. First, the constant term can be removed from the hologram using Fourier filtering. The constant-term-removed hologram is expressed as follows:

$$\bar{g}(x, y) = 2\mathrm{Re}\{O(x, y)\} + |O(x, y)|^2, \tag{15.5}$$

where we assume $R = 1$ for simplicity. The second assumption is that the nonlinear term is a model error $e(x, y)$ that should be eliminated through the reconstruction process (for more details, see [4], Sect. 5). Thus, we obtain the following linear model:

$$\bar{g}(x, y) = 2\mathrm{Re}\{O(x, y)\} + e = 2\mathrm{Re}\left\{\int h * s(x, y; z)dz\right\} + e. \tag{15.6}$$

We discretize this model to obtain the matrix–vector form. Let the image sensor have $N_x \times N_y$ pixels, and assume that the object consists of N_z image slices with $N_x \times N_y$ pixels. The matrix–vector form of Eq. (15.6) is expressed as follows:

$$\bar{\mathbf{g}} = 2\mathrm{Re}\{\mathbf{Hs}\} + \mathbf{e},$$

where $\bar{\mathbf{g}} \in \mathbb{R}^{N_x N_y}$, $\mathbf{s} \in \mathbb{C}^{N_x N_y N_z}$ and $\mathbf{e} \in \mathbb{R}^{N_x N_y}$ are the vector forms of $\bar{g}$, s, and e, and $\mathbf{H} \in \mathbb{C}^{N_x N_y \times N_x N_y N_z}$ is the matrix for computing Eq. (15.4), respectively. $\mathbf{s}$ is expressed as $\mathbf{s} = [\mathbf{s}_1^T, \mathbf{s}_2^T, \cdots, \mathbf{s}_{N_z}^T]^T$ where $\mathbf{s}_k$ is the kth slice.

The reconstruction of object $\mathbf{s}$ from hologram $\bar{\mathbf{g}}$ is an **ill-posed inverse problem** that does not have a unique solution. However, CS can solve this problem if the desired solutions are sparse in a certain space. We use the following regularization to infer image slices:

$$\min_{\mathbf{x}} \frac{1}{2}\|2\mathrm{Re}\{\mathbf{Hx}\} - \bar{\mathbf{g}}\|_2^2 + \tau G(\mathbf{x}), \tag{15.7}$$

where G is a regularizer that promotes the sparsity of the estimated value, and τ is the regularization parameter of G. This formula is the generalization of Eq. (15.3). If the object is spatially sparse, we can select ℓ_1 norm as a regularizer. For piecewise smooth objects, **total variation (TV)** is typically used for regularizers [25]. Assuming $\mathbf{u}$ is a 2D image with $M \times N$ pixels, the isotropic TV is defined as follows:

$$\|\mathbf{u}\|_{\mathrm{TV}} := \|\nabla\mathbf{u}\|_2 = \sum_{i,j}\sqrt{(\nabla_1\mathbf{u})_{i,j}^2 + (\nabla_2\mathbf{u})_{i,j}^2}.$$

Here, $\nabla\mathbf{u} := (\nabla_1\mathbf{u}, \nabla_2\mathbf{u})$ is the discrete gradient of $\mathbf{u}$ defined by

$$(\nabla_1\mathbf{u})_{i,j} := u_{i,j} - u_{i,j} \quad (1 \le i \le M - 1),$$
$$(\nabla_2\mathbf{u})_{i,j} := u_{i,j} - u_{i,j} \quad (1 \le j \le N - 1),$$

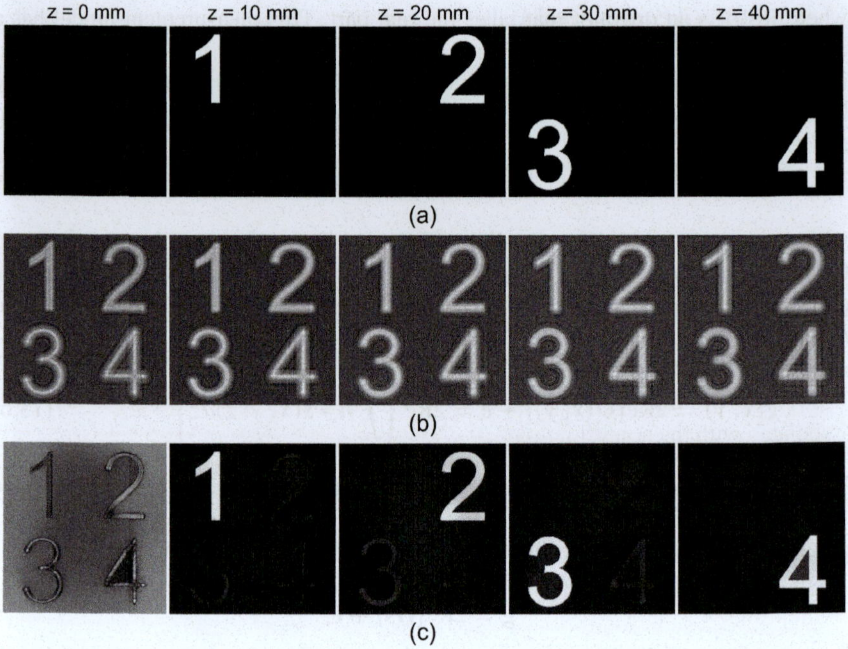

Fig. 15.2 **a** Three-dimensional object and reconstructed images by **b** backpropagation and **c** compressive holography

where $u_{i,j}$ denotes an element in the ith row and jth column. For objects composed of N_z slices, as $\mathbf{x} = [\mathbf{x}_1^T, \mathbf{x}_2^T, \cdots, \mathbf{x}_{N_z}^T]^T$ where $\mathbf{x}_k$ is the kth slice, the TV-based regularizer can be $G(\mathbf{x}) = \sum_{k=1}^{N_z} \|\mathbf{x}_k\|_{\mathrm{TV}}$. By solving this regularized optimization problem, image slices can be reconstructed from the Gabor hologram.

An illustrative example of compressive holography was presented using simulation. Figure 15.2a shows the 3D object used in the simulation. The 3D object space consists of $N_x \times N_y \times N_z = 512 \times 512 \times 5$ with 10.0μ m lateral pitch and 10.0 mm axial pitch, and characters (1, 2, 3, and 4) are at various distances along z axis. A Gabor hologram was obtained using a plane wave with a 632.8-nm wavelength, and the constant term was eliminated, as shown in Eq. (15.5). Zero pixels were padded around the hologram and an $N_x \times N_y = 1024 \times 1024$ image was created to avoid circular convolution. Figure 15.2b, c show reconstructed images by backpropagation and compressive holography, respectively. Here, the $z = 0$ plane was included for reconstruction to remove model error $\mathbf{e}$, which tends to concentrate on the $z = 0$ plane. Although out-of-focus images, twin images, and autocorrelation fields exist in the reconstructed images by backpropagation, compressive holography can suppress these unwanted images from the reconstructed images. The results of compressive holography were obtained using TV regularization, which required 181.4 s on the CPU and 4.2 s on the GPU (see Sect. 15.4 and Table 15.1 for the implementation details and evaluation environment).

15.4 GPU-Accelerated Compressive Holography

15.4.1 Signal Reconstruction Algorithm

The optimization problem (15.7) is generalized as follows:

$$\min_{\mathbf{x}} F(\mathbf{x}) + G(\mathbf{x}) \tag{15.8}$$

where F and G are differentiable and nondifferentiable functions, respectively. The basic gradient descent methods cannot be applied to this problem because the objective function has a nondifferentiable term. A basic algorithm for solving problem (15.8) is the proximal gradient method [23]. In this algorithm, the proximal operator associated with the nondifferentiable function G is used, which is defined as follows:

$$\mathrm{prox}_G(\mathbf{v}) := \operatorname*{argmin}_{\mathbf{x}} G(\mathbf{x}) + \frac{1}{2}\|\mathbf{x} - \mathbf{v}\|_2^2.$$

This operation is seen as a generalization of Euclidian projection. The proximal gradient method can be easily implemented when the proximal operator can easily be computed. The algorithm iteratively updates the approximate solution at the kth step, $\mathbf{x}^k$ as follows:

$$\mathbf{x}^{k+1} = \mathrm{prox}_{t^k G}\left(\mathbf{x}^k - t^k \nabla F(\mathbf{x}^k)\right),$$

where ∇F is the gradient of F, and $t^k > 0$ is the step size. When ∇F is Lipschitz continuous with constant L, this method converges at a rate of $O(1/k)$ when a fixed step size $t^k = t \in (0, 1/L]$ is used.

An accelerated version of the proximal gradient method, called the **fast iterative shrinkage-thresholding algorithm (FISTA)** [3], is used to solve problem (15.8), which improves the convergence rate using the last two iterations at each iteration step. The algorithm is described in Algorithm 1. FISTA incurs a small additional computational cost for the proximal gradient method but improves the convergence rate to $O(1/k^2)$.

Algorithm 1 Fast iterative-shrinkage thresholding algorithm.

1: **for** $k = 1$ to K **do**
2: $\quad \mathbf{x}^{k+1} = \mathrm{prox}_{tG}\left(\mathbf{z}^k - t\nabla F(\mathbf{z}^k)\right)$
3: $\quad a^{k+1} = \frac{1+\sqrt{1+4(a^k)^2}}{2}$
4: $\quad \mathbf{z}^{k+1} = \mathbf{x}^{k+1} + \frac{a^k}{a^{k+1}}\left(\mathbf{x}^{k+1} - \mathbf{x}^k\right)$
5: **end for**

In the optimization steps of the proximal gradient method, the gradient of F is first computed. Our differentiable cost function is $F(\mathbf{x}) = \frac{1}{2}\|2\mathrm{Re}\{\mathbf{Hx}\} - \bar{\mathbf{g}}\|_2^2$, and according to Wirtinger derivative [9], its complex gradient is computed by

$$\nabla F(\mathbf{x}) = 2\mathbf{H}^* \left(2\mathrm{Re}\{\mathbf{Hx}\} - \bar{\mathbf{g}}\right)$$

where $\mathbf{H}^*$ denotes the adjoint of $\mathbf{H}$. Because matrix multiplication with $\mathbf{H}$ and $\mathbf{H}^*$ is expressed as convolution, the gradient can be efficiently computed using fast Fourier transforms (FFTs).

We selected ℓ_1 norm or TV as a nondifferentiable regularizer G. The **proximal operator** for ℓ_1 norm is explicitly expressed as follows:

$$\mathrm{prox}_{\tau\|\cdot\|_1}(\mathbf{x})_i = \mathrm{sgn}(x_i)\max\{|x_i| - \tau, 0\},$$

where sgn denotes the sign function. This operation is called **soft thresholding** and can be easily computed element by element.

The proximal operator for TV is more difficult to compute than that for the ℓ_1 norm because it does not have an explicit form. Therefore, the subminimization problem should be solved numerically at each iteration as follows:

$$\mathrm{prox}_{\tau\|\cdot\|_{\mathrm{TV}}}(\mathbf{v}) = \min_{\mathbf{x}} \tau\|\mathbf{x}\|_{\mathrm{TV}} + \frac{1}{2}\|\mathbf{x} - \mathbf{v}\|_2^2.$$

This subproblem is solved by formulating its dual problem and solving it using **gradient projection** [2]. Algorithm 2 shows the algorithm, where the divergence operator is defined as follows:

$$\nabla \cdot (\mathbf{p}, \mathbf{q}) := p_{i,j} - p_{i-1,j} + q_{i,j} - q_{i,j-1} \quad (1 \le i \le M,\ 1 \le j \le N),$$

where we assume $p_{0,j} = p_{M,j} = q_{i,0} = q_{i,N} = 0$. The operator P is defined as follows:

$$P(\mathbf{p}, \mathbf{q}), \left(\frac{p_{i,j}}{\max\left\{1, \sqrt{p_{i,j}^2 + q_{i,j}^2}\right\}}, \frac{q_{i,j}}{\max\left\{1, \sqrt{p_{i,j}^2 + q_{i,j}^2}\right\}} \right),$$

which is the projection onto a set of 2D image pairs, $\{(\mathbf{p}, \mathbf{q}) \mid p_{i,j}^2 + q_{i,j}^2 = 1\}$.

Algorithm 2 Gradient-projection-based TV denoising.

1: **for** $k = 1$ to V **do**
2: $(\mathbf{p}^{k+1}, \mathbf{q}^{k+1}) = P\left\{(\mathbf{p}^k, \mathbf{q}^k) + \frac{1}{8\tau}\nabla \mathbf{x}^k\right\}$
3: $\mathbf{x}^{k+1} = \mathbf{v} - \tau\nabla \cdot (\mathbf{p}^{k+1}, \mathbf{q}^{k+1})$
4: **end for**

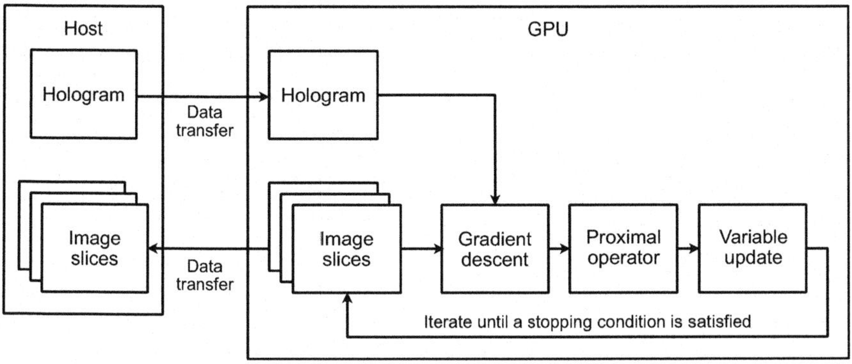

Fig. 15.3 Overview of GPU-accelerated compressive holography implementation

15.4.2 GPU Implementation

We implemented a FISTA-based signal reconstruction using a GPU. Our code is available on the book site and at GitHub.[1] Our implementation is based on CUDA, a parallel computing platform and programming model invented by NVIDIA. It allows developers to use a CUDA-enabled GPU for general-purpose processing. In the CUDA programming model, the computing system consists of a *host*, CPU, and one or more *devices*, GPUs. We write a function called *kernel* that describes the work of a single thread on a device and invokes it with numerous threads from a host. These threads are executed in parallel on thousands of cores on a GPU. This programming model fits data-parallel computing and supports developers in using GPUs as a massively parallel computing device.

Figure 15.3 illustrates an overview of GPU-accelerated reconstruction. The host first sends the hologram data to the device as input data. Subsequently, the device iteratively computes FISTA, which consists of gradient descent, proximal operator, and variable update, until a stopping condition is satisfied. A fixed number of iterations was used as the stopping condition. After the FISTA step, the device transferred the reconstructed results to the host. Kernels for FISTA are easily parallelized because most operations in those kernels are vector operations, where each element can be processed independently by a single thread.

The evaluation of the forward model and its adjoint (i.e., the matrix–vector multiplication with $\mathbf{H}$ and $\mathbf{H}^*$) is the most computationally intensive in FISTA. We implemented these operations using 2D FFTs, as the forward and adjoint models can be expressed as the sum of 2D convolutions, as shown in Eq. (15.4). The cuFFT library was used to execute 2D-FFTs on the GPU.

[1] https://github.com/ytkend/compressive-holography-cuda.

15.4.3 Performance Evaluation

We evaluated the performance of the GPU-based compressive holography implementation. Table 15.1 shows the evaluation environment. The CPU and GPU were AMD Ryzen 7 2700X and NVIDIA Geforce RTX 2080Ti, respectively. We compared the CPU-based implementation, in which FFTW is used for an FFT library, and OpenMP is used for multithreading.

Table 15.2 presents the computation times of the CPU and GPU-based implementations for ℓ_1 and TV regularizations. The number of slices $N_z = 10$ was fixed, and the hologram size $N_x \times N_y$ was changed. The number of iterations of FISTA was 300, which produced moderate reconstruction results in our cases. The total computation time of our implementations increased almost linearly with the number of iterations. For TV regularization, the number of iterations to solve the subminimization problem by Algorithm 2 was set to 10. The computation times are the averages of 10 runs on the GPU and CPU. The results revealed that GPU-based implementation is more than ten times faster than CPU-based implementation.

Table 15.1 Evaluation environment

CPU	AMD Ryzen 7 2700X (8 cores, 16 threads, and 3.7 GHz base clock)
Memory	DDR4 2667 MHz 32 GB
GPU	NVIDIA Geforce RTX 2080 Ti (4352 cores, 1350 MHz base clock, 11 GB GDDR6 RAM)
C++ compiler	GCC 9.4.0
CUDA	11.8

Table 15.2 Computation time of compressive holography using FISTA with 300 iterations

Hologram size	ℓ_1 regularization		TV regularization	
$N_x \times N_y$	CPU (s)	GPU (s)	CPU (s)	GPU (s)
256 × 256	4.963	0.217	7.764	0.788
512 × 512	23.36	0.510	35.91	2.449
768 × 768	40.48	0.950	68.53	5.111
1024 × 1024	152.4	1.517	224.5	7.806

References

1. Baraniuk, R.: Compressive Sensing [Lecture Notes]. IEEE Signal Processing Magazine **24**(4), 118–121 (2007). https://doi.org/10.1109/msp.2007.4286571
2. Beck, A., Teboulle, M.: Fast Gradient-Based Algorithms for Constrained Total Variation Image Denoising and Deblurring Problems. IEEE Transactions on Image Processing **18**(11), 2419–2434 (2009). https://doi.org/10.1109/TIP.2009.2028250
3. Beck, A., Teboulle, M.: A Fast Iterative Shrinkage-Thresholding Algorithm for Linear Inverse Problems. SIAM Journal on Imaging Sciences **2**(1), 183–202 (2009). https://doi.org/10.1137/080716542
4. Brady, D.J., Choi, K., Marks, D.L., Horisaki, R., Lim, S.: Compressive Holography. Optics Express **17**(15), 13040–13049 (2009). https://doi.org/10.1364/OE.17.013040
5. Candès, E., Romberg, J.: Sparsity and incoherence in compressive sampling. Inverse Problems **23**(3), 969–985 (2007). https://doi.org/10.1088/0266-5611/23/3/008
6. Candes, E., Romberg, J., Tao, T.: Robust uncertainty principles: Exact signal reconstruction from highly incomplete frequency information. IEEE Transactions on Information Theory **52**(2), 489–509 (2006). https://doi.org/10.1109/tit.2005.862083
7. Candes, E., Wakin, M.: An Introduction To Compressive Sampling. IEEE Signal Processing Magazine **25**(2), 21–30 (2008). https://doi.org/10.1109/msp.2007.914731
8. Candès, E.J.: The restricted isometry property and its implications for compressed sensing. Comptes Rendus Mathematique **346**(9-10), 589–592 (2008). https://doi.org/10.1016/j.crma.2008.03.014
9. Candes, E.J., Li, X., Soltanolkotabi, M.: Phase Retrieval via Wirtinger Flow: Theory and Algorithms. IEEE Transactions on Information Theory **61**(4), 1985–2007 (2015). https://doi.org/10.1109/TIT.2015.2399924
10. Candes, E.J., Tao, T.: Near-Optimal Signal Recovery From Random Projections: Universal Encoding Strategies? IEEE Transactions on Information Theory **52**(12), 5406–5425 (2006). https://doi.org/10.1109/tit.2006.885507
11. Cetin, M., Stojanovic, I., Onhon, O., Varshney, K., Samadi, S., Karl, W.C., Willsky, A.S.: Sparsity-Driven Synthetic Aperture Radar Imaging: Reconstruction, autofocusing, moving targets, and compressed sensing. IEEE Signal Processing Magazine **31**(4), 27–40 (2014). https://doi.org/10.1109/MSP.2014.2312834
12. Davenport, M.A., Duarte, M.F., Eldar, Y.C., Kutyniok, G.: Introduction to compressed sensing. In: Y.C. Eldar, G. Kutyniok (eds.) Compressed Sensing, pp. 1–64. Cambridge University Press, Cambridge (2012). https://doi.org/10.1017/CBO9780511794308.002
13. Donoho, D.: Compressed sensing. IEEE Transactions on Information Theory **52**(4), 1289–1306 (2006). https://doi.org/10.1109/TIT.2006.871582
14. Donoho, D., Huo, X.: Uncertainty principles and ideal atomic decomposition. IEEE Transactions on Information Theory **47**(7), 2845–2862 (2001). https://doi.org/10.1109/18.959265
15. Endo, Y., Shimobaba, T., Kakue, T., Ito, T.: GPU-accelerated compressive holography. Optics Express **24**(8), 8437–8445 (2016). https://doi.org/10.1364/oe.24.008437
16. Foucart, S., Rauhut, H.: A Mathematical Introduction to Compressive Sensing. Springer New York, New York (2013). https://doi.org/10.1007/978-0-8176-4948-7
17. Goodman, J.W.: Introduction to Fourier Optics, third edn. Roberts & Company Publishers, Englewood, Colorado (2005)
18. Kim, M.K.: Principles and techniques of digital holographic microscopy. Journal of Photonics for Energy 018005 (2010). https://doi.org/10.1117/6.0000006
19. Lim, S., Marks, D.L., Brady, D.J.: Sampling and processing for compressive holography [Invited]. Applied Optics **50**(34), H75–H86 (2011). https://doi.org/10.1364/AO.50.000H75
20. Lustig, M., Donoho, D., Pauly, J.M.: Sparse MRI: The application of compressed sensing for rapid MR imaging. Magnetic Resonance in Medicine **58**(6), 1182–1195 (2007). https://doi.org/10.1002/mrm.21391
21. Lustig, M., Donoho, D., Santos, J., Pauly, J.: Compressed Sensing MRI. IEEE Signal Processing Magazine **25**(2), 72–82 (2008). https://doi.org/10.1109/MSP.2007.914728

22. Marcia, R.F., Willett, R.M., Harmany, Z.T.: Compressive Optical Imaging: Architectures and Algorithms. In: G. Cristobal, P. Schelkens, H. Thienpont (eds.) Optical and Digital Image Processing, first edn., pp. 485–505. Wiley (2011). https://doi.org/10.1002/9783527635245.ch22
23. Parikh, N.: Proximal Algorithms. Foundations and Trends® in Optimization **1**(3), 127–239 (2014). https://doi.org/10.1561/2400000003
24. Rivenson, Y., Stern, A., Javidi, B.: Overview of compressive sensing techniques applied in holography [Invited]. Applied Optics **52**(1), A423–A432 (2013). https://doi.org/10.1364/AO.52.00A423
25. Rudin, L.I., Osher, S., Fatemi, E.: Nonlinear total variation based noise removal algorithms. Physica D: Nonlinear Phenomena **60**(1-4), 259–268 (1992). https://doi.org/10.1016/0167-2789(92)90242-F
26. Wu, J., Zhang, H., Zhang, W., Jin, G., Cao, L., Barbastathis, G.: Single-shot lensless imaging with fresnel zone aperture and incoherent illumination. Light: Science & Applications **9**(1), 53 (2020). https://doi.org/10.1038/s41377-020-0289-9
27. Zhang, W., Cao, L., Brady, D.J., Zhang, H., Cang, J., Zhang, H., Jin, G.: Twin-Image-Free Holography: A Compressive Sensing Approach. Physical Review Letters **121**(9), 093902 (2018). https://doi.org/10.1103/PhysRevLett.121.093902

Chapter 16
Sparse CGH and the Acceleration of Phase-Added Stereograms

David Blinder

Abstract Sparse CGH algorithms encode the wavefield in a certain transform space where the holographic signals to be computed are "sparse", i.e., require a small number of coefficient updates to be accurate. This principle can be leveraged to achieve high-speed CGH needing only a fraction of the calculations that are used in conventional CGH. We detail several examples and focus on Phase-Added Stereograms (PAS) in this chapter. Thereafter, we elaborate on a cache-friendly data structure consisting of "lozenge" cells, which can speed up PAS CGH by another order of magnitude.

16.1 Sparsity

Sparsity is a notion in signal processing, where some class of signals can be accurately represented by a small number of coefficients in a well-chosen transform basis. This property is useful for many different applications, such as data compression, filtering, compressed sensing, and accelerated calculations.

We will illustrate this with a few examples. Natural images and photographs predominantly consist of low frequencies and have local features such as edges. This makes multi-resolution wavelets a good candidate for efficiently encoding images, and is, e.g., why they serve as the basis for the JPEG 2000 image compression standard [1].

When transforming an exemplary image using the Cohen-Daubechies-Feauveau wavelet with a 4-level Mallat decomposition (cf. Fig. 16.1), we can observe that most coefficients are near-zero. Smooth features will be captured by the lowpass

---

Supplementary Information The online version contains supplementary material available at https://doi.org/10.1007/978-981-99-1938-3_16.

D. Blinder (✉)
Vrije Universiteit Brussels, 1050 Brussels, Belgium
e-mail: david.blinder@vub.be

imec, 1050 Brussels, Belgium

T. Shimobaba and T. Ito (eds.), *Hardware Acceleration of Computational Holography*,
https://doi.org/10.1007/978-981-99-1938-3_16

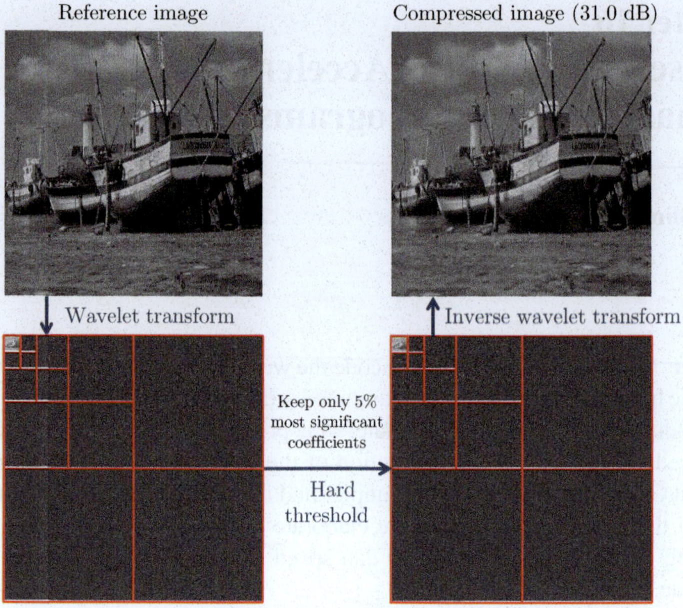

Fig. 16.1 Illustration of the relationship between sparsity and compression. By wavelet-transforming a typical photographic image and keeping only 5% of the most significant coefficients, most of the important details are preserved. This is akin to compression and is also the core idea behind sparse CGH

band, while edges will be captured by a few high-pass band coefficients. By only keeping 5% of the most significant coefficients, i.e., those with the highest absolute value, we obtain a distorted version still having 31.0 dB PSNR. This is the basic principle behind compression.

Another useful application is removing noise from signals, i.e., denoising. Noise is per definition random and thus uncorrelated to any set of signals, so its energy will be spread out over all coefficients no matter the chosen transform basis. This means that if we choose the right basis, large coefficients will likely chiefly be part of the signal and small coefficients will likely mostly correspond to noise. Significant parts of the noise can thus be removed through thresholding algorithms, cf. Fig. 16.2. In [2], an adaptive soft-thresholding technique is used for this purpose.

Sparsity is highly useful in CGH as well. This is what will be covered in the remainder of this chapter.

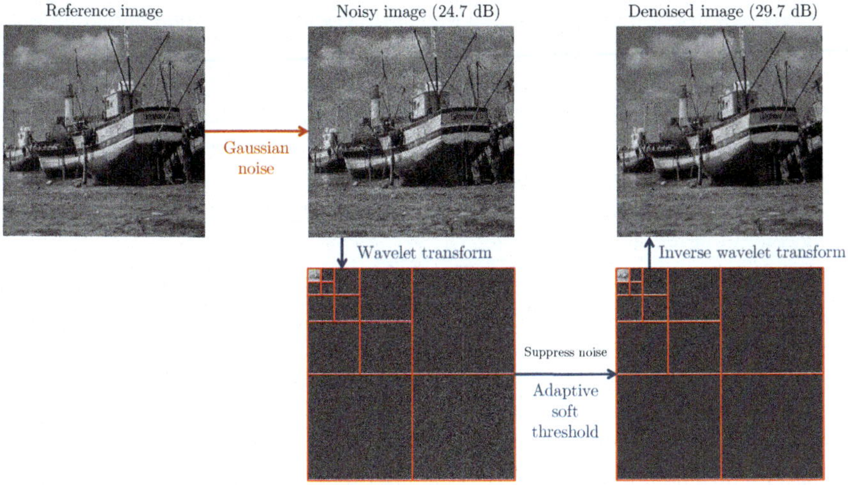

Fig. 16.2 Illustration of denoising images with wavelets. Because noise is uncorrelated with linear transforms, one can isolate a lot of its energy in the otherwise near-zero coefficients of a sparsifying transform. After an adaptive soft-thresholding operation, the PSNR is improved by 5 dB

16.1.1 Sparse CGH

In general, due to the nature of wave-based diffraction, all pixels of the hologram can be affected by every scene element. This is apparent from the Huygens–Fresnel principle, where luminous points create spherical waves, emitting light in all directions. But by expressing the holographic signal in the right basis, we can compute only a small fraction of the total number of transform coefficients, thereby considerably speeding up calculations over the default CGH calculation in the spatial/hologram domain. Consider a collection of N elements $E_j(x, y), \forall j \in \{1, .., N\}$. These E_j can represent any kind of objects such as point emitters, polygons, line or curve segments, surface pieces, etc. For a chosen transform $\mathcal{T}$, a linear combination of these E_j can be expressed as

$$H(x, y) = \mathcal{T}^{-1}\big\{\mathcal{T}\{H(x, y)\}\big\} = \mathcal{T}^{-1}\left\{\sum_{j=1}^{N} \mathcal{T}\{E_j(x, y)\}\right\}. \tag{16.1}$$

The different $\mathcal{T}\{E_j(x, y)\}$ can directly be computed or copied from a precomputed look-up table rather than evaluated in the spatial domain, which we refer to as **sparse CGH** [3].

However, several conditions need to be met for sparse CGH to be effective, i.e., significantly faster than the conventional spatial domain computation. These are [3]:

1. *High sparsity*; the ratio of needed transform coefficients to total hologram pixel count must be small so that the target signal can be accurately approximated with only a few coefficient updates.
2. *Efficient computation or insertion* of the coefficient values. It should be computationally efficient to compute and/or copy the coefficients for the different elements $\mathcal{T}\{E_j(x, y)\}$ directly in transform space. It should not be significantly more costly than evaluating $E_j(x, y)$ values in the hologram plane; otherwise, acceleration will not be possible.
3. *Efficient inverse transform* $\mathcal{T}^{-1}$. The computational complexity should be relatively low compared to having a non-sparse CGH algorithm. This condition is often met if N is sufficiently large.

Note that although the individual E_j should be sparse in $\mathcal{T}$, their sum does not have to be sparse. Moreover, the transform $\mathcal{T}$ can be applied multiple times, even combining different transforms, so long as the previously mentioned constraints are satisfied.

Several different candidates have been proposed for the transform space $\mathcal{T}$. In [4, 5], parts of the object were computed in selected batches to affect a limited number of coefficients in virtual planes, making them sparse as inputs for the subsequent FFT needed for calculating the light propagation. Using the "sparse FFT", significant acceleration was achieved.

Another approach is to use **coefficient shrinking**, whereby only some fraction of the most significant coefficients of the transformed E_j are kept, whereas the rest of the coefficients with values below some chosen threshold are set to 0. This threshold can be set depending on the quality and speed requirements. It is a trade-off between accuracy and calculation speed. The WAvelet ShrinkAge-Based superposItIon (WASABI) method [6] uses this approach with the Daubechies-4 wavelet transform for $\mathcal{T}$, cf. Fig. 16.3a. The thresholded coefficients are pre-computed, and applied in wavelet space for every transformed point-spread function $\mathcal{T}\{E_j(x, y)\}$.

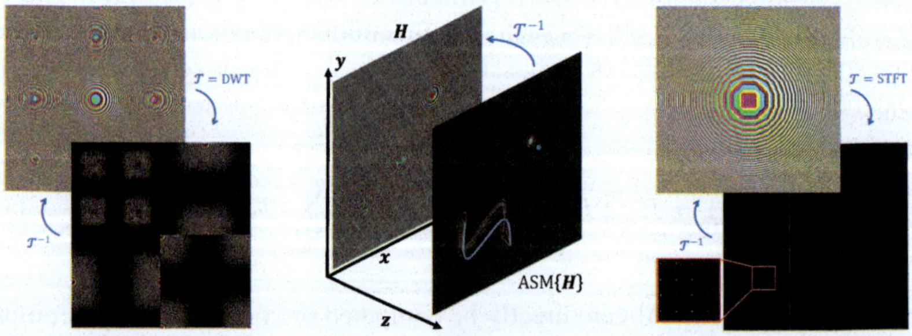

(a) WASABI (b) WRP (c) PAS

Fig. 16.3 Examples of sparse CGH methods. **a** A point-spread function and its corresponding sparse 2-level Daubechies-4 wavelet transform [6]. **b** The blue 3D curve and point are close to the WRP, only affecting nearby coefficients delineated in the red regions. This contrasts with the hologram plane H, where all pixels are affected. **c** Accurate PAS on 16 × 16 coefficient blocks with redundancy 2. (Based on [3], Fig. 16.6)

This method was shown to be about 30 times faster than the reference point-cloud CGH method while retaining acceptable visual quality. Later, a similar shrinkage method was proposed using the **short-time Fourier transform (STFT)** for $\mathcal{T}$ instead of wavelets. Because of higher average sparsity and frequency symmetry, a 2 dB PSNR quality gain was achieved over wavelet-based methods, with better off-axis view quality [7]. This method was further accelerated by analytically computing coefficients [8] rather than using precomputed values from look-up tables.

We can also use a convolutional diffraction operator for $\mathcal{T}$, such as the Fresnel transform or the angular spectrum method (cf. Chap. 1). By backpropagating the hologram close to the virtual objects in the 3D scene, the energy of the point-spread functions will be spatially concentrated. This will make them spatially sparse in $\mathcal{T}$, which is the principle leveraged in **wavefront recording planes** [9] (WRP): thanks of the proximity of scene elements to the WRP plane, the $\mathcal{T}\{E_j(x, y)\}$ can be calculated using only a small number of pixels close to the virtual objects, cf. Fig. 16.3b.

In this chapter, we will focus on a specific sparse CGH method called **"Phase-added stereograms" (PAS)**. The sparsifying transform $\mathcal{T}$ is also the STFT, where the hologram is subdivided into small spatial regions such as blocks, which are called **hogels**. But these techniques will only update a single coefficient per hogel per element E_j, see Fig. 16.3c; we will elaborate on this method in the section hereinbelow.

16.2 The Phase-Added Stereogram (PAS)

Holographic stereograms perform hologram calculations by approximating them by a discrete light field. **Light fields** can be represented by a four-dimensional **plenoptic function** $L(x, y, \theta, \phi)$ on some surface, describing the radiance in every point (x, y) of that surface emitted along angles (θ, ϕ) w.r.t. to the surface normal. This principle is illustrated in Fig. 16.4a. These light fields can be sampled along every dimension, resulting in a discrete light field utilized e.g. in light field cameras and displays.

These discrete light fields can be mapped to holograms by associating a small plane wave segment to each light field sample, cf. Fig. 16.4b. The center of the plane wave segment will coincide with the sample position (x, y), and its frequency components are proportional to the incidence angle (θ, ϕ), given by the **grating equation**

$$\sin(\theta) = \lambda\nu \tag{16.2}$$

where λ is the hologram wavelength and ν is the spatial frequency. This collection of plane wave segments can be modeled in general by a **short-time Fourier transform (STFT)** for $\mathcal{T}$. The two-dimensional STFT transform is formally defined as a family of apodized functions S with different combinations of spatial translations (τ, υ) and frequency modulations (ω, η), namely

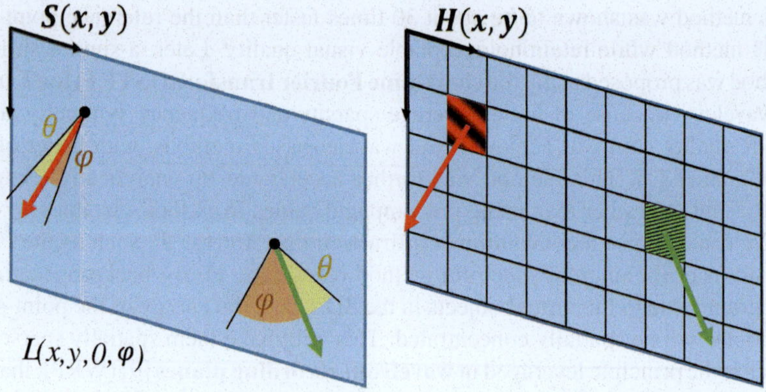

(a) Light field (b) Holographic stereogram

Fig. 16.4 Mapping between light field rays and stereogram plane wave segments. **a** the light field L is parameterized by a surface S and ray angles (θ, ϕ), with two exemplary rays shown in red and green, respectively. **b** The corresponding stereogram is shown on the right, subdivided into blocks, where plane wave segments are associated with the exemplary rays. Their diffraction angles are proportional to their frequencies

$$\begin{aligned} &\text{STFT}\{H(x, y)\}(\tau, \upsilon, \omega, \eta) \\ \equiv S(\tau, \upsilon, \omega, \eta) \quad &= \iint_{-\infty}^{\infty} H(x, y) w(x - \tau, y - \upsilon) e^{-i(\omega x + \eta y)}\, dx\, dy \end{aligned} \tag{16.3}$$

where w is a **window function**. Typically, a rectangular window function w is used for stereogram CGH, but it can also be a smooth shape, such as the **Hamming window**, the **Gaussian window**, or the **Hann window** used, e.g., in [12]. Once the plane wave coefficients have been computed from the light field samples, we can invert the STFT process, and obtain the CGH from the inverse-STFT-transformed coefficients.

16.2.1 Point Cloud CGH

The way plane wave coefficients are determined depends on the CGH algorithm. We will focus on the "phase-added stereogram" (PAS) method, which computes the CGH from **point cloud** data. Point clouds approximate objects by a discrete set of points in 3D space. In point-based CGH, every point with coordinates $(\delta, \epsilon, \zeta)$ will create a diffraction pattern called a "**point spread function**" (**PSF**) or "**zone plate**", given by

$$P(x, y) = a \cdot \exp\left(\frac{\pi i}{\lambda \zeta}\left[(x - \delta)^2 + (y - \epsilon)^2\right]\right) \tag{16.4}$$

where i is the imaginary unit and $a \in \mathbb{C}$ is the point amplitude. This expression can be used to calculate a hologram H by summing over a collection of Q PSFs P_j, $j \in \{1, ..., Q\}$, each with their respective coordinates $(\delta_j, \epsilon_j, \zeta_j)$ and amplitudes a_j:

$$H(x, y) = \sum_{j=1}^{Q} P_j(x, y) = \sum_{j=1}^{Q} a_j \cdot \exp\left(\frac{\pi i}{\lambda z_j}\left[(x - \delta_j)^2 + (y - \epsilon_j)^2\right]\right). \quad (16.5)$$

This expression can be evaluated directly, but this is highly computationally demanding because every hologram pixel needs to be updated Q times. That is why we use the STFT as a sparse transform for PAS.

16.2.2 Optimizing PAS Coefficients

In phase-added stereograms, we subdivide holograms into blocks of $B \times B$ pixels, matching the plane wave segment sizes. Rather than calculating the plane wave coefficients directly in the spatial domain, we express them in Fourier space so that we only have to update a single coefficient per plane wave segment. This principle is illustrated in Fig. 16.5.

Every block is represented by a $S \times S$ coefficient in FFT space, corresponding to plane wave pieces with different frequencies. As we choose larger values for S, we sample Fourier space more finely, providing higher precision for specifying the carrier frequency, i.e., propagation angle of each plane wave segment. Whenever $S > B$, we have a variant of PAS which is also called the "**accurate PAS**" method [11].

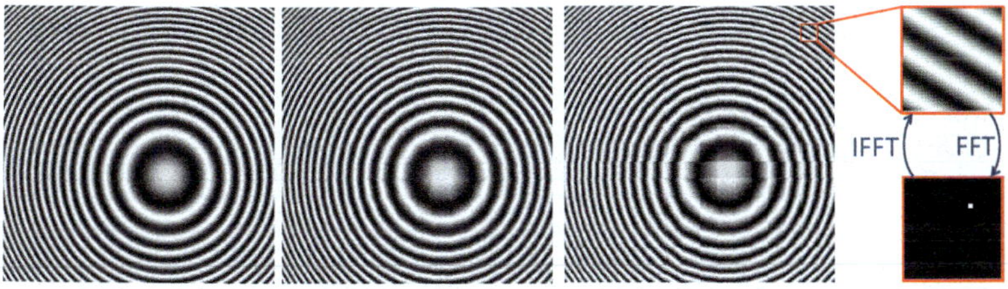

(a) Reference PSF (b) Full PAS (c) Accurate PAS

Fig. 16.5 Real part of the computed PSF signal for different CGH algorithms: **a** the reference ray-tracing equation (using (16.4)), **b** the fully computed PAS and the **c** Accurate PAS. The latter subdivides the holograms into small blocks and assigns a plane wave piece with a quantized frequency per block. A zoomed-in example of such a block is shown next to it, and its associated discrete Fourier transform. There is only one non-zero FFT coefficient per block for every PSF. This is a reprint of Fig. 16.1 from [13]

For a given PSF, how do we find out what coefficients we should update with what value in each block? This problem amounts to minimizing the energy difference between the target PSF (16.4) and the planar wave segment found within the block boundaries $[-A, +A]$, where $A = \frac{Bp}{2}$ [10]. Formally, we have

$$\operatorname*{argmin}_{m,n,\varphi} \iint_{-A}^{+A} \left|\exp\left(\frac{\pi i}{\lambda\zeta}[(x-\delta)^2 + (y-\epsilon)^2]\right) \exp\left(2\pi i(mx+ny+\varphi)\right)\right|^2 dxdy \tag{16.6}$$

solving for the best frequencies m, n and phase delay φ.

This expression can be simplified by using the identities

$$\begin{aligned} |\exp(i\phi_1) - \exp(i\phi_2)|^2 &= \sin(\phi_1 - \phi_2)^2 + (1 - \cos(\phi_1 - \phi_2))^2 \\ &= 2(1 - \cos(\phi_1 - \phi_2)) \end{aligned} \tag{16.7}$$

making (16.6) equivalent to

$$\operatorname*{argmin}_{\varphi_x} \iint_{-A}^{+A} 1 - \cos\left(\frac{\pi}{\lambda\zeta}[(x-\delta)^2 + (y-\epsilon)^2] - 2\pi(mx+ny+\varphi)\right) dxdy. \tag{16.8}$$

To solve this expression, we will make use of the following ansatz; when m, n are chosen to closely match the local frequency of the PSF within the purview of the block, we can assume that the phase difference between the plane wave and the target PSF chirp will be small. We can therefore use **Taylor approximation** $\cos t \approx 1 - \frac{t^2}{2}$ valid for small values of $|t|$. This makes (16.8) equivalent to solving

$$\frac{\partial}{\partial\varphi} \iint_{-A}^{+A} \left(\frac{\pi}{\lambda\zeta}[(x-\delta)^2 + (y-\epsilon)^2] - 2\pi(mx+ny+\varphi)\right)^2 dxdy = 0 \tag{16.9}$$

resulting in the sought phase delay

$$\varphi = \frac{1}{2\lambda\zeta}\left(\frac{2}{3}A^2 + \delta^2 + \epsilon^2\right) \tag{16.10}$$

where the term containing A^2 can be ignored, because it will cause the same phase delay across all blocks. For a general block centered at coordinates (M_x, M_y), we get

$$m = \left\lfloor \frac{(M_x - \delta) p S}{\lambda \zeta} \right\rceil \tag{16.11}$$

$$n = \left\lfloor \frac{(M_y - \epsilon) p S}{\lambda \zeta} \right\rceil \tag{16.12}$$

$$\varphi = \frac{(M_x - \delta)^2 + (M_y - \epsilon)^2}{2\lambda\zeta} - \frac{B}{2S}(m+n) \tag{16.13}$$

where $\lfloor \cdot \rceil$ is the **rounding operator**.

16.2.3 Example Code

This section includes exemplary code for computing Fresnel PAS in **Python**. The `Hologram_Settings` helper object describes the hologram and PAS properties, the `accurate_fresnel_stereogram` procedure calculates the STFT coefficients for a given point cloud, and the `inverse_stft` procedures computes the inverse STFT transform, returning the resulting CGH.

Listing 16.1 Core code for computing Fresnel accurate phase-added stereograms in Python.

```
import math
import numpy as np

# Hologram settings object
class Hologram_Settings:
    def __init__(self, res, pp, wlen, B, F):
        self.res = res # hologram resolution (in pixels)
        self.pp = pp # pixel pitch (in m)
        self.wlen = wlen # wavelength (in m)
        self.B = B # block size
        self.F = F # scaling factor

    def pas_dimensions(self):
        blockdim = (self.res[0]/self.B, self.res[1]/self.B) # block dimensions
        assert blockdim[0].is_integer() and blockdim[1].is_integer()
        SS = self.B * self.F # segment size
        return (int(blockdim[0]), int(blockdim[1]), SS, SS) # coefficient tensor dimension

# computes exp(1j*phase), for a real-valued input 'phase'
def expi(phase):
    return np.complex(np.math.cos(phase), np.math.sin(phase))

def accurate_fresnel_stereogram(hs, pcloud, ampl = None):
    """
    Returns the coefficients of the accurate phase-added stereogram
    quadratic Fresnel approximation in a 4D tensor.
    INPUTS:
        hs (Hologram_Settings)
        pcloud (Nx3): point list of N points in (x, y, z) coordinates
```

```
        ampl (Nx1): array of amplitudes [optional, default: all
            amplitudes = 1]
    """
    SS = hs.B * hs.F # segment size
    cdim = hs.pas_dimensions() # coefficient tensor dimension
    wk = 2/hs.wlen # double reciprocal of the wavelength

    # PAS coefficient matrix
    C = np.empty(cdim, np.complex64)

    # block center coordinates
    block_centers = lambda i: (np.arange(0, hs.res[i], hs.B, dtype=np.
        float32) + hs.B/2)*hs.pp
    ucenters = block_centers(0)
    vcenters = block_centers(1)

    # iterate over every block
    for u in range(cdim[0]):
        for v in range(cdim[1]):
            blockdata = np.zeros((SS, SS), np.complex64)
            center = np.array([ucenters[u], vcenters[v]]);

            # iterate over every point
            for p in range(np.shape(pcloud)[0]):
                pos = center - pcloud[p, 0:2]
                f = pos / (pcloud[p,2]*hs.wlen)
                fi = np.rint(f*hs.pp*SS).astype(int)
                fr = fi + SS//2

                # are the target Fourier coefficient coordinates
                    within block bounds?
                if np.all(fr>=0) and np.all(fr<SS):
                    coeff = expi(math.pi * (wk * pcloud[p,2] + np.sum(f*pos)
                        - fi.sum()/hs.F))
                    if ampl is not None: coeff *= ampl[p]
                    blockdata[fr[0],fr[1]] += coeff

            # assign computed block to tensor
            C[u,v,:,:] = blockdata
    # result
    return C

# Inverse STFT of PAS coefficients, with optional FFT cropping
def inverse_stft(C, B = None):
    cdim = np.shape(C)

    # inverse Fourier transform, slice
    C = np.fft.ifft2(np.fft.ifftshift(C, (2,3)))

    # crop frequency blocks, if applicable
    if B: C = C[:, :, 0:B, 0:B]
    else: B = cdim[2]

    # output hologram
    H = np.empty(np.array(cdim[0:2])*B, np.complex64)

    # reorder samples into hologram
    for u in range(cdim[0]):
        for v in range(cdim[1]):
            H[B*u:B*(u+1), B*v:B*(v+1)] = C[u,v,:,:]
    #result
    return H
```

To run it, another code snippet is provided below, computing the PAS for a toy example made out of three points. The resulting hologram is shown in Fig. 16.6.

Listing 16.2 Example code on how to calculate a PAS.

```
import pascode as pas
import matplotlib.pyplot as plt
import numpy as np

# Hologram settings, with typical parameter values
hs = pas.Hologram_Settings((2048, 2048), 4e−6, 633e−9, 32, 2)

# Point cloud data; toy example consisting of 3 points.
pcloud = np.array([
    [6e−3, 4e−3, 7e−2],
    [3e−3, 3e−3, 8e−2],
    [2e−3, 6e−3, 9e−2],
], dtype = np.float32)

## Compute Phase-added stereogram
H = pas.accurate_fresnel_stereogram(hs, pcloud)
H = pas.inverse_stft(H, hs.B)

## Display results (real part of the hologram)
fig, ax = plt.subplots()
plt.imshow(np.real(H), cmap='gray')
ax.set_title('Computed PAS')
plt.show()
```

16.3 Acceleration Structures for PAS

The **sparsity** of the PAS algorithm is quite high, since we only update a single coefficient per block and per point. The sparsity is equal to B^{-2}; so for a typical block size of $B = 32$, this amount to less than 0.1%. Despite this fact, when PAS are implemented on a GPU, the calculation time reduction w.r.t. reference point-based CGH method is more limited than what may be expected solely from the sparsity. Rather than a 1000-fold speedup, the GPU implementation was only about 3 times faster.

What explains this discrepancy? The main reason is due to **memory caching** limitations. Computation times are not only determined by the execution time of the mathematical instructions but also by memory access patterns. For a $N \times N$ pixel hologram, we need to store $\frac{S^2}{B^2N^2}$ FFT coefficients in memory, which can in principle all be accessed depending on the different positions of the point cloud elements. This is not conducive to caching, as these data structures do not fit in local memory.

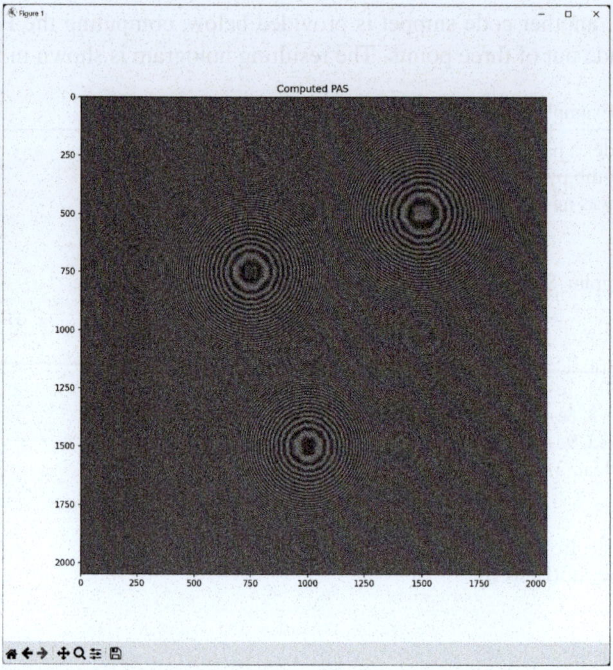

Fig. 16.6 Resulting CGH from the PAS algorithm by running the Python code in Listing 16.2 (real part)

16.3.1 Lozenge Cell Lattices

To address this problem, we should find a way to access the same small number of coefficients in every block. We start from the following observation: suppose we take a small $K \times K$ sub-block of FFT coefficients within a single $S \times S$ PAS block. These coefficients correspond to a restricted set of acceptance angles in which points will only affect selected the $K \times K$ coefficients, cf. Fig. 16.7. However, acceptable points should not only lie in the acceptance angles of one PAS block but also of all PAS blocks simultaneously. The trick is to carefully choose different $K \times K$ sub-blocks in every PAS block to have a non-empty intersection in space, as shown in Fig. 16.8.

Because these cell shapes resemble a rhombus, we call them "**lozenge cells**". We need to use multiple of these cells to fully cover space. This can be done systematically with the following construction; we start with the full hologram bandwidth, that will dictate the allowed acceptance angles of the incoming light rays as large cone. Incidence angles beyond that limit cannot be resolved as they will cause frequency aliasing; this is illustrated with the "**aliasing-free cone**" [14]. Given the subset of frequencies $F = \frac{K}{S}$, we can partition the cone into a fan of smaller adjacent cones. By carefully choosing their offsets at every point, we can make them intersect to form a lozenge cell lattice. This is illustrated in Fig. 16.9a, where the fans are drawn

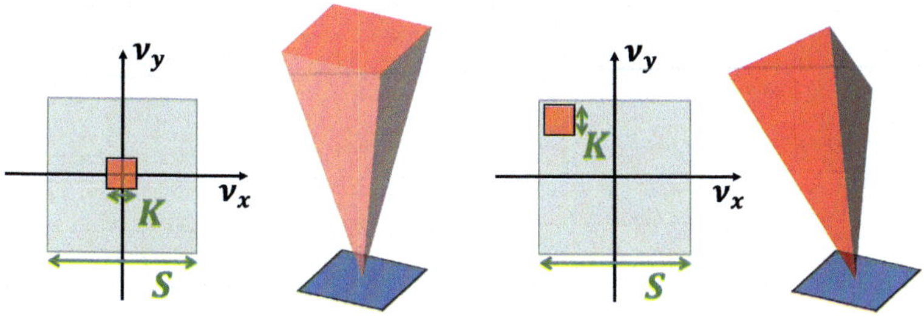

(a) Angle bundle (low frequencies) (b) Angle bundle (high frequencies)

Fig. 16.7 Diagram showing how frequency bands map to angle bundles in 3D space. In both figures, the left parts represent the $S \times S$ Fourier coefficients of a PAS block, where only a $K \times K$ sub-block is highlighted in red. The right parts show the corresponding (blue) PAS block located in space, with a (red) pyramid covering the spatial region whose points would only affect the designated sub-block of FFT coefficients. The actual pyramid region extends infinitely far away from the block center. **a** For low frequencies, the covered angles are close to the hologram plane normal, **b** while the high frequencies will cover more oblique angles. This is a reprint of Fig. 16.2 from [13]

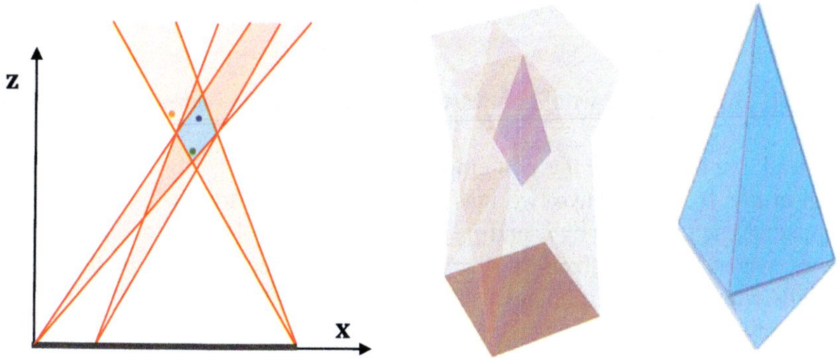

(a) Lozenge cell geometry (2D) (b) Lozenge cell geometry (3D) (c) Lozenge cell shape

Fig. 16.8 Diagram of lozenge cell shapes. **a** For well-chosen active sub-blocks in every PAS block, all cones intersect into a lozenge cell. **b** This principle is extended to 3D, showing the resulting cell at the center. **c** It's shaped as a distorted octahedron, which cannot be used to tile 3D space without overlaps. Adapted with permission from [10] ©The Optical Society

for the extremities of the hologram in 2D. The colored regions correspond to the cell volumes where points can be present, while the white region is forbidden, lying outside of the aliasing-free cone.

Unfortunately, we cannot extend this principle directly to 3D space because it is mathematically impossible to seamlessly tile space with octahedrons. This means there will inevitably be some redundancy, where cells overlap, if we want to cover the entirety of the aliasing-free cone. We could take the Cartesian product of all cell combinations in the $x - -z$ and $y - -z$ planes, respectively, as shown in Figs. 16.9a

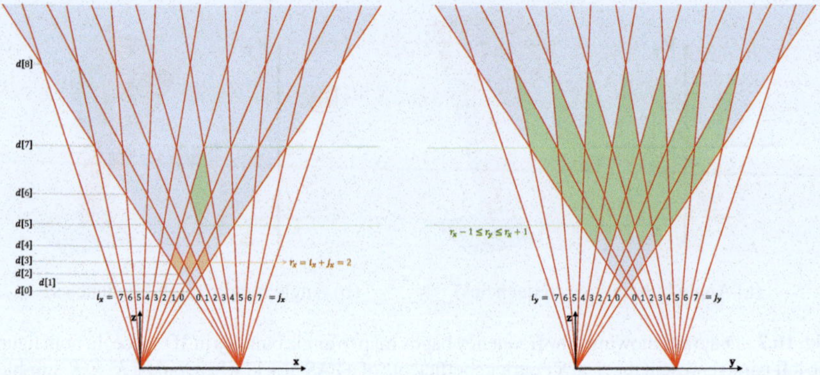

(a) x-z plane (b) y-z plane

Fig. 16.9 Diagram of the 3D point cloud cell partitioning by combining two 2D lozenge cell lattices. Every cell in the x–z plane is characterized by its coordinates i_x and j_x (**a**). The row index is given by the sum of the coordinates, as shown for the orange cells for the example $r_x = 2$. Every lozenge cell in (**a**) can have an intersection with a cell in (**b**), so long as they have an overlap along z, as shown by the green cells. Reprinted with permission from [10] ©The Optical Society

and 16.9b. However, this would be wasteful, as many of the intersections would be empty.

Instead, we only select the non-empty combinations based on their depths. As shown in Fig. 16.9a, we have two fans of cones at each extremity consisting of F components each, indexed by i_x and j_x, respectively. These lozenge cells are stacked in rows, whose index is given by $r_x = i_x + j_x$, consisting of $r_x + 1$ cells per row. The cells within a row are bounded in z by the array entries $d[r_x]$ and $d[r_x + 2]$, respectively, given by

$$d[r_x] = \frac{p^2 N}{\lambda(1 - r_x F)}. \tag{16.14}$$

provided that $r_x F < 1$. Because of (16.14), one can deduce that every 2D lozenge cell in the $x - z$ plane extruded along the y-axis will only intersect three rows in the $y - z$ plane (assuming that the same construction is used for both planes). This can also be observed in Fig. 16.9 for the exemplary cells marked in green.

This concept can be used to partition the input point cloud. Any point with coordinates $(\delta, \epsilon, \zeta)$ can be assigned lozenge cell coordinates

$$i_x = f(\delta); \quad i_y = f(\epsilon); \quad j_x = f(Np - \delta); \quad j_y = f(Np - \epsilon). \tag{16.15}$$

using the mapping $f : t \mapsto \left\lfloor \frac{1}{2F} - \frac{tp}{F\lambda\zeta} \right\rfloor$, where $\lfloor \cdot \rfloor$ is the **floor operator**. This can be used to linearize every valid tuple (i_x, i_y, j_x, j_y) uniquely into a single index

$$\ell = r_x \cdot (r_x^2 + r_x + 1) + 3i_x \cdot (r_x + 1) + \tfrac{1}{2} r_y \cdot (r_y + 1) + i_y \tag{16.16}$$

where $r_x = i_x + j_x$ and $r_y = i_y + j_y$ are the lozenge cell row indices. This simplifies the point storage structure into a linear array of point lists, which can be processed sequentially using a small amount of memory, benefiting computational performance. More details on the concrete implementation follow in the remainder of this chapter.

16.3.2 Implementation and Results

The proposed algorithm lends itself to massively parallel processors, such as GPUs, FPGAs, or ASICs. In this section, we will focus on a concrete implementation for NVIDIA GPUs using **CUDA**.

In the CUDA programming model, as for most architectures, it is important to distinguish between the different types of memory. As covered in Chap. 6, the main GPU memory is called "global memory", which typically consists of several gigabytes. This off-chip memory is relatively slow compared to the other parts of the CUDA memory hierarchy. On the other hand, the CUDA multiprocessor consists of many cores which each have their own very fast registers, and share a common on-chip cache called "local memory", which is partitioned into L1 cache and shared memory. Since the PAS algorithms are much more memory-bound than compute-bound, it is beneficial for performance to minimize memory usage and to prioritize faster memory whenever possible.

For the base reference PAS algorithm, we allocate one thread per PAS block, as they can operate independently. Each thread loops over the entire point cloud, independently updating a single coefficient as described earlier in this chapter. The amount of memory per thread is relatively large: S^2 coefficients per block, each taking up 8 bytes when represented by complex-valued single-precision floating-point numbers. For a typical value of $S = 64$, this amounts to 32KB per thread, which is too large for local memory. Therefore, all updates should happen in global memory instead.

This memory bottleneck will hamper computational performance. This can be partially addressed in CUDA by using some optimizations. We can allocate multiple threads per block if we use **atomic** additions; these ensure that when two or more threads operate concurrently on the same variable, they do not incorrectly overwrite each other's results. Atomic additions are treated as completing in a single step w.r.t. other threads; though they tend to perform slower than regular additions. This can be combined with a randomization of the input point cloud pressing order, to reduce the probability of coincident access to the same coefficients by different threads. More details on these optimizations, their parameterizations, and effects are found in [13].

Despite these optimizations, the reference method will still be several times slower than the proposed algorithm. Moreover, supporting these optimizations requires more complex hardware and parameter tuning. This overhead will make FPGA or ASIC implementations difficult.

For the proposed algorithm, we first bin all the points in the point cloud into the different cells based on (16.15) and (16.16). We process each cell one by one,

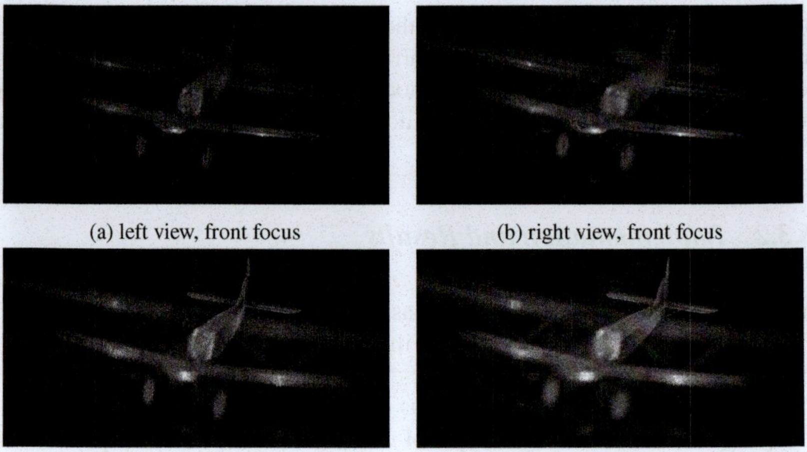

(a) left view, front focus (b) right view, front focus

(c) left view, back focus (d) right view, back focus

Fig. 16.10 Several numerical reconstructions of the hologram calculated with the PAS algorithm, showing left and right views, each refocused at the front and back of the plane model. This is a reprint of Fig. 16.8 from [13]

guaranteeing that only the coefficients in a known sub-stereogram will be affected for every PAS block. This ensures that they all fit in the local memory for every thread. For the typical case of $K = 4$, we need only 16 8-byte coefficients or 128 bytes in total per thread. Like in the base reference algorithm version, one thread is allocated per PAS block, ensuring that the threads do not mutually interfere. When all points in a lozenge cell are processed, the resulting coefficients are transferred to global memory, making room for the next (non-empty) lozenge cell to be processed. Since these global transfers only happen once per lozenge cell, the needed amount of global memory transfers is much smaller (Fig. 16.10).

To test the algorithm, we used a gray-scale version of the "Bi-plane" point cloud, consisting of 1 million points with associated intensities for the point color. The virtual plane object was axially placed at 20 cm from the hologram plane, and laterally centered to match the hologram's optical axis. The hologram was calculated with a wavelength of $\lambda = 633$ nm and a pixel pitch of $p = 2\mu m$ along both the x and y axes. Its resolution was 16384×16384 pixels, totalling to $2^{28} \approx 2.56 \cdot 10^8$ pixels. The PAS algorithm was parameterized using $B = 32$ for the hologram subdivision block size, PAS block coefficient size of $S = 64$, and sub-stereogram block size of $K = 4$.

The algorithm was tested on a machine configured as described in Table 16.1. The reference base PAS implementation took 711.7 s, while the proposed solution only took 20.6 s, resulting in a 34.5-fold speedup. Although the proposed algorithm also requires some overhead for distributing the points in the various lozenge cells, the impact is negligible, as it required only about 29 ms. Its impact may be reduced

Table 16.1 Implementation environment. Adapted from [10]

OS	Windows 10 Pro
CPU	Intel Xeon E5-2687W v4
Memory	256 GB
Programming language	C++17 with CUDA 10.1
Compiler	Visual studio 2019
GPU	NVIDIA TITAN RTX

further still by pipelining. Since the proposed algorithm is a more memory-efficient version of the reference base PAS algorithm, they both essentially produce identical and thus have no difference in quality.

16.4 Conclusions

The complete process is summarized in Fig. 16.11. The PAS algorithm is a sparse CGH technique, requiring only one coefficient update per $B \times B$ block and per point. To further enhance computational performance, we first partition the object point cloud into different lozenge cells, which are then processed on a cell-per-cell basis. Because all points within the same cell will only affect a known small subset of the total coefficients, the memory requirement is significantly reduced, which is conducive to memory caching. Only when a cell is completely processed, are the results copied to the larger but slower global memory. After computing a final IFFT for every PAS block, we obtain the final hologram wavefield, which can be used for a holographic display.

David Blinder

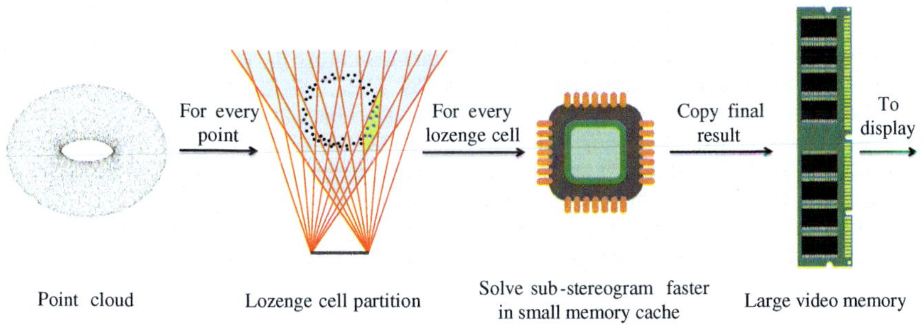

Fig. 16.11 Graphical summary of the PAS algorithm pipeline

PAS calculations can be sped up by a factor of about 30, or are about 100 times faster than the reference point-based CGH algorithm. The algorithm is especially suited for customized hardware solutions such as FPGA or ASIC because of its much lower memory requirements and dependencies, and could potentially realize even higher CGH speeds.

Sparse CGH is a versatile algorithmic principle, which may be key in realizing high-resolution real-time video holographic display.

Fundings The Research Foundation—Flanders (FWO), Senior postdoctoral fellowship (12ZQ223N); the Japan Society for the Promotion of Science (JSPS), International research fellowship (P22752).

References

1. Schelkens, P., et al.: The JPEG 2000 Suite, Wiley Publishing (2009). https://doi.org/10.1002/9780470744635
2. Chang, S. G., et al.: Adaptive wavelet thresholding for image denoising and compression. IEEE transactions on image processing (2000) https://doi.org/10.1109/83.862633
3. Blinder, D., et al.: The state-of-the-art in computer generated holography for 3D display. Light: Advanced Manufacturing (2022) https://doi.org/10.37188/lam.2022.035
4. Kim, H. G. and Ro, Y. M.: Ultrafast layer based computer-generated hologram calculation with sparse template holographic fringe pattern for 3-D object. Opt. Express (2017) https://doi.org/10.1364/OE.25.030418
5. Jia, J., et al.: Fast two-step layer-based method for computer generated hologram using sub-sparse 2D fast Fourier transform. Opt. Express (2018) https://doi.org/10.1364/OE.26.017487
6. Shimobaba, T., and Ito, T.: Fast generation of computer-generated holograms using wavelet shrinkage. Opt. Express (2017) https://doi.org/10.1364/OE.25.000077
7. Blinder, D., and Schelkens, P.: Accelerated computer generated holography using sparse bases in the STFT domain. Opt. Express (2018) https://doi.org/10.1364/OE.26.001461
8. Blinder, D.: Direct calculation of computer-generated holograms in sparse bases. Opt. Express (2019) https://doi.org/10.1364/OE.27.023124
9. Shimobaba, T., et al.: Simple and fast calculation algorithm for computer-generated hologram with wavefront recording plane. Opt. Lett. (2009) https://doi.org/10.1364/OL.34.003133
10. Blinder, D., and Schelkens, P.: Phase added sub-stereograms for accelerating computer generated holography. Opt. Express (2020) https://doi.org/10.1364/OE.388881
11. Kang, H., et al.: Accurate phase-added stereogram to improve the coherent stereogram. Appl. Opt. (2008) https://doi.org/10.1364/AO.47.005784
12. Padmanaban, N., et al.: Holographic Near-Eye Displays Based on Overlap-Add Stereograms. ACM Trans. Graph. (2019) https://doi.org/10.1145/3355089.3356517
13. Blinder, D., and Schelkens, P.: Accelerating phase-added stereogram calculations by coefficient grouping for digital holography. Proc. SPIE (2020) https://doi.org/10.1117/12.2553918
14. Blinder, D., et al.: Signal processing challenges for digital holographic video display systems. Signal Processing: Image Communication (2019) https://doi.org/10.1016/j.image.2018.09.014

Chapter 17
Efficient and Correct Numerical Reconstructions

Tobias Birnbaum

Abstract This chapter concerns itself with visual quality assessment and the numerical reconstruction of holograms. Both topics are essential when designing, tuning, or evaluating any component in the signal processing chain of a holographic 3D imaging system. The chapter will cover the fundamental requirements, best practices, and considerations for a correct and efficient implementation.

17.1 Introduction

Digital holography has many applications in metrology as well as 3D imaging [1]. However, certainly, 3D imaging applications are more attractive to the mainstream as only holography holds the promise of being able to provide the ultimate 3D viewing experience. Because a hologram can reproduce the amplitude and phase of the light field over a given surface, ideal 3D holograms are visually indistinguishable from reality.

In practice, however, artifacts from the recording or display setups and the various processing steps render even the best holograms still discernible from reality at present. Conceptually, the end-to-end pipeline can be visualized as shown in Fig. 17.1. To improve over the state of the art and approach ultimate realism in 3D imaging, it is mandatory to assess and quantify the quality of the final reconstructed hologram. This means also assessing the effects and potential implications of any of the signal processing components. In particular, when designing a new algorithm for any component of the pipeline—unless it is guaranteed to be lossless, for example, lossless compression—then the various design choices and parametrizations have to be

Supplementary Information The online version contains supplementary material available at https://doi.org/10.1007/978-981-99-1938-3_17.

T. Birnbaum (✉)
Vrije Universiteit Brussels, 1050 Brussels, Belgium
e-mail: tobias.birnbaum@vub.be

IMEC, 1050 Brussels, Belgium

T. Shimobaba and T. Ito (eds.), *Hardware Acceleration of Computational Holography*,
https://doi.org/10.1007/978-981-99-1938-3_17

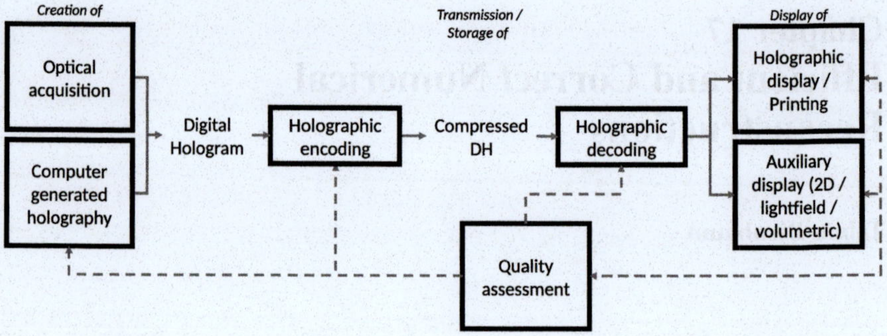

Fig. 17.1 Simplified signal processing pipeline for digital holography for 3D imaging

weighed against each other. In this chapter, some of the intricacies of this weighing will be discussed.

Ideally, any given algorithm tries to maximize the perceived **visual quality** under a given hardware resource, real-time, and various other constraints. Visually perceived quality is subjective and non-numeric. It is comparatively difficult to obtain sufficient statistical significance. Thus, often mathematical functions are used to yield objective scores which approximate a score of the perceived visual quality. These functions have of course to be tuned beforehand to best model subjective quality and are most frequently used during the design stage of signal processing components. For final evaluation of the design choices, subjective quality with a strong statistical basis remains as the ultimate criterion. The research domain of visual objective and subjective quality assessment is called **visual quality assessment (VQA)**. It is part of the bigger field of quality assessment, which itself is a branch of signal processing rooted in the mathematical field of measures.

A natural question for VQA in the context of holography is, what data shall be used as the basis for the assessments? In a perfect world, holograms would always be consumed on an ideal holographic display modality and subjective quality scores would be immediately accessible. In practice, holographic displays remain limited in resolution. Currently, spatial light modulators—the essential component in any true holographic display—have resolutions of ~ 8 mega-pixel. They severely lack behind the resolutions of high-quality holograms, which have several hundred mega-pixel to hundreds of giga-pixel [1]. Because subjective quality assessments (with statistical significance) can not be obtained without delays, objective quality assessment is often a mandatory aid. Digital holograms are frequently numerically reconstructed by reversing the direction of propagation in the common diffraction propagation kernels. Numerical reconstructions are used for quality assessment as well as for the actual consumption of high-quality holograms on provisional display solutions such as regular 2D, volumetric, or light field displays [2, 3]. Various methods exist for reconstructing digital holograms. But, attention needs to be paid to the efficient implementation of the reconstruction methods, as well as their correctness and potential further constraints due to VQA best practices or file formats, etc. In Sect. 17.3, several numerical reconstruction methods along with their mentioned areas of special attention will be discussed.

A concise overview of key concepts relevant to holograms will be presented in Sect. 17.2.

17.2 Visual Quality Assessment

Visual quality assessment (VQA) can be classified as either subjective or objective. This section will provide an overview of both types and largely follow the best practices outlined in the common test conditions [4] specified within the scope of **JPEG Pleno Holography** [5, 6]. JPEG Pleno Holography is the first international standardization effort for the compression of holographic content and contains the most recent consensus from multiple leading teams in the domain on the VQA of holographic content.

The reason why VQA for holograms differs from well-understood modalities such as classic images and video is deeply connected to the signal characteristics of holograms [7]. The key difficulties are:

Non-locality The **non-locality** of information in the hologram plane makes it difficult to compare the visual content of any two holograms directly without reconstructions. In addition, the non-locality also changes fundamentally the signal characteristics and the sensitivity of holograms on various distortions. For example, a low-pass filtered image is still well recognizable, while a low-pass filtered Fresnel hologram is missing higher viewing angles; or while missing information in an image is lost, missing information in the hologram plane is usually recoverable upon reconstruction. See, for example, Fig. 17.2.

Plenoptic data The fact that each hologram supports a continuous spectrum of viewpoints (gaze angles, foci) and the fact that there exist distortion types, which affect the set of supported viewpoints not equally, means that per hologram, in

(a) Natural image (b) Lossy image (c) Hologram rec. (d) Lossy hologram rec.

Fig. 17.2 In contrast to a natural image, (**a**), the reconstructed hologram from a diffuse surface with the same texture (**c**) shows speckle noise. However, information loss in the image can not be recovered for images, but the same loss in the hologram plane is barely impacting the reconstructed hologram, see (**b**) versus (**d**)

theory, an infinite set of scores would be required. Thus, a sampling of the viewpoint space and a subsequent pooling of scores are required to draw meaningful conclusions. This problem is shared among all **plenoptic imaging** modalities, such as light-field displays and volumetric displays, but it is the most severe for holography.

Speckle noise Reconstructed holograms typically suffer from a multiplicative, signal-dependent noise called **speckle noise**. It is due to the reconstruction of scene objects with a natural surface roughness using coherent illumination. This leads to point-wise constructive and destructive interferences in the reconstruction. In practice, this makes any reconstruction appear to be polluted by a salt-and-pepper type of noise, compare (Fig. 17.2a) versus (Fig. 17.2c). This is a problem both for subjective VQA as well as for objective metrics.

Display/Printing limitations Because of limitations of current holographic displays (e.g., limited resolution, large pixel pitch, and slow response time, ...) no direct rendering of a large number of high-quality holograms is possible. Trade-offs on holographic printers, holographic display devices, or alternative displays have to be made. Each modality places additional constraints on the VQA. For example, printing of holograms is currently limited to real-valued holograms and expensive/time-intensive. Thus, printing all holograms contained in a test dataset is not feasible due to the large numbers even for small studies. Alternative display modalities, such as 2D, volumetric, or light field displays enforce restrictions on the explorable degrees of freedom, supported bitdepth, and in part resolution, too.

To better understand how these problems are tackled, we will describe next the essence of subjective VQA and thereafter of objective VQA.

17.2.1 Subjective Visual Quality Assessment

A subjective VQA experiment consists of data preparation, a survey over a sufficiently large pool of participants, and a statistical analysis of the results.

17.2.1.1 Data Preparation

Typically, the data preparation involves an expert curating a versatile and large enough dataset to be subsequently evaluated by test subjects. Thereby, it is important that meaningful levels of distortion are selected and the display constraints are accounted for. That is provided a selected display modality, the distortions should be neither exclusively indistinguishable nor exclusively extremely poor. The size of the dataset has to be chosen with care. A large dataset implies the need for a large group of test subjects and/or long test sessions, to draw statistically relevant conclusions from the results. A small dataset on the other hand may not be representative of all content scenarios considered. Thus, the dataset and the question(s) to be answered by a specific survey need to be defined as a function of one another.

17.2.1.2 Survey Design

For the survey itself, a large list of possible design choices exists. As mentioned already before the display modality represents one choice. In [3], the VQA performance and the comparability of holographic displays were compared with the same content shown on a light field and a conventional 2D display for the first time. As a result of that study, regular displays were found to be the most sensitive to artifacts. Based on this finding and for reasons of limited availability of high-end holographic display setups, subsequent studies were thus far conducted using conventional displays.

Another design choice is the precise form of the survey. Decisions need to be taken on, for example, how long subjects may inspect the content; how long does each test session run; if and how test subjects are to be trained beforehand; is a reference presented, if so when; how is the scoring implemented, e.g., discrete or continuous numeric scores or discrete classes. For regular 2D displays, multiple well-established experimental VQA designs exist. For example, single, sequential double, or simultaneous double stimulus experiments with a binary, discrete, or continuous scoring per image can be studied. Most of the current studies on holographic VQA, see [3, 4, 8] and references therein used a simultaneous double stimulus for continuous evaluation as per ITU-R BT.500 recommendation. Irrespective of which test design is ultimately selected, care needs to be taken to make the study as independent from local test environment constraints as possible and ensure reproducibility. For this reason, any subjective survey is typically conducted in at least 2 independent test labs.

For the JPEG Pleno Holography-related experiments, one more component was essential in the experimental design, which is the diffraction-limited reconstruction of perspective reconstructions [10, 11]. It allows for the reconstruction of holograms at their true intrinsic resolution. For example, the direct reconstruction from the "Dices16K" hologram from the b<>com dataset (https://hologram-repository.labs.b-com.com) using a 2048 × 2048 px spatial aperture has a resolution of 16384 × 16384 px. It can be shown [11] using phase space arguments and ab initio considerations that the diffraction-limited resolution, in that case, would be only 1680 × 1680 px. This differs from naive downsampling strategies, such as bilinear downsampling, which provides no such guarantees or guides on which downsampling factors are acceptable and may thus not be used for VQA but only for scene previews. Figure 17.3 shows an example of a region of interest crop of size 2048 × 2048 px from the high-resolution reconstruction, whose intrinsic resolution is solely 210 × 210 px. The bilinear resized and diffraction-limited reconstruction of sizes 2048 × 2048 px and 1680 × 1680 px, respectively, are shown as well.

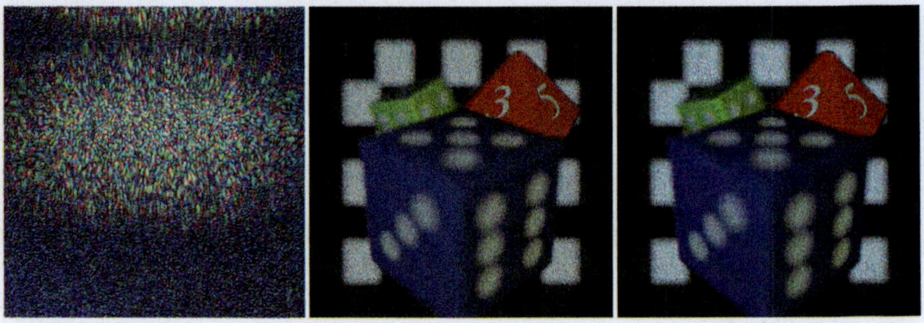

(a) Crop (b) Bilinear downsampling (c) Diff. Lim. Rec.

Fig. 17.3 Reconstruction of "Dices16K" using a 2048 × 2048 px square aperture. Shown are **a** a 2048 × 2048 px region-of-interest crop; **b** a reconstruction bilinearly downsampled to 2048 × 2048 px; **c** a diffraction-limited perspective reconstruction of resolution 1680 × 1680 px obtained from a 2048 × 2048 px spatial aperture applied in the hologram plane before propagation

17.2.1.3 Statistical Analysis

For the statistical analysis, mean-pooling across the set of test participants yields the so-called mean-opinion scores. Further pooling such as overall viewpoints per hologram or a subset thereof depends on the question of interest and no general rule of thumb can be provided here. Though the "best" assignment of a unique score (vector) for all viewpoints per distorted hologram is still an open research question and positive synergies can be expected in the interaction with VQA of other plenoptic imaging modalities.

17.2.1.4 Best Practices

In practice, subjective test surveys on regular 2D screens with holographic content and as published in [3, 4, 8] thus took the following form: in a test environment according to ITU-R BT.500-11 recommendations a color and contrast calibrated TX-65AX800e display of resolution 3840 × 2160 px was used to display the reference and a distorted hologram side-by-side. Test proponents had to pass a short training session showcasing the extremes under supervision. Thereafter, they had to rate each hologram reconstruction on a discrete scale of 1–5 in sessions of 20–40 min with a limited viewing time per image and from a fixed viewing position.

Another survey type that was evaluated was dynamic VQA [9] in the form of 3D pseudo-video sequences generated from viewpoint tours through a single static hologram. Unfortunately, video tours are thus far ill-suited for double stimulus experiments, due to the informational overload. Another complication of pseudo-video tours is that view-dependent noise may not be sufficiently weighted in the final score and a strong dependence on the viewpoint path design exists. Furthermore, it was found that sensitivity to artifacts is lowered due to the informational overload in speckle noise-polluted video sequences. The removal of speckle noise through numerical filtering on the other hand, although practical, is generally ill-advised. Again there is a risk that targeted artifacts could be removed by chance.

17.2.1.5 Subjective VQA Summary

To summarize, it is important to note that albeit its obstacles only subjective quality assessment can serve as an ultimate judge of 3D imaging quality. Especially, because only digital holography holds the promise of providing the ultimate 3D imaging experience. Therefore, it will always remain important in the final design steps of any potentially lossy/proximal component of the signal processing pipeline involved in the holographic display of 3D scenes. But exactly because of its biggest drawback, the test duration approximated VQA in terms of mathematical functions is highly sought after and an active research field. We will discuss the current state of the art on **objective quality assessment** in the following section.

17.2.2 Objective Visual Quality Assessment

Objective VQA includes any method that provided a distorted hologram returns one or multiple numerical scores which relate to the visual quality of any/all viewpoint reconstructions of the hologram. Common are reference-based methods, but few no-reference exist as well.

An ideal method yields accurate quality predictions based on calculations in the hologram domain directly, instead of requiring costly reconstructions at ambiguously sampled viewpoints. Naturally, such a method needs to account for the general signal characteristics of digital holograms, which are oscillatory patterns that are, in the most general case, signed, complex-valued, and of unbound dynamic range. This is in contrast to non-negative intensity recordings with natural image characteristics. Any suitable metric will also need to account for the specific regime/display setup geometry a given hologram was acquired/created for. Solely consider the signal differences in Fresnel, Fourier, and image-plane holograms.

17.2.2.1 Classic Image Metrics

Before, any new metrics were designed multiple studies compared the suitability of existing quality metrics (potentially with minor modifications). For example, compression-related distortions were studied in [12–14]. Especially notable is thereby [13] which presents one of the most comprehensive studies of 11 conventional metrics as well as 2 holography-aware metrics, including a summary of the underlying principles and the utilized parametrizations. The underlying dataset was a mix of computer-generated and optically captured Fourier holograms of $\approx 2k \times 16k$ px

Table 17.1 Objective visual quality assessment metric recommendations reproduced from [13] and extended by current recommendations within JPEG Pleno Holography for VQA [15]. **Legend**: Positive, if use is advised; Zero, if the study results are inconclusive; Negative, if use is discouraged; Starred, if currently used within JPEG Pleno Holography. Italic metrics were not studied in [13]

Quality metric	Fourier hologram plane	Fresnel hologram plane	Reconstruction	Speckle denoised reconstruction
SNR	*	*	*	—
Renormalized SSIM	*	*	—	—
MSE	0	0	1	1
NMSE	1	0	1	1
PSNR	0	0	0*	0
SSRM	−1	−1	1	1
SSRMt	−1	−1	0	0
SSIM	1*	1*	−1*	−1
IWSSIM	0	0	0	0
MS-SSIM	0	0	−1	−1
UQI	−1	0	−1	−1
GMSD	1	0	0	0
FSIM	−1	−1	−1	0
NLPD	0	0	0	−1
VIFp	−1	1	−1*	−1

resolution. The metrics were evaluated in the hologram plane of a Fourier and a Fresnel representation, as well as after numerical reconstruction and after reconstruction and speckle denoising. A recommendation table reproduced in Table 17.1 is the outcome of that study. The table was extended with the current recommendations from the common test conditions ver. 9 [15] and quality assessment pipeline developed for JPEG Pleno Holography [4]. Specifically, the implemented metrics of the common test conditions software ver. 7 and recommended were used. For more information on the metrics and implementation details, we refer to [13, 15] and references therein.

Other metrics that are used in the scope of deep neural networks are often regularized combinations of ℓ_2-norm and ℓ_1-norm errors, see for example, [19].

One additional metric requires explicit introduction as it is the de facto standard for comparing any two compression schemes, also within the holography context. The **Bjøntegaard-Delta peak signal-to-noise ratio (BD-PSNR)** [20] is computed from a series of PSNR scores obtained from a single hologram compressed at different rates. The scores are proportional to the signed area between two log(bitrate)-distortion curves over a chosen bitrate range, see Fig. 17.4. This accounts for PSNR scores for low bitrates being more important than for higher bitrates. Given the bitrate-distortion points of two methods, the BD-PSNR is evaluated as follows:

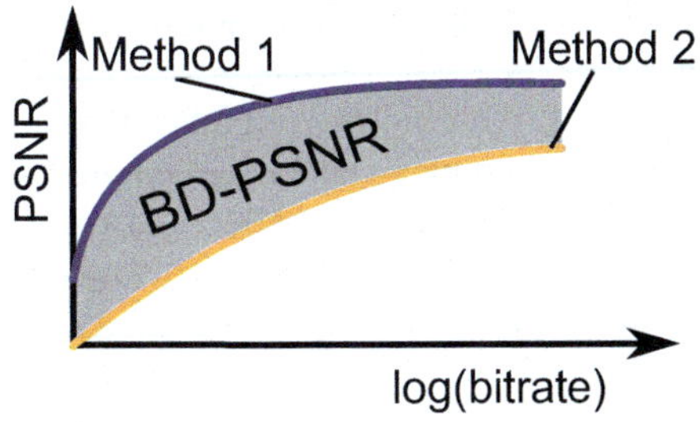

Fig. 17.4 Geometric meaning of BD-PSNR of method 1 versus method 2

1. Perform a cubic interpolation of the PSNR scores as a function of the log($bitrate$).
2. Calculate the area under the interpolated curve for each method.
3. The BD-PSNR equals the area of method 1 minus the area of method 2.

Note, by convention, the natural logarithm is used for bitrates. Furthermore, if method 1 is better performing than method 2, it can be a negative number. Although not broadly used yet, whenever the PSNR is not available the SNR may be used to express the gains. It functions the same way as the BD-PSNR; however, the range is scaled and therefore only comparisons between one error measurement type, either SNR or PSNR, are possible.

17.2.2.2 Holography-Aware Metrics

Going beyond classic image quality metrics, the **versatile similarity metric (VSM)** [13, 16] and the **sparseness significance ranking measure** (SSRM) [13, 17, 18] were proposed and tested. See also Table 17.1. Starting with SSRM, new research on holographic VQA targets some abstract transform domain for quality evaluation. Another example is the latent space of a neural network which was shown as a proof of concept for defocused 2D images in [21]. But more research is needed here.

The dynamic focal stack is yet another metric, that was demonstrated successfully for smooth phase holograms [22].

17.2.2.3 Remarks on Speed

If objective VQA should be used in a parameter search or algorithmic optimization with many iterations, simple metrics, such as (P)SNR, (N)MSE, applied to the hologram plane are preferable over, e.g., SSIM. Unfortunately though, (P)SNR and (N)MSE are extremely sensitive to some errors like pixel shifts or different instantiations of random parts of the algorithms. Therefore, metrics such as SSIM applied to the hologram plane or even focal stack analyses and viewpoint sampling are not always avoidable.

17.2.2.4 Remarks on Objective Quality Metrics in the Hologram Plane

Typically, the following modifications are required to use conventional metrics on unbound, complex-valued digital holograms: averaging scores from real and imaginary parts, consistent rescaling, and ideally skipping any quantization. When utilizing conventional visual quality metrics on holograms two important facts should be kept in mind. First, the behavior of those metrics such as score ranges will be differing substantially from other modalities and any prior knowledge from those modalities should not be used during evaluation. Second, scores are rarely meaningful as absolute numbers. Often only relative scores within a dataset or even only per hologram for varying distortion levels bear meaning.

17.2.2.5 Remarks on Objective Quality Metrics for Numerical Reconstructions

When evaluating quality metrics on the numerical reconstructions of digital holograms multiple choices exist. Metrics may be applied to the absolute values of the reconstructed wavefields, i.e., typically to floating point precision data, or to 8 bit, 16 bit quantized data. The metrics behavior may differ considerably provided the large dynamic range of digital hologram reconstructions sourced in the presence of speckle noise. For VQA within a standardization process, speckle denoising methods are currently not permitted. Therefore, rescaling and clipping of the dynamic range at fixed thresholds (obtained from a "ground truth") before quantization to 16 bit is recommended by JPEG Pleno Holography [4, 15]. Despite any efforts to reduce the dynamic range of the reconstructions, most of the remarks valid for metrics evaluated in the hologram plane remain applicable. One degree of freedom that has been neglected in VQA research thus far is the effect of any camera model on numerical reconstructions. Starting from the simple perspective or orthographic reconstructions, [11] also more complete camera models approaching the human eye model have been proposed [23, 24] but remain yet to be analyzed in the context of VQA.

17.2.2.6 Objective VQA Summary

Up until now, modified classic image metrics are the de-facto standard in VQA for holograms. Only few new proposals have been presented in literature and more research in their performance over a wide range of holographic content as well as research on alternative methods is urgently needed. Fortunately, because research on deep neural networks in many contexts of holography is picking up, more researchers focus on the construction of a proximal visual objective loss function while processing vast amounts of data. These are perfect conditions of the creation of new metrics.

Nonetheless, until convincing new metrics have reached wide acceptance the most common and the best-interpretable, albeit slow way of objective VQA will remain the evaluation of classic image metrics on numerical reconstructions. In the following section, we discuss therefore among others the effects of various camera models on numerical reconstructions.

17.3 Numerical Reconstructions from Digital Holograms

When digital holograms can or shall not be optically reconstructed, the diffraction of light can be reversed numerically. For this, the respective propagation kernels (e.g., in part I, Chap. 1) are conjugated thereby reversing the sense of the complex phase exponentials. The back-propagation of light is then facilitated by an application of the conjugated kernels to the hologram, and part III, Chap. 8. In this section, we will first discuss several ways of performing numerical reconstructions, which may be used in visual quality assessment. We will use a basic form of the **angular spectrum method** for these discussions. Thereafter, we provide some remarks on memory efficient implementation, correctness, and close by providing a more complete implementation example of the angular spectrum method.

17.3.1 Types

Provided a specific choice of display setup, multiple ways of reconstructing a digital hologram exist. Here, we will discuss five scenarios in decreasing popularity before comparing them in terms of computational complexity as well as geometric and signal characteristics. For the convenience of the notation, we will base the discussion on a complex-valued, monochrome, on-axis Fresnel hologram with planar reference wave $\exp(0\pi i) = 1$ and with an even number of rows N and columns M. All **MATLAB** code samples will make use of the angular-spectrum method described in earlier chapters. For convenience, it will be briefly re-introduced again below.

17.3.1.1 Full-Field

The most common and trivial reconstruction type is achieved by back-propagating the entire hologram from its hologram plane to some in-focus image plane located in the scene. This is also referred to as **full-field propagation**. Any suitable propagation kernel may be used. Without loss of generality, we will use the angular spectrum method Listing 17.1 for the discussion.

Listing 17.1 Simple angular spectrum method.

```
function X = asm(X,p,z,wlen)
  % function X = asm(X,p,z,wlen)
  %
  % Returns the angular spectrum propagated wavefield
```

```
  % with distance z, pixel pitch p, and wavelength wlen.
  %
  % INPUT:
  % X@numeric(N, M)... digital hologram at z'
  % p@numeric(1,2)... pixel pitch in meters
  % z@numeric(1)... propagation distance in meters
  % (z<0 for back-propagation)
  % wlen@numeric(1)... wavelength in meters
  %
  % OUTPUT:
  % X@numeric(N, M)... digital hologram at z'+z

  % Early exit
  if(z == 0), return; end

  if(isscalar(p)), p = p * [1,1]; end
  res = size(X);
  pad = max(res/2);

  % Zero-padding in the spatial domain
  X = padarray(X, [pad, pad]);
  resPad = size(X);
  X = fft2(X);
  [x, y] = meshgrid( (wlen/p(2)*(-resPad(2)/2:resPad(2)/2-1)/resPad(2)), ...
                     (wlen/p(1)*(-resPad(1)/2:resPad(1)/2-1)/resPad(1)));
  X = X .* ifftshift(exp(2i*pi*real(sqrt(1 - x.^2 - y.^2))*z/wlen));
  X = ifft2(X);

  % Undo zero-padding through center cropping
  X = centercrop(X, res);
end
```

The function "centercrop" is defined in Listing 17.2.

Listing 17.2 Centercrop.

```
function X = centercrop(X, S)
  % function Y = centercrop(X, S)
  %
  % Crops at the center of image X (i.e. first 2 dims) with size S. If the image
  % is nD array, shape will be preserved, apart from cropping the first two dimensions.

  s = size(X);
  col_beg = max(floor((s(2)-S(2))/2), 0); % Correct also for odd numbers
  row_beg = max(floor((s(1)-S(1))/2), 0);
  row_end = min(s(1), row_beg+S(1));
  col_end = min(s(2), col_beg+S(2));
  s(1:2) = [S(1), S(2)];
  X = reshape(X( row_beg+1:row_end, col_beg+1:col_end, :), s);
end
```

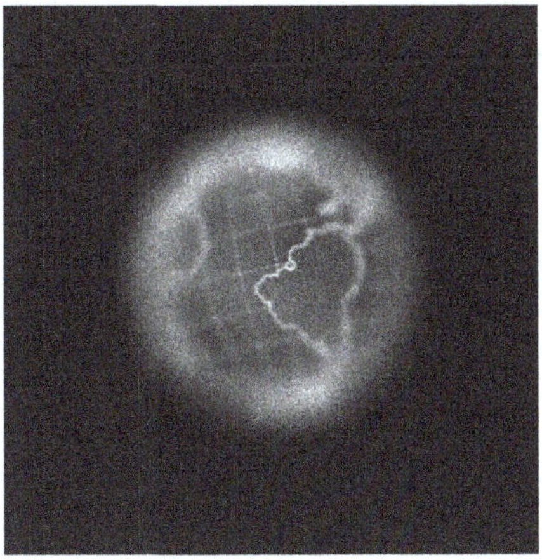

Fig. 17.5 Full-field reconstruction of the diffuse earth

A full-field reconstruction from a hologram X a distance z away from the scene can then be obtained simply as:

Listing 17.3 Full-field reconstruction.

```
X = abs(asm(X, p, −z, wlen));
```

An advantage of this propagation type is that it preserves the maximum amount of information from the hologram. Thus, any quality impairments of the hologram will also be contained in the full-field reconstruction. However, due to the non-local nature of diffraction most artifacts will appear as a global increase in speckle noise and therefore a lower global SNR.

As the full-field reconstruction does not constrict the limiting aperture of the hologram any further, the **depth of field** will be minimal. The depth range of scene parts in focus is roughly inversely proportional to the aperture size given by the physical extent of the hologram. This is put to the extreme with an ideal pinhole camera which has an infinite depth of field. The preservation of the maximal spatial aperture also results in a minimal speckle grain size, see e.g., [25].

The associated camera model with this type of propagation depends solely on the chosen propagation kernel and the reference wave shape. In the case of the angular spectrum method and a planar reference wave, it is orthographic. This means all camera rays cast toward the scene are parallel to the optical axis. An exemplary reconstruction of the diffuse earth of resolution 8192×8192, a pixel pitch of 1μm, and a wavelength of 633 nm reconstructed at 1.2 cm is shown in Fig. 17.5. The hologram is publically available as part of the Interfere-II dataset at http://erc-interfere.eu.

The full-field propagation is often used to explore a given hologram with respect to its scene depth, e.g., while producing scene focal stacks. To explore the parallax present in a hologram any of the following four propagation types may be used.

17.3.1.2 Perspective

The most common way to obtain perspective reconstructions from a digital hologram using a perspective camera model with view frustum, see for example, [11], is the so-called "**perspective reconstruction**". Conceptually, it corresponds to the application of a spatial filter in the hologram plane before reconstruction. Much like peeking through a key-hole dramatically increases the number of different observable perspectives of the inside of a room, as opposed to viewing the room through an open door. It may be implemented as in Listing 17.4.

Listing 17.4 Perspective reconstruction.

```
X = apply_aperture(X, hpos, vpos, apsize);
X = abs(asm(X, p, −z, wlen));
```

The function "apply_aperture" is defined in Listing 17.5.

Listing 17.5 Aperture application.

```
function Y = apply_aperture(X, hpos, vpos, apsize)
  % function Y = apply_aperture(X, hpos, vpos, apsize)
  %
  % Applies an aperture of size apsize at the relative positions
  % hpos, vpos from [-1, 1].
  %
  % INPUT:
  % X@numeric(N, M)... digital hologram
  % hpos@numeric(1)... normalized position within the hologram
     horizontally
  % vpos@numeric(1)... normalized position within the hologram
     vertically
  % apsize@numeric(1,2)... aperture size in pixel
  %
  % OUTPUT:
  % Y@numeric(N, M)... digital hologram with applied aperture


  [N, M]=size(X);
  vpos = −vpos;

  % Calculate aperture corners in pixel
  N_beg = max(1, round(N/2 + (N/2 − apsize(1)/2 ) ∗ vpos − apsize(1)/2) + 1);
  N_end = min(N, round(N/2 + (N/2 − apsize(1)/2 ) ∗ vpos + apsize(1)/2 ) );

  M_beg= max(1, round(M/2 + (M/2 − apsize(2)/2 ) ∗ hpos − apsize(2)/2 ) + 1);
  M_end= min(M, round(M/2 + (M/2 − apsize(2)/2 ) ∗ hpos + apsize(2)/2 ) );

  Y = zeros(N, M);
  Y(N_beg:N_end, M_beg:M_end) = X(N_beg:N_end, M_beg:M_end);
end
```

To better understand the effects of this and the following propagation types, we utilize visualizations of the phase space of our practical example the diffuse earth.

Phase space is a concept widely used in signal processing and vital to understand holograms, see [26] and references therein. For our purposes, it is sufficient to understand that the phase space of static holograms is spanned by the dimensions space and spatial frequencies. Since a hologram is two-dimensional, its full phase space has four dimensions. To develop an understanding, it is often sufficient to consider one-dimensional cross sections of the hologram, resulting in a two-dimensional phase space. For the visualizations in this section, a row through the center of the aperture was evaluated as representative.

Before we continue with the analysis of the perspective propagation type, let us shortly develop a basic intuition about **phase space**. Let us start by considering two edge cases: a pure plane wave along a line and a single (in-focus) point on a line. A plane wave will be represented by a horizontal line whose vertical offset is given by its inclination angle in proportion to the maximal diffraction angle given by the grating equation (17.1).

$$\sin(\Theta) = \frac{\lambda}{2p} , \tag{17.1}$$

where Θ is the diffraction angle, λ is the wavelength, and p is the pixel pitch or grating period, respectively. For a sub-half wavelength pixel pitch, a spatial frequency of ± 1 thus corresponds to $\pm 90°$, respectively. A single non-zero in focus point on a one-dimensional hologram cross-sectional results in a horizontal line in phase space. The pixel position is indicated by the sample index within the digital hologram.

We conclude the intuition building about phase space with the following few statements. For a Fresnel hologram, typically, a parallelogram is obtained in 2D phase space for any 1D hologram cross section. Any subsequent processing, such as Fourier transforms, translations, aperture application, modulations, and multiplications with propagation kernels results in shifts, shearing, rotations, filtering,...of the signal. For the highest information content in the reconstruction, typically as much as possible of the phase space should be filled with signal, provided that no form of white noise is introduced, e.g., through interpolation or quantization errors.

And now, let us continue with the analysis of each processing step of a perspective reconstruction up to the absolute value calculation. In Fig. 17.6, it is shown how the input hologram is first filtered by the spatial aperture and then sheared by the propagation operator. Note that the final processing step of computing the absolute value was omitted as it is highly non-linear and not educational. As can be seen, the spatial aperture applied in the hologram plane preserves all frequencies for a part of a hologram. Because spatial frequencies are linked to gaze angles by the grating equation, see [26], the following conclusion can be drawn. The described reconstruction corresponds to a perspective camera model with the limiting aperture applied in the hologram plane and camera rays diverging towards the scene as all diffraction angles are potentially still contained.

Because this type of reconstruction relies on the application of a spatially limiting aperture, the depth of field and speckle grain size are increased compared to the full-field reconstruction. This means that aside from being able to explore the parallax of

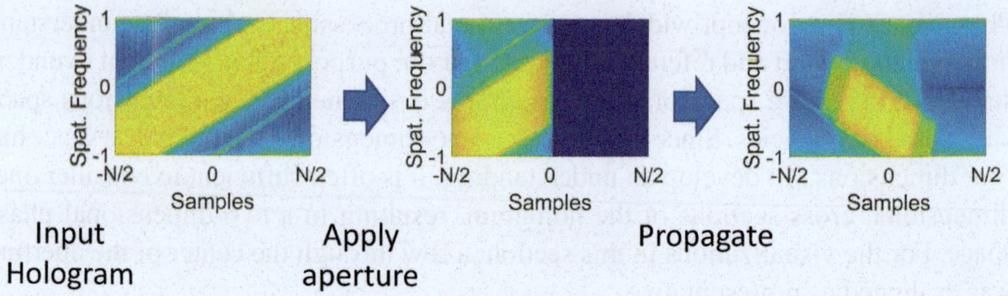

Fig. 17.6 Conceptual visualization of the phase space of the steps a hologram undergoes upon perspective reconstruction with an aperture of half the hologram resolution at high-gaze angle

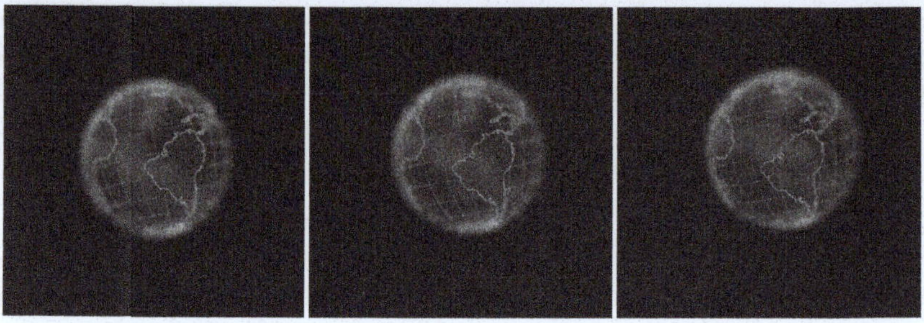

(a) hpos=vpos=0 (b) hpos=vpos=0.5 (c) hpos=vpos=1

Fig. 17.7 Perspective reconstructions of the diffuse earth with increasing gaze angle

the hologram, more of the scene can also be explored in focus at once. A series of reconstructions with increasing gaze angle is shown in Fig. 17.7.

17.3.1.3 Orthographic

The second most common view-dependent reconstruction type is the **orthographic view reconstruction**. It always uses the orthographic camera model after propagation. That means, in the case of spherical wavefronts and/or, for example, a Fresnel propagation based on a single Fourier transform, which both can introduce perspective distortions upon propagation, the (perspectively distorted) propagated scene may be only analyzed orthographically, or approximately so. The implementation is trivial. As was explained for the perspective reconstruction, viewing angles correspond to spatial frequencies. Therefore, it should come as no surprise that a selection of only a few spatial frequencies via aperture application onto the back-propagated wavefield yields orthographic views. In MATLAB, we have Listing 17.6.

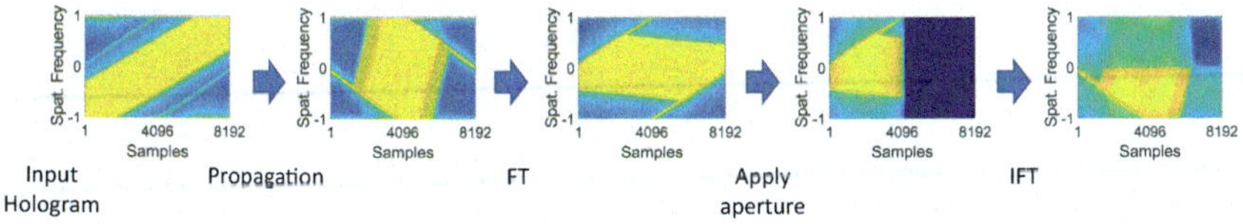

Fig. 17.8 Conceptual visualization of the phase space of the steps a hologram undergoes upon orthographic reconstruction with an aperture of half the hologram resolution at high-gaze angle

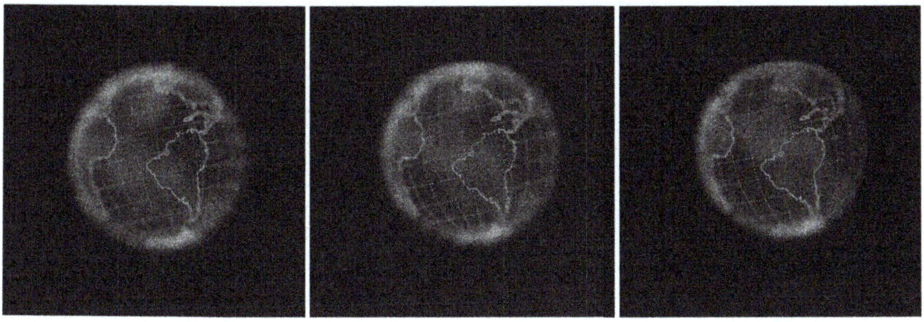

(a) hpos=vpos=0 (b) hpos=vpos=0.5 (c) hpos=vpos=1

Fig. 17.9 Orthographic reconstructions of the diffuse earth with increasing gaze angle

Listing 17.6 Orthographic reconstruction

```
X = asm(X, p, −z, wlen);
X = ifftshift( fft2( fftshift(X)));
X = apply_aperture(X, hpos, vpos, apsize);
X = ifftshift( ifft2( fftshift(X)));
X = abs(X);
```

Figure 17.8 shows the processing steps again in phase space. What can also be seen well is the effect of a Fourier transform and its inverse, which correspond to $\pm 90°$ rotations in phase space—as samples are mapped to frequencies and vice versa. Another imminent observation from the phase space footprints is that there is no further constriction of the limiting spatial aperture of the hologram. Comparing the second and the fifth step, both can be seen to cover the same range of samples. Also, comparing the fifth step of Fig. 17.8 with the third step of Fig. 17.6, which shows that although the same aperture sizes were used, the results are different.

An exemplary reconstruction is shown in Fig. 17.9. Because orthographic reconstructions do not narrow the aperture of a digital hologram, the speckle grain sizes and depth of field remain the same as in the full-field reconstruction. When compared directly to perspective reconstructions, orthographic reconstructions appear often less speckle-noise polluted because of their smaller speckle grain sizes. Another feature of orthographic reconstructions is, that any additional gazing angle comes at the price of just one- or two two-dimensional fast Fourier transforms with only one propagation. In contrast, a perspective reconstruction always requires a propagation,

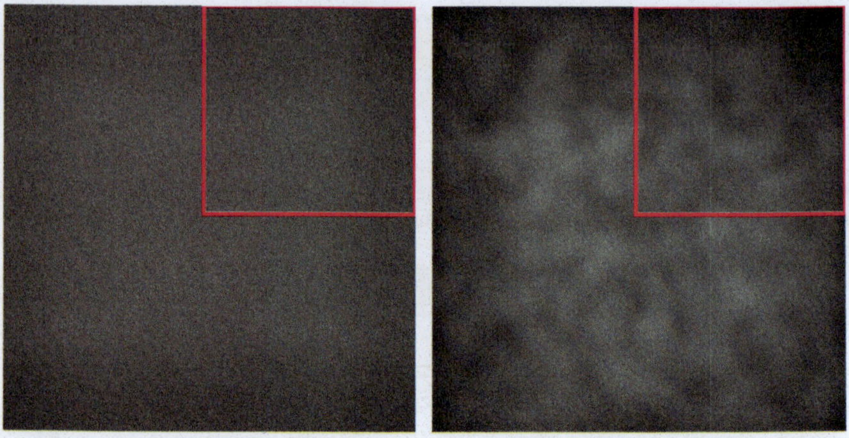

(a) Spatial domain (b) Frequency domain

Fig. 17.10 Spatial domain of the hologram plane and Fourier domain of the back-propagated wavefield. Outlined in red are the applied apertures in the perspective and orthographic reconstruction case, respectively. Figures copied from [11]

which is typically more expensive. One drawback of orthographic reconstructions, especially for VQA, is that the average intensity within a reconstruction is often view-dependent. This is because, similar to natural images, the intensity distribution of back-propagated wavefields is often non-homogeneous, see Fig. 17.10. Thus, the application of a fixed aperture in the Fourier domain of the back-propagated wavefield leads to the drift of mean intensity in the reconstructions. This, too, is unlike perspective reconstructions which are based on spatial windows in the hologram plane. This is especially pronounced for shallow scenes.

17.3.1.4 Spherical Lensing

Another, though not (yet) widely used, way of obtaining a view-dependent reconstruction from a hologram is conceptually the application of a windowed, focusing Fresnel-zone plate (that is an ideal spherical lens) in the hologram plane. This is exemplary for the application of any focusing lens in or imaged into the hologram plane. The focusing distance of the lens may be used to select a scene depth of interest. Subsequent aperture application selects a gaze angle and the propagation to the far field of the lens can be modeled by a Fourier transform—see Listing 17.7.

Listing 17.7 Spherical lensing reconstruction.

```
X = spherical_demodulate(X, p, z, wlen);
X = apply_aperture(X, hpos, vpos, apsize);
X = fftshift( fft2( ifftshift(X)));
X = abs(X);
```

The function "spherical_demodulate" is defined in Listing 17.8.

Listing 17.8 Spherical lens application.

```
function X = spherical_demodulate(X, p, z, wlen)
  % function X = spherical_demodulate(X, p, z, wlen)
  %
  % Applies an ideal spherical lens, as a focusing point-spread function
  % to a hologram. Applies lasting zero-padding to avoid aliasing.
  %
  % INPUT:
  % X@numeric(N, M)... digital hologram
  % p@numeric(1,2)... pixel pitch in meters
  % z@numeric(1)... propagation distance in meters
  % (z<0 for back-propagation)
  % wlen@numeric(1)... wavelength in meters
  %
  % OUTPUT:
  % X@numeric(NN, MM)... demodulated and padded hologram

  [N, M] = size(X);
  % Compute maximal padding frequencies
  tanpsf = max([N, M].*p/2/z);
  fratio = 2*max(p)*tanpsf/wlen/sqrt(tanpsf^2+1);

  % Sinc-interpolation of the hologram with a new pixel pitch
  % == Padding in Fourier domain
  H = ifft2(ifftshift( padarray(fftshift(fft2(X)), ceil(fratio*[N, M]/2)) ));
  p_new = p./(1+fratio);

  % Multiplication with point-spread function in the spatial domain
  res = size(X);
  [x, y] = meshgrid( p_new(2)*(-res(2)/2:res(2)/2-1), ...
                    p_new(1)*(-res(1)/2:res(1)/2-1));
  r = sqrt(z.^2 + x.^2 + y.^2);
  X = X .* exp(-2i*pi/wlen * r)./r;
end
```

The processing and ordering of steps become obvious when studying the propagation for a high-gaze angle in phase space as shown in1 Fig. 17.11. The idea of using a focusing point-spread function is similar to using a de-magnifying spherical reference wave, as used frequently in Fourier holography. It is also similar to the compact space-bandwidth representation explained in [27] as it produces a very space-bandwidth efficient representation first before applying the aperture.

Because of the included lensing function, this reconstruction type is only in parts perspectively distorted, see Fig. 17.12 versus Fig. 17.7. In addition, the lens enlarges the effective aperture at the cost of spatial resolution. Therefore, the depth of field is even shallower than in the full-field reconstruction but the lateral resolution is the lowest. See also entendue trade-offs as described, for example, in [28].

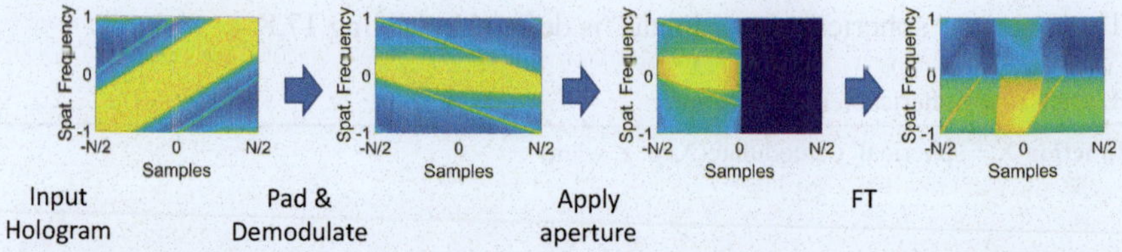

Fig. 17.11 Conceptual visualization of the phase space of the steps a hologram undergoes upon view-dependent reconstruction with spherical lensing and an aperture of half the hologram resolution at high-gaze angle

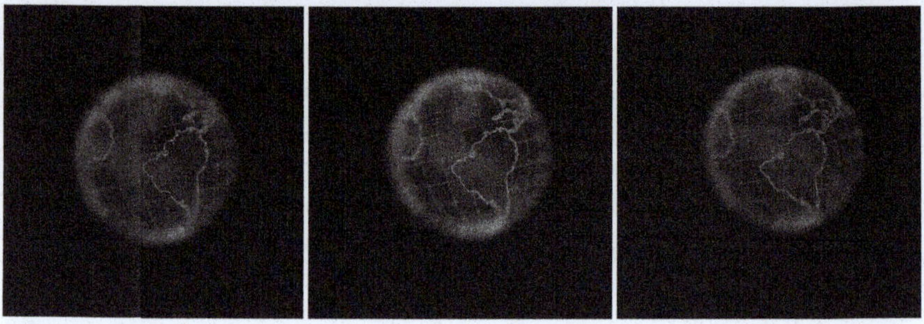

(a) hpos=vpos=0 (b) hpos=vpos=0.5 (c) hpos=vpos=1

Fig. 17.12 View-dependent reconstructions with spherical lensing of the diffuse earth with increasing gaze angle

The camera model is that of a lens superposed to the hologram and with a limiting spatial aperture. But it may as well be extended to more complete camera models approaching the human eye model [23, 24]. Though a study of those is beyond the scope of this short overview. Ultimately, the full modeling of the optical system is however the only way to build an accurate simulator for any experimental results.

Note that spherical lensing can be potentially implemented very efficiently by cropping the demodulated hologram before the final Fourier transform instead of only applying the aperture. And different gaze angles can be achieved simply by, cropping in different spatial positions after demodulation.

17.3.1.5 Hologram Rotation

A final, though more theoretical than practical, way to extract views from a hologram is obtained by tilting the hologram plane before propagation. While the exact computation of a tilted hologram is possible in theory, see Chap. 13 in [23]; in practice, several errors degrade the quality at angles larger than $30 - -60°$. Furthermore, the computational complexity is significant. This method is an example, that intuition can be a false friend. We also illustrate the use of phase space representations to analyze this problematic situation. To begin with, the "best" implementation can be written as follows:

Listing 17.9 Hologram plane rotation and reconstruction.

```
res = size(X);

% FT because tilt expects the FT domain
X = fftshift(fft2(fftshift(X)));
% Zeropad
X = padarray(X, res(1:2)/2);
X = rot_2d(X, p, wlen, rotx, roty);
% Undo zeropad & inverse FT
X = centercrop(X, res);
X = ifftshift(ifft2(ifftshift(X)));

X = abs(asm(X, p, −z, wlen));
```

The function "rot_2d" is defined in Listing 17.10.

Listing 17.10 Two-dimensional hologram plane rotation.

```
function X = rot_2d(X, p, wlen, rotx, roty)
  % function X = rot_2d(X, p, z, wlen)
  %
  % Rotates a hologram around its two lateral axes: x and y.
  %
  % INPUT:
  % X@numeric(N, M)... digital hologram
  % p@numeric(1,2)... pixel pitch in meters
  % wlen@numeric(1)... wavelength in meters
  % rotx@numeric(1)... rotation angle around x-axis in radian
  % roty@numeric(1)... rotation angle around y-axis in radian
  %
  % Note: z-axis is the optical axis.
  %
  % OUTPUT:
  % X@numeric(N, M)... rotated hologram

  [N, M] = size(X);
  L = N*p(1); K = M*p(2);

  % Early exit
  if((rotx == 0) && (roty == 0)); return; end

  % Assemble 3D rotation matrix
  RotZ = [1 0 0 ; 0 1 0 ; 0 0 1 ]; %Rz
  RotY = [cos(roty) 0 sin(roty); 0 1 0 ; −sin(roty) 0 cos(roty)]; %Ry
  RotX = [ 1 0 0 ; 0 cos(rotx) −sin(rotx); 0 sin(rotx) cos(rotx)]; %Rx
  R = RotZ*RotY*RotX;

  % Prepare rotation
  [V,U] = meshgrid([−M/2 : M/2−1] / K, [−N/2 : N/2−1] / L); % Re-sampled
      frequencies
```

```
    D = round([R(7)/wlen, R(8)/wlen]);
    W = real(sqrt(wlen^(-2) - (U+D(1)).^2 - (V+D(2)).^2)); % z-component k-vector
    J = ((roty*R(6)-R(3)*R(5))*(U+D(1)) + (R(3)*R(4)-rotx*R(6))*(V+D(2))) ./W + (rotx*R
        (5)-roty*R(4)); % Jacobian
    J(isinf(J)) = 0; % Sanitize Jacobian

    % Resample frequencies
    X = sqrt(J).*interp2(U,V,X, R(1)*(U+D(1))+R(2)*(V+D(2))+R(3)*W, R(4)*(U+D(1))+R
        (5)*(V+D(2))+R(6)*W, 'cubic',0);
end
```

For Fig. 17.13, the chosen maximal diffraction angles were given as half of the supported field of view as $0.5\sin^{-1}(\frac{\lambda}{2p})$ rad.

Intuitively, one might expect that this method has a higher quality than the perspective or orthographic reconstructions, as no windowing is applied. Instead, the direction of propagation and the wavefield are manipulated. As can be seen, the quality degrades quite fast with increasing gaze angle, see 17.14. How is that? To answer this question we turn to phase space representations—in this case the S-method see [26, 29]. Careful, study of the last sub-figure in Fig. 17.14 reveals the reason. Due to interpolation errors, in-focus scene points (e.g., the brightest lines contained

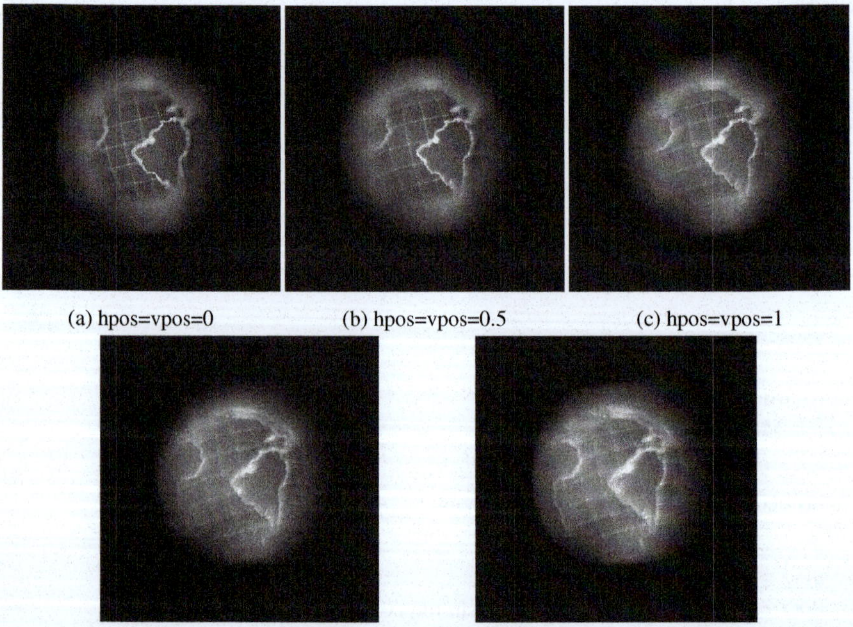

(a) hpos=vpos=0 (b) hpos=vpos=0.5 (c) hpos=vpos=1

(d) hpos=vpos=1; No padding (e) hpos=vpos=1; Spat. padding

Fig. 17.13 View-dependent reconstructions through rotation of the diffuse earth hologram with increasing gaze angle (using Fourier domain padding). Also included are high-gaze angle reconstructions without zero-padding or padding only in the spatial domain, respectively

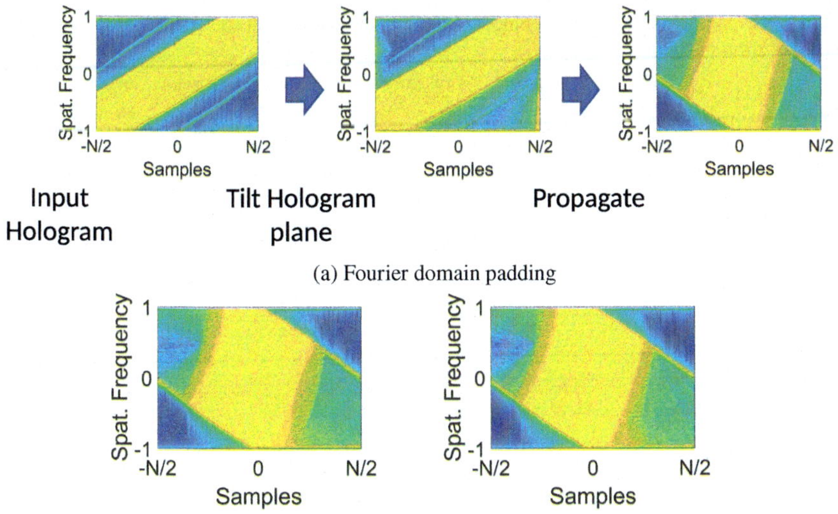

(a) Fourier domain padding

(b) No padding

(c) Spatial padding

Fig. 17.14 Conceptual visualization of the phase space of the steps a hologram undergoes upon view-dependent reconstruction with prior hologram rotation at high-gaze angle. Also included are the last steps without zero-padding or padding only in the spatial domain, respectively

within the signals footprint) are not taking the shape of a vertical straight line but remain curved. This means, that their signal is smeared over a small area, which gives the reconstruction its fuzzy appearance at large gaze angles. This can be verified, by studying padding in the spatial instead of the Fourier domain thus increasing the dependence on the correctness of the spatial frequency re-sampling and interpolation. In the bottom of Figs. 17.13 and 17.14, the resulting worse reconstruction at the highest gaze angle and its phase space footprints are shown, respectively, next to the same information for reconstructions without any padding.

17.3.2 Comparison

We have discussed several types of numerical holographic reconstructions. Their differences with respect to effective limiting aperture size, computational complexity as well as geometric aspects are listed and summarized in Table 17.2.

Table 17.2 Summary of various computational and geometric characteristics of the discussed reconstruction methods

	Full-field	Perspective	Orthographic	Hologram rotation	Spherical lensing
Effective aperture spatial size	Large	Small	Large	Large	Medium
Logic	P	A→P	P→FT→A→IFT	FT→Pad→Rot →Unpad→IFT→P	Demod→A→FT
Computational complexity	Low	Medium	Medium	High	Lowest
Reuse possible for multiple gaze angles?	n/a	No	Yes	No	Yes
Depth-of-field	Small	Large	Small	Variable	Smallest
Speckle grain size	Small	Large	Small	Small	Small
Lateral resolution	High	Medium	High	Medium-High	Low
Intensity view-dependent?	n/a	No	Yes	No	No
Geometric deformation?	No	Yes	No	No	Yes

Legend: **P** = Propagation; **A** = Aperture application; **Rot** = Hologram rotation along X, Y; **Demod** = Spherical lens demodulation. Note that spherical lensing can be implemented even more efficiently by cropping the demodulated hologram before the final Fourier transform instead of only applying the aperture

17.3.3 Efficient Implementation

To reduce the computational complexity for any of the presented reconstruction methods with an arbitrary, fixed propagation kernel, any zero-padding should be as small as possible; a pre-assembled kernel, or parts thereof, should be kept in memory to avoid re-computation; and a cheaper propagation kernel should be used otherwise.

For a more memory-efficient implementation of numerical reconstructions, however, the following measures exist:

1. First and foremost, it should be ensured, that the implementation does not have any redundant copies of the data and large matrices have a lifetime limited to their actual use instead of the entire program.
2. Use minimal amounts of zero-padding. Provided a specific signal, often the amount of zero-padding can be lowered based on its and the propagation kernels bandwidth.
3. In case of multiple color frames, avoid storing unused color channels in memory during computation. Compute one color for all frames first, to avoid repeated computation of kernels. With this strategy, also only one color channel needs to be ever accessed in memory at any time.

4. Limit the numerical precision to floating point precision or even integer precision. But be aware of the extremely sensitive modulo operations in the exponential of most propagation kernels, because of z/λ. Eventually, the recurrence algorithm part I, Sect. 2.3, should be used.
5. If the padded hologram is too big to fit in memory, the two-dimensional fast Fourier transforms may be decomposed and applied to single rows/columns or groups thereof at a time, while most of the data stays out of the random access memory on a storage disk, see [30]. For kernel multiplication, the recomputation of blocks or even per element are good strategies to lower memory consumption. Unfortunately, both techniques are impairing substantially also the computation speed.

We will show some of these concepts in application in the end of the next subsection, after discussing some aspects for ensuring the correctness and comparability of the numerical reconstructions. Those are especially important in the context of VQA.

17.3.4 Correctness

Aside from numerical errors of under-/overflow, evanescent frequencies or aliasing are obvious candidates that require corrections of the simple propagation methods. An exemplary corrected and optimized implementation of Listing 17.1 is given in Listing 17.11. It is meant for memory-efficient processing of color holograms.

Listing 17.11 Optimized angular spectrum method.

```
function X = asm_full(X,p,z,wlen,pad_)
  % function X = asm_full(X,p,z,wlen,pad_)
  %
  % Returns the angular spectrum propagated wavefield
  % with distance z, pixel pitch p, and wavelength wlen
  % and zero-padding of size ~min(min(size(X)/2), pad).
  %
  % INPUT:
  % X@numeric(N, M)... digital hologram at z'
  % p@numeric(1,2)... pixel pitch in meters
  % z@numeric(1)... propagation distance in meters
  % (z<0 for back-propagation)
  % wlen@numeric(1)... wavelength in meters
  % pad@numeric(1)... preferred one-sided padding size per dimension in px
  %
  % OUTPUT:
  % X@numeric(N, M)... digital hologram at z'+z

  % Early exit
  if(z == 0), return; end

```

```
% Initialization
if(isscalar(p)), p = p * [1,1]; end
size_X_ = size(X(:,:,1));
ncolors_ = numel(wlen);
if(ncolors_ ~= size(X,3))
    error('asm_full:Channel mismatch between hologram and wlen.');
end

% Prepare zero-padding
if(nargin < 5)
    thetaX = real( asin(max(wlen)*size_X_(1)/(2*size_X_(1)*p(1))) );
    pad_ = abs(round(abs(tan(thetaX)*z)/p(1)));
    pad_ = double(pad_-mod(pad_,16)+16);
    pad_ = min([pad_, size_X_(1:2)/2]); % , (2.^ceil(log2(res)+1)-res)/2
end

disp(['Pad by: ' num2str(pad_)])

% Loop over color channels
for color_id_ = 1:ncolors_
    X(:,:,color_id_) = subfun(X(:,:,color_id_), p, z, wlen(color_id_), pad_);
end

function Y = subfun(Y, p, z, wlen, pad_)
    % Zero-padding in the spatial domain
    Y = padarray(Y, [pad_, pad_]);

    % Initialize
    si_y = size(Y);
    rhalf = ceil(si_y(1)/2);
    chalf = ceil(si_y(2)/2);
    rquart = ceil(rhalf/2);
    cquart = ceil(chalf/2);

    % Operation: Fourier transform
    % X = fft2(X);
    for cBlk = 1:4
        Y(:, (cBlk-1)*cquart+1:min((cBlk)*cquart, si_y(2))) = fft(Y(:, (cBlk-1)*cquart+1:
                min((cBlk)*cquart, si_y(2))));
    end
    Y = transpose(Y);
    for rBlk = 1:4
        Y(:, (rBlk-1)*rquart+1:min((rBlk)*rquart, si_y(1))) = fft(Y(:, (rBlk-1)*rquart+1:
                min((rBlk)*rquart, si_y(1))));
    end
    Y = transpose(Y);

    % Operation: FFTshift
    Y = fftshift(Y);

    % Operation: Prepare quadrant of spatial frequency grid
    % [x, y] = meshgrid( (wlen/p(1)*(-size(X,2)/2:size(X,2)/2-1)
        /size(X,2)).^2, ...
```

```
    % (wlen/p(2)*(-size(X,1)/2:size(X,1)/2-1)/size(X,1)).^2);
    [xhalf, yhalf] = meshgrid( double(wlen/p(2)*[0:chalf]/(2*chalf)).^2, ...
                    double(wlen/p(1)*[0:rhalf]/(2*rhalf)).^2);

    % Apply Kernel: TopLeft [-N/2:1:1], [N/2:-1:-1]
    x = flipud(yhalf(2:rhalf+1,2:chalf+1)) + fliplr(xhalf(2:rhalf+1,2:chalf+1));
    x = get_kernel(x);
    Y(1:rhalf,1:chalf) = Y(1:rhalf,1:chalf) .* x;

    % Apply Kernel: BottomLeft [0:1:N/2-1], [N/2:-1:-1]
    x = yhalf(1:rhalf,2:chalf+1) + fliplr(xhalf(2:rhalf+1,2:chalf+1));
    x = get_kernel(x);
    Y(rhalf+1:min(2*rhalf, si_y(1)),1:chalf) = Y(rhalf+1:min(2*rhalf, si_y(1)),1:chalf) .* x;

    % Apply Kernel: TopRight [N/2:-1:-1], [0:1:N/2-1]
    x = flipud(yhalf(2:rhalf+1,2:chalf+1)) + xhalf(2:rhalf+1,1:chalf);
    x = get_kernel(x);
    Y(1:rhalf,chalf+1:min(2*chalf, si_y(2))) = Y(1:rhalf,chalf+1:min(2*chalf, si_y(2))) .* x;

    % Apply Kernel: BottomRight [0:1:N/2-1], [0:1:N/2-1]
    x = yhalf(1:rhalf,2:chalf+1) + xhalf(2:rhalf+1,1:chalf);
    x = get_kernel(x);
    Y(rhalf+1:min(2*rhalf, si_y(1)),chalf+1:min(2*chalf, si_y(2))) = Y(rhalf+1:min(2*rhalf,
        si_y(1)),chalf+1:min(2*chalf, si_y(2))) .* x;

    % Operation: Clean up
    clear x xhalf yhalf;

    % Operation: Inverse FFTshift
    Y = ifftshift(Y);

    % Operation: Inverse Fourier transform
    % X = ifft2(X);
    Y = transpose(Y);
    for rBlk = 1:4
        Y(:, (rBlk-1)*rquart+1:min((rBlk)*rquart, si_y(1))) = ifft(Y(:, (rBlk-1)*rquart+1:
            min((rBlk)*rquart, si_y(1))));
    end
    Y = transpose(Y);
    for cBlk = 1:4
        Y(:, (cBlk-1)*cquart+1:min((cBlk)*cquart, si_y(2))) = ifft(Y(:, (cBlk-1)*cquart+1:
            min((cBlk)*cquart, si_y(2))));
    end

    % Operation: Undo zero-padding through center cropping
    Y = centercrop(Y, size_X_);

    %% Auxiliary functions
    function Y = transpose(Y)
        % Performs block-wise transposition if the data is square
        % otherwise full transposition from Matlab is used.s
        if(rhalf==chalf) % if square
            Y(1:rhalf,1:chalf) = Y(1:rhalf,1:chalf).';
```

```
                x = Y(rhalf+1:min(2*rhalf, si_y(1)),1:chalf).';
                Y(rhalf+1:min(2*rhalf, si_y(1)),1:chalf) = Y(1:rhalf,chalf+1:min(2*chalf, si_y(2))).';
                Y(1:rhalf,chalf+1:min(2*rhalf, si_y(1))) = x;
                Y(rhalf+1:min(2*rhalf, si_y(1)),chalf+1:min(2*chalf, si_y(2))) = Y(rhalf+1:min(2*rhalf, si_y(1)),chalf+1:min(2*chalf, si_y(2))).';
            else
                Y = Y.';
            end
        end

        function x = get_kernel(x)
            % Unrolled propagation, avoid having multiple temporary copies in
            % memory
            x = real(sqrt(1 - x));
            x = 2i*pi/wlen*z*x;
            x = exp(x);
            x = cast(x, 'like', Y); % Convert to single eventually % Peak usage
        end
    end
end
```

Beyond the numerical correctness and anti-aliasing measures, the visual correctness of numerical reconstructions is also a sensitive topic — especially for VQA. It should always be considered when conclusions are drawn from numerical reconstructions. Potential pitfalls are the treatment of the large dynamic range, the consideration of amplitudes versus intensities, accidental artifact removal through post-processing steps, or ringing from applied apertures.

The large dynamic range of the absolute value of a reconstructed hologram is very unequally used. This is shown in Fig. 17.15a where a uniform 8 bit quantization was assumed. The first 18 bins contain more than 50% of the non-zero pixels, leading to a low contrast upon direct quantization. Therefore, clipping the dynamic range, e.g., at the 99%-quantile is advised. The effect of such a clipping is shown in Fig. 17.15c. Clipping each image individually is not a good idea either as this can lead to or amplify mean drift and reduce the comparability of results considerably. Therefore, it is recommended to calculate per hologram—eventually also per viewpoint—an absolute clipping threshold using a quantile from the respective reference reconstruction and re-use this clipping threshold for all subsequent reconstructions.

As explained in part I, recorded holograms and captured images from a holographic display setup, are in the visible light regime always intensity-based. An understandable question is: if either amplitude or intensity should be considered for evaluating numerical reconstructions. Surprisingly, the answer is that amplitude reconstructions should be used for 2D or light-field displays. The reason is the implicit gamma correction in all modern 2D display devices, which are based on the sRGB color space. The gamma correction factor that is applied to any given pixel value is about 2.2. Therefore, an amplitude distribution before gamma correction is approximately a distribution of intensities after gamma correction. For volumetric displays

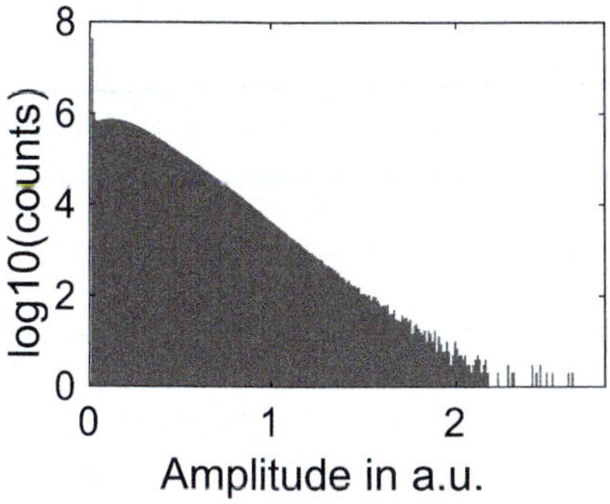

(a) Unclipped

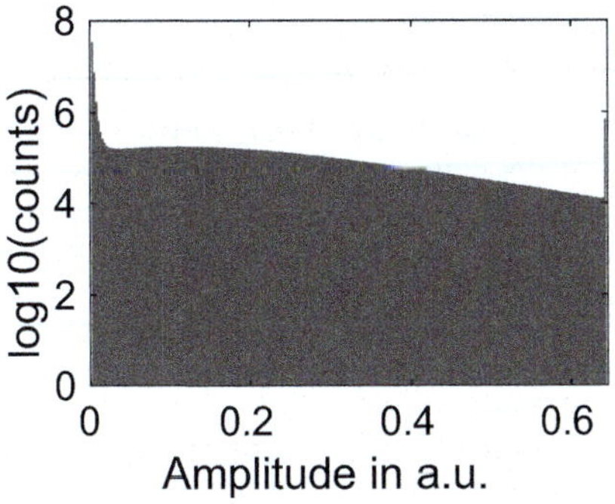

(b) Clipped at 99% quantile

Fig. 17.15 The histogram of the full-field reconstruction shown in Fig. 17.5 computed for 255 bins, shows that bin counts differ by more than 10^5 without clipping. With clipping at the 99% quantile only a difference of about 10 exists between all bins $\geq 0 + \epsilon$ with small ϵ

and other displays that may use a linear color space, the intensity should be explicitly signaled.

The accidental removal of artifacts was already mentioned in the previous section on VQA. More generally, if the characteristics of the potential artifacts from a processing step are unclear a sensitivity analysis should be conducted first to assess the ability of the VQA rendering pipeline to show those artifacts.

If apertures are applied in any domain before using a(n inverse) Fourier transform, the aperture profiles should not follow the rectangle function. Because the Fourier transform of a rectangle function leads to massive signal ringing (sinc-function) in its Fourier domain. Therefore, better-behaving window functions should be used to improve the SNR considerably at the expense of a slightly smaller footprint of the window being significantly larger than zero. The Hann window is a popular choice as its side lobes fall off very fast while providing still a decent window size. See also the excellent reference for more information [31].

In general, also the order of processing steps and parametrizations should be maintained for the comparability of results. Because currently most researchers working on digital holography have created their own rendering pipeline and comparability is rarely guaranteed. For this reason, JPEG Pleno Holography created the numerical reconstruction software for holograms v. 11 [11, 32], which is publically available as wg1n100417-098-PCQ-Numerical_Reconstruction_Software_for_Holography_v11_0.zip at: https://ds.jpeg.org/documents/jpegpleno. It provides the most recent consensus of leading experts in the field and was validated within the scope of the JPEG Pleno standardization effort which aims at the development of a first compression standard for digital holograms [5, 6].

References

1. Schelkens, P., et al.: Compression strategies for digital holograms in biomedical and multimedia applications. Light: Advanced Manufacturing (2022) https://doi.org/10.37188/lam.2022.040
2. Symeonidou, S., et al.: Three-dimensional rendering of computer-generated holograms acquired from point-clouds on light field displays. Proc. SPIE (2016) https://doi.org/10.1117/12.2237581
3. Ahar, A., et al.: Suitability analysis of holographic vs light field and 2D displays for subjective quality assessment of Fourier holograms. Opt. Express (2020) https://doi.org/10.1364/OE.405984
4. Pinheiro, A. M. G., et al.: Definition of common test conditions for the new JPEG Pleno Holography standard. SPIE Optics, Photonics and Digital Technologies for Imaging Applications VII (2022) https://doi.org/10.1117/12.2624499
5. Schelkens, P., et al.: JPEG Pleno: a standard framework for representing and signaling plenoptic modalities. SPIE Applications of Digital Image Processing XLI (2018) https://doi.org/10.1117/12.2323404
6. Muhamad, R. K., et al.: JPEG Pleno holography: scope and technology validation procedures. Applied Optics (2021) https://doi.org/10.1364/AO.404305
7. Blinder, D., et al.: Signal processing challenges for digital holographic video display systems. Signal Processing: Image Communication (2019) https://doi.org/10.1016/j.image.2018.09.014
8. Prazeres, J., et al.: Quality evaluation of the JPEG Pleno Holography Call for Proposals response. IEEE 14th International Conference on Quality of Multimedia Experience (QoMEX) (2022) https://doi.org/10.1109/QoMEX55416.2022.9900913
9. Ahar, A., et al.: Validation of dynamic subjective quality assessment methodology for holographic coding solutions. IEEE 13th International Conference on Quality of Multimedia Experience (QoMEX) (2021) https://doi.org/10.1109/QoMEX51781.2021.9465388
10. Birnbaum, T., Blinder, D., Schelkens, P.: Diffraction limited perspective, numerical reconstruction of macroscopic DH. Optica Digital Holography and 3-D Imaging (2021) https://doi.org/10.1364/dh.2021.dw6c.2
11. Birnbaum, T., et al.: JPEG Pleno Holography presents the numerical reconstruction software for holograms - An excursion in holographic views. Applied Optics. (submitted)
12. Corda R., et al.: An exploratory study towards objective quality evaluation of digital hologram coding tools. Applications of Digital Image Processing XLII (2019) https://doi.org/10.1117/12.2528402
13. Ahar, A., et al.: Comprehensive performance analysis of objective quality metrics for digital holography. Signal Processing: Image Communication (2021) https://doi.org/10.1016/j.image.2021.116361
14. Amirpour, H., et al.: Quality Evaluation of Holographic Images Coded With Standard Codecs. IEEE Transactions on Multimedia (2022) https://doi.org/10.1109/TMM.2021.3096059
15. Muhamad, R. K., et al.: wg1n100341 - JPEG Pleno Holography Common Test Conditions 9.0. JPEG Pleno Holography (2022)
16. Ahar, A., et al.: A new similarity measure for complex amplitude holographic data. Optica Digital Holography and 3-D Imaging (2017) https://doi.org/10.1117/12.2274761
17. Ahar, A., Barri, A., Schelkens, P.: From Sparse Coding Significance to Perceptual Quality: A New Approach for Image Quality Assessment. IEEE Transactions on Image Processing (2017) https://doi.org/10.1109/TIP.2017.2771412
18. Ahar, A., et al.: Performance Evaluation of Sparseness Significance Ranking Measure (SSRM) on Holographic Content. Optica Digital Holography and 3-D Imaging (2018) https://doi.org/10.1364/3D.2018.JTu4A.10
19. Lee, J., et al.: Deep neural network for multi-depth hologram generation and its training strategy. Opt. Express (2020) https://doi.org/10.1364/OE.402317
20. Bjontegaard, G.: Calculation of average PSNR differences between RD-curves. ITU-T, VCEG–M33 (2001)

21. Eybposh, M. H., et al.: Perceptual Quality Assessment in Holographic Displays With a Semi-Supervised Neural Network. Optica Digital Holography and 3-D Imaging (2022) https://doi.org/10.1364/DH.2022.Th1A.6
22. Shi, L., et al.: Towards real-time photorealistic 3D holography with deep neural networks. Nature (2021) https://doi.org/10.1038/s41586-020-03152-0
23. Matsushima, K.: Introduction to Computer Holography. Springer (2020)
24. Higashida, R., et al.: Simulation of Reconstructed Image Quality for Designing Ideal Holographic Near-eye Display. Optica Digital Holography and 3-D Imaging (2022) https://doi.org/10.1364/DH.2022.W5A.26
25. Bianco, V., et al.: Strategies for reducing speckle noise in digital holography. Light: Science & Applications (2018) https://doi.org/10.1038/s41377-018-0050-9
26. Birnbaum, T., Kozacki, T., Schelkens, P.: Providing a Visual Understanding of Holography Through Phase Space Representations. Applied Sciences (2020) https://doi.org/10.3390/app10144766
27. Kozacki, T., Falaggis, K.: Angular spectrum method with compact space-bandwidth: generalization and full-field accuracy. Applied Optics (2016) https://doi.org/10.1364/AO.55.005014
28. Haist, T., Osten, W.: Holography using pixelated spatial light modulators – Part 1: theory and basic considerations. Journal of Micro/Nanolithography, MEMS, and MOEMS (2015) https://doi.org/10.1117/1.JMM.14.4.041310
29. Stankovic, L.: A method for time-frequency analysis. IEEE Transactions on Signal Processing (1994) https://doi.org/10.1109/78.258146
30. Blinder, D., Shimobaba, T.: Efficient algorithms for the accurate propagation of extreme-resolution holograms (2019) https://doi.org/10.1364/OE.27.029905
31. Stankovic, L.: Digital Signal Processing. CreateSpace Independent Publishing Platform (2015)
32. Birnbaum, T., et al.: A standard way for computing numerical reconstructions of digital holograms. Proc. SPIE (2022) https://doi.org/10.1117/12.2625521

Chapter 18
Digital Holography Techniques and Systems for Acceleration of Measurement

Tatsuki Tahara

Abstract Digital holography (DH) [1, 2] is a 3D image sensing technique for conducting single-shot holographic 3D measurement with an image sensor. An image sensor records a digital hologram that contains the 3D information of a measured object, and a computer reconstructs a holographic 3D image through an image reconstruction algorithm. High-speed recording and image reconstruction are highly required to accelerate the measurement in DH, and single-shot holographic measurement is essential for high-speed recording. The developments of an algorithm and a hardware architecture are the keys to achieve high-speed image reconstruction. This chapter presents techniques for high-speed recording and image reconstruction in DH and spatially incoherent digital holography [3–9].

18.1 Acceleration of Measurement in Single-Shot Full-Color Digital Holography with Spatial Frequency-Division Multiplexing [16, 17]

Classically, holography and DH adopt an off-axis configuration [10] to conduct single-shot 3D imaging. The single-shot recording of a full-color analog/digital hologram has been conducted with off-axis analog/digital holography to conduct single-shot full-color 3D imaging even in the case where a monochrome image sensor is used to record a **full-color hologram** [11–14]. On the other hand, the acceleration of image reconstruction procedures is important for real-time full-color 3D measurement. An algorithm and the use of a graphical processing unit (GPU) system for the acceleration of holographic image reconstruction are introduced.

T. Tahara (✉)
Radio Research Institute, National Institute of Information and Communications Technology (NICT), 4-2-1 Nukuikitamachi, Koganei, Tokyo 184-8795, Japan
e-mail: tahara@nict.go.jp

T. Shimobaba and T. Ito (eds.), *Hardware Acceleration of Computational Holography*,
https://doi.org/10.1007/978-981-99-1938-3_18

18.1.1 Principle of Advanced Image-Reconstruction Algorithm

Figure 18.1 schematically illustrates the recording geometries of the conventional [11–15] and advanced [16] algorithms in **single-shot full-color off-axis DH** with **spatial frequency-division multiplexing**. Figure 18.1a indicates that holograms at different wavelengths are multiplexed in the space domain and a multiplexed hologram is recorded. The spatial frequency modulation of a hologram at a wavelength is conducted by tilting the optical axis of a reference wave against that of an object wave [11–13]. Another approach is to use the difference in wavelength to generate a different type of spatial frequency modulation [14]. Object-wave spectra at respective wavelengths are separated in the spatial frequency domain by setting different optical axes for different reference waves. Therefore, as shown in Fig. 18.1b, object-wave information at multiple wavelengths is selectively extracted using the Fourier transform (FT) method [15] (see Chap. 3). FT and inverse FT (IFT) are usually required, as shown in Fig. 18.1b, to reconstruct object waves from the multiplexed hologram. The computational cost in time for the conventional image reconstruction algorithm is $O(N \log N)$, where N is the number of pixels. This is because 2D FT and IFT are required.

In contrast, the advanced algorithm [16] shown in Fig. 18.1c does not require FT or IFT when extracting the desired complex amplitude information separately from the multiplexed hologram. Let $H(x, y)$ be a recorded image, $I_m(x, y)$ be an intensity distribution containing a complex amplitude distribution $U_{om}(x, y)$ and a reference amplitude distribution $U_{rm}(x, y)$, where m as an integer from 1 to M (M as the number of wavelengths measured), A be the amplitude, ϕ be the phase, i be the imaginary unit, and $*$ be the complex conjugate, a spatially multiplexed image $H(x, y)$ is expressed as

$$H(x, y) = \sum_{m=1}^{M} I_m(x, y) \tag{18.1}$$

$$\begin{aligned} I_m(x, y) = |U_{om}(x, y)|^2 + |U_{rm}(x, y)|^2 + U_{om}(x, y)U_{rm}(x, y)^* \\ +U_{om}(x, y)^* U_{rm}(x, y), \ (m = 1, \ldots, M) \end{aligned} \tag{18.2}$$

$$U(x, y) = A(x, y) \exp[i\phi(x, y)]. \tag{18.3}$$

In DH, the spatial frequency of each object wave is modulated by introducing different spatial carrier frequencies. The spatial carrier frequency depends on the angle between the object and the reference wave [11–13], and the wavelength is used for recording λ [14]. When Eq. 18.2 is multiplied by $\exp[i\phi_{rm}(x, y)]$, to remove the spatial carrier frequency from the third term on the right-hand side of Eq. 18.2, only $U_{om}(x, y)$ is localized in the low-spatial-frequency region. In the same manner, when $\exp[i\phi_{r1}(x, y)]$, $\exp[i\phi_{r2}(x, y)]$, and $\exp[i\phi_{r3}(x, y)]$ are multiplied by Eq. 18.1 to remove each spatial carrier generated by each reference beam, $U_{o1}(x, y)$, $U_{o2}(x, y)$, and $U_{o3}(x, y)$ are respectively moved to the low-spatial-frequency region. Then, a smoothing process, such as mean filtering or smoothing based on the sinc function, is applied to each multiplexed image with the removed spatial carrier. As a result, only the desired information, $U_{o1}(x, y)$, $U_{o2}(x, y)$, and $U_{o3}(x, y)$, on the image sensor plane is selectively extracted in the space domain by smoothing. The order of the advanced algorithm is estimated as $O(N)$ for the multiplexed hologram.

A single smoothing procedure is performed to extract the desired information from the multiplexed image after the removal of the spatial carrier. Figure 18.1d illustrates its principle. We use zero points in the spatial frequency domain, which are generated by smoothing in the space domain. When we use $p \times q$ mean filtering, zero points appear at the spatial frequencies $f_x = \pm K/(pd)$, $f_y = \pm K/(qd)$, where d denotes the pixel pitch of an image sensor, K is an arbitrary nonzero integer, and $-1/(2d) \leq K/(pd), K/(qd) \leq 1/(2d)$. Therefore, by setting the distance between the complex amplitude images, the conjugate ones, and the summation of zero-order diffraction ones as $\pm 1/(3d)$ on the x- and y-axes in the spatial frequency domain, a single 3×3 mean filtering procedure enables the extraction of the desired image from the multiplexed image. Note that there is a trade-off between the filter size and the available spatial bandwidth for a complex amplitude image. The advanced algorithm can also freely arrange the object-wave spectra in the spatial frequency domain by mean filtering iteratively [17]. Furthermore, measurement accuracy is improved by filtering several times at an increased calculation cost.

18.1.2 Numerical Simulation

The validity of the advanced algorithm has been investigated by numerical simulation [16, 17]. In the numerical simulation, image DH was assumed, and the wavelengths of the light sources were $\lambda_1 = 671$ nm, $\lambda_2 = 532$ nm, and $\lambda_3 = 473$ nm. The pixel pitch d and the number of pixels of the monochrome image sensor were assumed to be 2.2 μm and 2048×2048, respectively. The intensity distributions at the three wavelengths and the phase information of the object assumed are shown in Fig. 18.2a–e. Reference beams at the three wavelengths illuminate the sensor from different directions simultaneously, and a monochrome image sensor records a spatially multiplexed hologram shown in Fig. 18.2f. The multiplexed hologram has the spatial frequency distribution in Fig. 18.2g. The spatial carrier frequencies were set as $f_x = 1/(3d)$ and $f_y = 0$ at λ_1, $f_x = 1/(6d)$ and $f_y = 1/(3d)$ at λ_2, and $f_x = 1/(6d)$ and $f_y = -1/(3d)$ at λ_3. 3×3 mean filtering was carried out to selectively extract an object wave at the desired wavelength from the image hologram. Figure 18.3 shows the results of the numerical simulation. Figure 18.3c–g indicate the following. Faithfully reproduced object images are successfully obtained at the different wavelengths by the advanced algorithm. Figure 18.3f shows that a clear color object image was reconstructed owing to the strong suppression of undesired waves by smoothing. The root-mean-square errors (RMSEs), signal-to-noise ratios (SNRs), and cross-correlation coefficients (CCs) were calculated to clarify the image

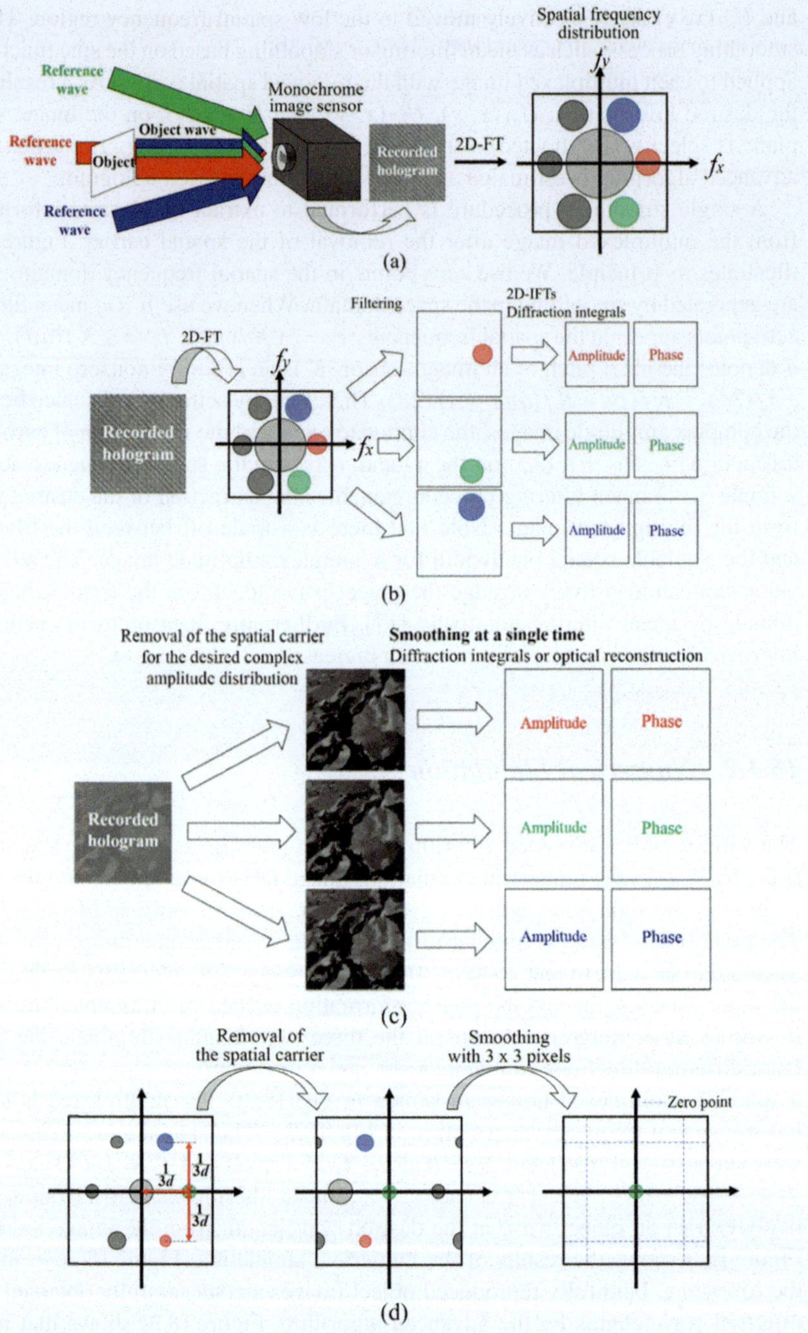
Reference wave
Reference wave
Reference wave
Object wave
Object
Monochrome image sensor
Recorded hologram
2D-FT
Spatial frequency distribution
fy
fx
(a)
2D-FT
Recorded hologram
fy
fx
Filtering
2D-IFTs
Diffraction integrals
Amplitude
Phase
Amplitude
Phase
Amplitude
Phase
(b)
Removal of the spatial carrier for the desired complex amplitude distribution
Smoothing at a single time
Diffraction integrals or optical reconstruction
Recorded hologram
Amplitude
Phase
Amplitude
Phase
Amplitude
Phase
(c)
Removal of the spatial carrier
Smoothing with 3 x 3 pixels
1/3d
1/3d
1/3d
Zero point
(d)

◀**Fig. 18.1** Schematic of off-axis digital holography with spatial frequency-division multiplexing. **a** Geometry for recording a multiplexed hologram. **b** Conventional [15] and **c** advanced [16, 17] image reconstruction algorithms. **d** Spatial spectra arranged for the advanced algorithm and the extraction of the desired object-wave spectra by a single smoothing procedure. In the figure, $p = q = 3$ is used as an example. Reprinted with permission from [16] © The Optical Society

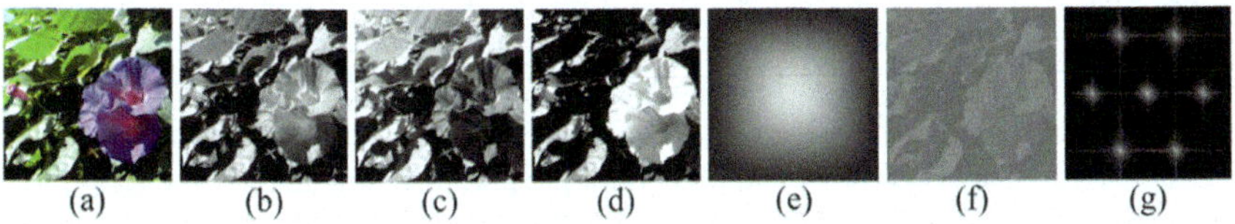

Fig. 18.2 Complex amplitude distribution of the object used for the numerical simulation and its numerically generated hologram. **a** Color-synthesized image and its intensity distributions at **b** λ_1, **c** λ_2, and **d** λ_3. **e** Phase distribution of the object. **f** Multiplexed hologram of the object and **g** its spatial frequency distribution. Adapted from Reprinted with permission from [16] © The Optical Society

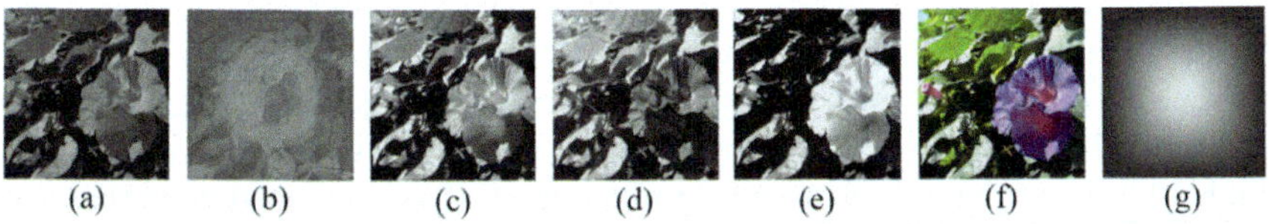

Fig. 18.3 Numerical results. **a** Color-synthesized intensity and **b** phase images reconstructed with the calculations of spatial carrier removal, diffraction integrals, and no smoothing. Intensity images at **c** λ_1, **d** λ_2, and **e** λ_3 retrieved by the advanced algorithm. **f** Color reconstructed image obtained from (**c**)–(**e**) and (**g**) reconstructed phase image. Reprinted with permission from [16] © The Optical Society

Table 18.1 RMSEs, SNRs, and CCs of the reconstructed images. Reprinted with permission from [16] © The Optical Society

	Red	Green	Blue	Phase
RMSE	2.85	3.31	2.34	0.118 rad
SNR	32.6	31.8	34.6	27.2
CC	0.999	0.999	1.00	0.996

quality quantitatively. Table 18.1 shows the small RMSEs and high SNRs and CCs of the amplitude and phase images reconstructed by the advanced algorithm. The relationship between the quality of the reconstructed image and the iterative use of mean filtering was investigated in [17]. Numerical and experimental comparisons between the conventional and advanced algorithms were reported in [16, 17].

18.1.3 Acceleration of Image Reconstruction with Advanced Algorithm and GPU System

The degree of acceleration has been investigated quantitatively [16, 17]. Although the image quality is slightly lower than that for the FT method, as reported in [17], high-speed image reconstruction is the main feature of the advanced algorithm. Further acceleration can be achieved by using a GPU in DH [18]. Three different procedures were compared: the advanced algorithms with 3×3 mean filtering and 5×5 mean filtering, and a conventional algorithm [15]. To investigate the time required to reconstruct images, $512^2, 1024^2, 2048^2$, and 4096^2 were set as the numbers of pixels. A three-wavelength-multiplexed image hologram was assumed, and no diffraction integral was calculated. These object waves were repeatedly reconstructed at least 100 times for each number of pixels, and the average times were regarded as the calculation times.

Table 18.2 shows the implementation environment used for the calculation time measurement. A fast FT library (Fastest FT in the West; FFTW [19]) was used for the calculation of 2D FTs in the conventional algorithm. Tables 18.3 and 18.4 indicate the calculation times to reconstruct three-wavelength object waves from the multiplexed hologram with a central processing unit (CPU) and a GPU, respectively. Table 18.3 shows that the advanced algorithm markedly accelerated multiple image reconstructions as the number of pixels, N, increased when using a commercially available computer with a CPU. A throughput of 10 times that of the FT method can be achieved when using a CPU and an image sensor with 4 megapixels. This is because the order of the advanced algorithm is $O(N)$ for a 2D image, whereas a 2D FFT algorithm has an order of $O(N \log N)$. Table 18.3 shows the results supporting these theoretical estimations. Furthermore, Table 18.4 clarifies that threefold acceleration compared with the FT method is possible when a GPU is used. For the GPU, the acceleration for multiple image reconstructions did not change with the number of pixels, N, because of its architecture. However, the threefold acceleration and the data in Table 18.4 imply that the advanced algorithm can achieve real-time, multicolor, and holographic motion-picture image reconstruction with 16 megapixels, which was not possible with the FT method.

An image reconstruction acceleration for single-shot full-color DH with spatial frequency-division multiplexing was introduced. Acceleration can be achieved with a high-performance processor for general DH [18]. The degree of acceleration has

Table 18.2 Implementation environment. Reprinted with permission from [16] © The Optical Society

OS	Windows 7 Professional 64 bit
CPU	Intel Core i5-4690
Memory	8 GB
Compiler	Visual Studio 2015
GPU	GeForce GTX960

Table 18.3 Calculation time to reconstruct three waves with a CPU. Reprinted with permission from [16] © The Optical Society

	Proposed algorithm [ms]		
Number of pixels	3 × 3 mean filter	5 × 5 mean filter	FT method [ms]
512^2	5.90	18.0	23.8
1024^2	23.7	73.0	151
2048^2	94.5	294	958
4096^2	381	1,257	5,481

Table 18.4 Calculation time to reconstruct three waves with a GPU. Reprinted with permission from [16] © The Optical Society

	Proposed algorithm [ms]		
Number of pixels	3 × 3 mean filter	5 × 5 mean filter	FT method [ms]
512^2	0.319	0.368	0.852
1024^2	1.10	1.38	3.19
2048^2	4.19	4.78	12.4
4096^2	17.3	19.4	49.3

been investigated and reviewed [20, 21]. The results presented in this section were derived with a GPU system produced more than three years ago [16]. Therefore, higher performance is expected with the use of state-of-the-art processors.

18.2 Incoherent Digital Holography Techniques for Acceleration of Measurement by Simultaneous Recording of Multiple Holograms

The acceleration of image reconstruction in DH has been realized by introducing a high-performance processing system such as a GPU system, as presented in the previous section. The acceleration of the recording of a digital hologram is another important factor that increases the measurement speed. Off-axis [15] and **single-shot phase-shifting (SSPS)** [22–24] configurations for single-shot 3D measurement with DH have been researched. In **incoherent digital holography (IDH)** [25–29], off-axis [30, 31] and SSPS [32–36] configurations have also been adopted. Most single-shot IDH systems have been combined with in-line SSPS configurations because coherence length is critically small in many IDH systems. Therefore, an in-line configuration is essential for high-speed 3D measurement. In this section, single-shot IDH systems that adopt SSPS are presented. In addition, a methodology [37–39] for accelerating the measurement of spectroscopic information of a 3D object using IDH

[40–46] is introduced. Furthermore, the quantitative phase information of transparent specimens is obtained with **self-reference DH (SRDH)**. The combination of SSPS and SRDH is described.

18.2.1 Single-Shot Phase-Shifting Incoherent Digital Holography [32–36]

Figure 18.4 illustrates a schematic of **single-shot phase-shifting incoherent digital holography (SSPS-IDH)** [32–36], which is a combination of SSPS and IDH. Figure 18.4 indicates that an SSPS-IDH system consists of an SSPS interferometry system and an IDH system. A digital hologram of a 3D object illuminated with spatially incoherent light is generated on the basis of IDH. Self-interference is frequently utilized to generate an incoherent digital hologram. Multiple phase-shifted incoherent holograms are simultaneously recorded with a single-shot exposure of an image sensor by the space-division multiplexing of incoherent holograms. Various types of configurations, which are briefly summarized in [25–29], have been proposed. Today, one can construct an SSPS-IDH system with commercially available optical elements

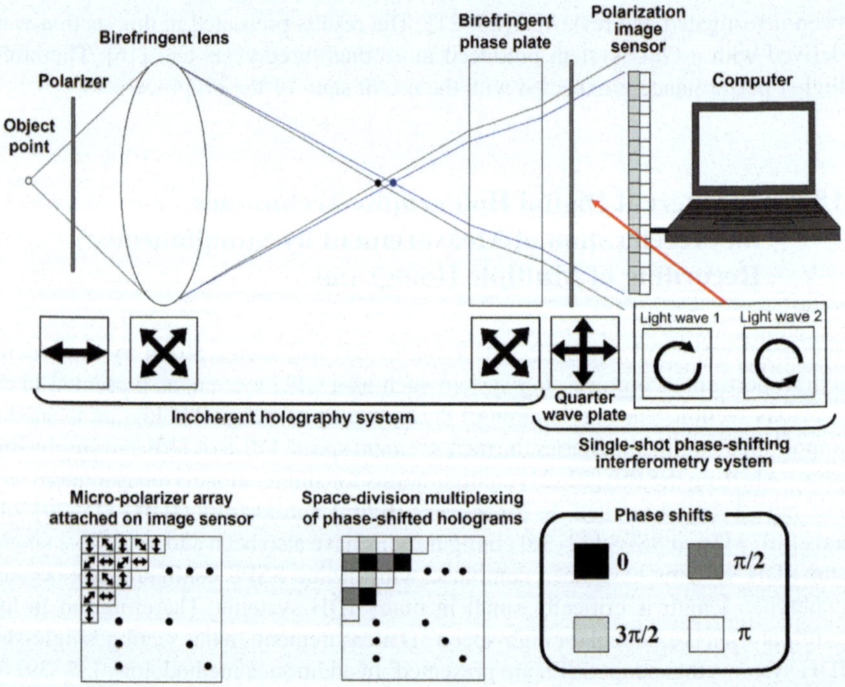

Fig. 18.4 Schematic of SSPS-IDH

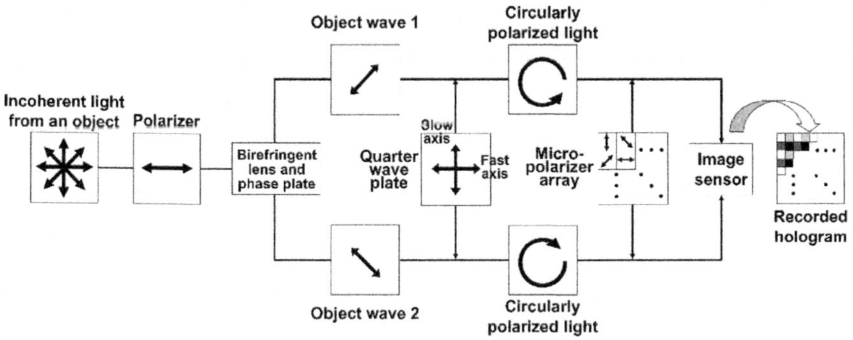

Fig. 18.5 Polarization transition of the SSPS-IDH system in Fig. 18.4

and image sensors [32–36]. Figure 18.5 shows an implementation of SSPS-IDH. A polarizer is initially set to align the polarization direction of an incoherent object wave. Then, a birefringent dual-focus lens, such as a crystal lens, and a birefringent plate are set to generate two incoherent object waves with different wavefront curvature radii and polarization directions [46]. A quarter-wave plate is then set to induce the circular polarization of the two object waves, whose rotation directions are mutually orthogonal. Each polarizer in a polarization image sensor aligns the polarization directions of the two object waves. The phase difference between the two object waves differs along the transmission axis of each polarizer. Using the implementation setup in Fig. 18.5, four phase-shifted holograms with phase shifts of 0°, 90°, 180°, and 270° are simultaneously recorded by the polarization image sensor in the case where the transmission axes of the four polarizers are 0°, 45°, 90°, and 135° relative to the horizontal direction. An incoherent 3D image of the measured object is reconstructed from the recorded image that contains four phase-shifted holograms by applying an image reconstruction algorithm of SSPS. The main point resulting in the acceleration of the measurement with SSPS-IDH is the single-shot recording of multiple phase-shifted incoherent holograms. The multiple phase-shifted incoherent holograms are sequentially recorded with sequential changes in the phase of one of the two object waves in ordinary phase-shifting IDH [7, 26, 47]. The synchronization of the exposure of an image sensor and the movement of a phase shifter is complicated, and the movement of a phase shifter is generally time-consuming. In contrast, the recording speed of SSPS-IDH is equal to the frame rate of the image sensor owing to the single-shot acquisition of multiple phase-shifted holograms. As a result, the recording speed of SSPS-IDH is more than four times higher than that of phase-shifting IDH. For example, the 3D motion-picture recording of incoherent holograms at a rate of more than 100 fps has been experimentally performed in holographic fluorescence microscopy with SSPS-IDH [25]. Research on the downsizing of SSPS-IDH systems has also been conducted, and portable [25] and finger-sized [48] optical systems have been proposed.

18.2.2 Multidimensional IDH

SSPS-IDH as a single-shot incoherent 3D imaging technique is presented in the previous section. Single-shot incoherent full-color 3D imaging is also achieved by introducing a color polarization image sensor [36]. This is because full-color and 3D information is obtained with a color-filter array and SSPS with a micropolarizer array. However, both spectroscopic information and 3D information are recorded on a monochrome image sensor and retrieved from the recorded multiplexed image(s) through a phase-encoding scheme, termed the **computational coherent superposition (CCS)** scheme [37–39], and **multidimension-multiplexed full-phase-encoding holography (MPH)** [46]. The CCS scheme is also combined with SSPS-IDH to conduct single-shot measurement [43]. These spectroscopic IDH techniques are introduced.

18.2.2.1 Multiwavelength-Multiplexed Incoherent Digital Holography with Computational Coherent Superposition Scheme

Figure 18.6 shows a schematic of CCS [37–39]. Holograms at two wavelengths are multiplexed and simultaneously recorded when a monochrome image sensor is used. An image sensor with a color-filter array is frequently used to capture a multiwavelength image. A color filter introduces wavelength-dependent intensity modulation, and wavelength information is separated by the modulation. In holography, phase-shift information affects the intensity distribution of an interference fringe image.

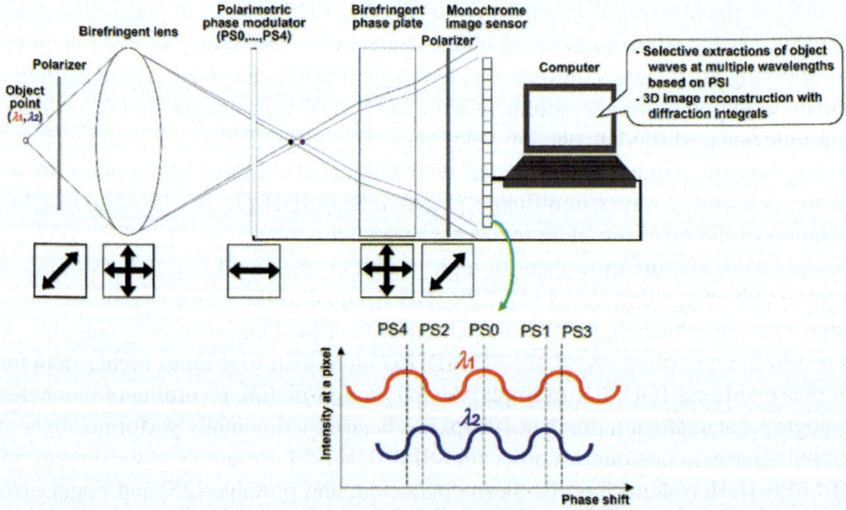

Fig. 18.6 Schematic of CCS

It is straightforward to introduce different phase shifts to holograms at different wavelengths as illustrated in Fig. 18.6. Optical implementations of two-wavelength phase-shifting DH and IDH are applicable as examples. Wavelength-dependent intensity modulation is derived from wavelength-dependent phase shifts generated by a mirror with a **piezo actuator** [37–39] or a **liquid crystal phase modulator** (LC-PM) [46, 49]. Wavelength-multiplexed phase-shifted holograms are obtained with wavelength-dependent phase shifts. Such modulation results in the encoding of wavelength information through phase shifts. Object waves at multiple wavelengths are selectively extracted from the recorded wavelength-multiplexed phase-shifted holograms using the phase-shift information and the principle of phase-shifting interferometry (PSI) [37–39, 46, 50, 51]. As a result, spectroscopic 3D information is retrieved only using phase-shift information. CCS was initially demonstrated for laser DH [37–39, 49] and then applied to IDH [41–45]. Then, the concept of CCS was extended to both obtain multidimensional information and identify the variety of light from spatially and temporally incoherent holograms. The extended concept is termed MPH [46].

Figure 18.7 illustrates an example of experiments with CCS-IDH adopting **Fresnel incoherent correlation holography (FINCH)**, termed **multiwavelength-multiplexed phase-shifting incoherent color digital holography (MP-ICDH)**. FINCH is a well-known IDH technique applicable for 3D imaging with an LED [7, 26]. Exploiting the principle of FINCH, a liquid crystal on a silicon spatial light modulator (LCoS-SLM) works as both a polarization-sensitive dual-focus lens

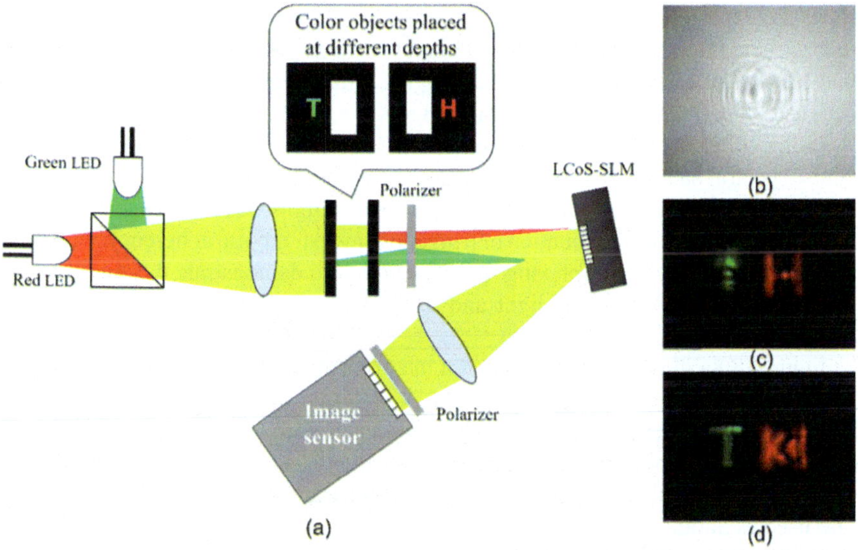

Fig. 18.7 Experimental results. **a** Schematic of the constructed setup of MP-ICDH for transparent color objects. **b** One of the wavelength-multiplexed phase-shifted incoherent holograms. **c** and **d** Images reconstructed by MP-ICDH. The depth difference between **c** and **d** was 36 mm. Reprinted with permission from [41] © The Optical Society

and a wavelength-dependent phase shifter. A simple, compact, and single-path self-interference interferometer was constructed using FINCH, as shown in Fig. 18.7a. FINCH with polarization multiplexing [26] was adopted to MP-ICDH in [41]. An LCoS-SLM (X10468-01, fabricated by Hamamatsu Photonics K.K.) to generate wavelength-dependent phase shifts is set and a wide phase-modulation range is obtained. The phase-modulation range of this model of the SLM was extended twofold, then both a diffractive lens and phase shifts were simultaneously generated. The nominal wavelengths of the LEDs used as light sources were $\lambda_1 = 625$ and $\lambda_2 = 530$ nm, and the **full widths at half maximum (FWHMs)** of these LEDs were 18 and 33 nm, respectively. The LCoS-SLM set phase shifts at the wavelengths of (λ_1, λ_2) as $(-100\pi/123, -100\pi/99)$, $(-40\pi/123, -40\pi/99)$, $(0, 0)$, $(-40\pi/123, -40\pi, 99)$, and $(-100\pi/123, -100\pi/99)$. A monochrome sCMOS image sensor of 6.5 μm pixel size and 12 bits recorded five wavelength-multiplexed in-line phase-shifted incoherent holograms with 2048 × 2048 pixels. A lens was set in front of the image sensor to collect wavelength-multiplexed light.

Color objects were prepared using transparent films and a color inkjet printer, and set in the optical path of the single-path interferometer. A green 'T', a red 'H', and a black background were drawn on the films using a color inkjet printer. Each character was 7 pt. Rectangular transparent areas were set to these objects as shown in Fig. 18.7a. Two colored objects were 36 mm apart from each other in the depth direction. An image reconstruction algorithm [39] was applied to the recorded holograms and two-wavelength object waves were retrieved. Two-wavelength focused images of objects were reconstructed by calculating diffraction integrals. Figure 18.7b–d show the experimental results. Color object images were clearly and successfully reconstructed by MP-ICDH, as shown in Fig. 18.7c and d. Thus, it has been clarified that MP-ICDH can perform color 3D imaging without an imaging lens or crosstalk between object waves at multiple wavelengths.

Using MPH, which is an extension of the CCS scheme, varieties of light waves whose spectral bandwidths are different are distinguished. MPH exploits the temporal coherence difference between different varieties of light waves and separates these waves with PSI and the attenuation of fringes owing to temporal coherency, which can be termed coherence multiplexing. An experiment to demonstrate the simultaneous 3D sensing of self-luminous light and diffraction light of illumination light was conducted.

Figure 18.8a illustrates a schematic of the experimental setup. In this experiment, the case where the wavelength bandwidths of the two varieties of light overlap was examined. A self-luminous object was illuminated with an ultraviolet LED whose central wavelength was 365 nm to generate blue fluorescence light, and another object was illuminated with a blue LED to obtain the blue transmission light of a 1 mm aperture used as another object. A block of tin halide perovskite nanocrystal containing metal complex molecules was set as the blue luminescent material.

The spectral intensity distribution of its fluorescence light was obtained with a spectrometer in advance and is shown in Fig. 18.8b. The peak wavelength of the material was 451 nm. The FWHM of the luminescence was 65 nm and the luminescence wavelength ranged from 401 to 521 nm; at these two values, the inten-

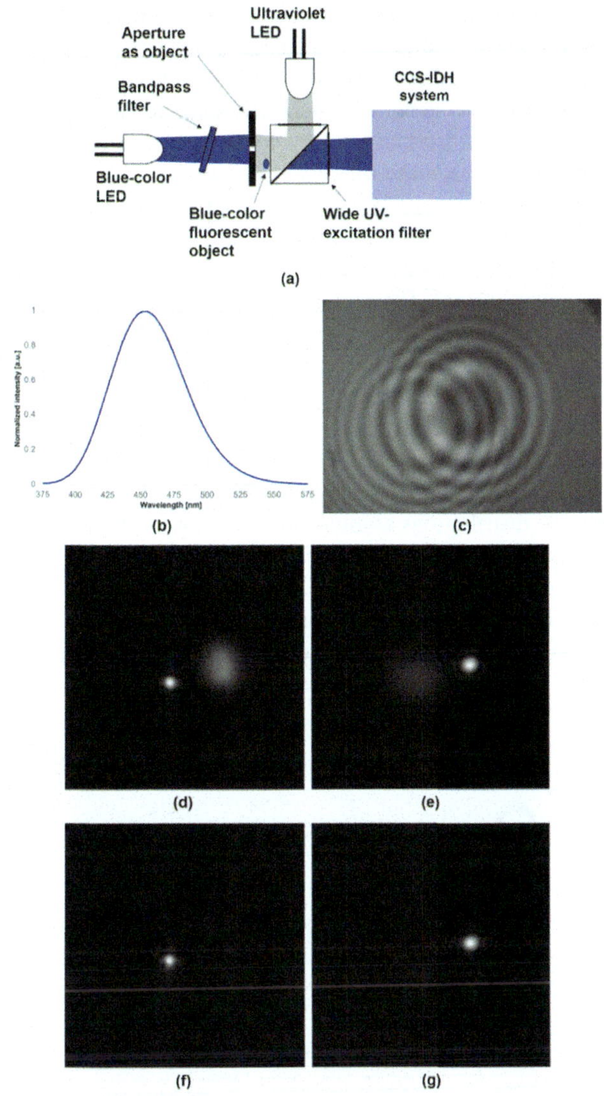

Fig. 18.8 Experimental results for light-multiplexed 3D imaging with the developed hologram recorder. **a** Schematic of experimental setup. **b** Spectral intensity of the self-luminous object. **c** One of the recorded holograms. Left and right Gabor zone plate patterns were generated with LED light and fluorescence light, respectively. **d**, **e** Images reconstructed by commonly applied four-step PSI. The depth difference between **d** and **e** corresponded to 35 mm in the object plane. **f** LED light and **g** fluorescence light images of the objects. The numerical propagation distances of (**f**) and (**g**) were the same as those of (**d**) and (**e**), respectively. Reprinted with permission from [46] © The Optical Society

sity was one-tenth of that at 451 nm, giving a luminescence wavelength width of 120 nm. A blue LED with a nominal wavelength of 455 nm, which was mounted in a four-wavelength LED head (LED4D201, Thorlabs), was used as the illumination light source. The peak wavelength of the LED was between 450 and 455 nm and the FWHM was 18 nm. A bandpass filter whose transmission bandwidth was 446–468 nm was inserted between the blue LED and the aperture to improve the temporal coherency of the illumination light. Each variety of light contained the same wavelength, and the wavelength bandwidth of the LED light was fully overlapped with that of the fluorescence light. Furthermore, the difference between their peak wavelengths was within 5 nm. The self-interference multiplexed hologram shown in Fig. 18.8c was recorded, and then multiple phase shifts and recordings were repeated to obtain the 3D information of each variety of light and to distinguish them. In this experiment, eight exposures were used to perform MPH.

Figure 18.8d–g show the experimental results. The 3D information of each variety of light was retrieved with the commonly used four-step PSI technique, as shown in Fig. 18.8d and e. PSI can also be applied to the recorded single image, and then complex amplitude distributions at multiple wavelengths are retrieved using a CCS algorithm [39, 46]. A multiwavelength 3D image is reconstructed by calculating diffraction integrals (Fig. 18.9).

A WPP array was developed to combine the two IDH techniques. Figure 18.10 shows a schematic of the WPP array and a photograph of the image sensor with

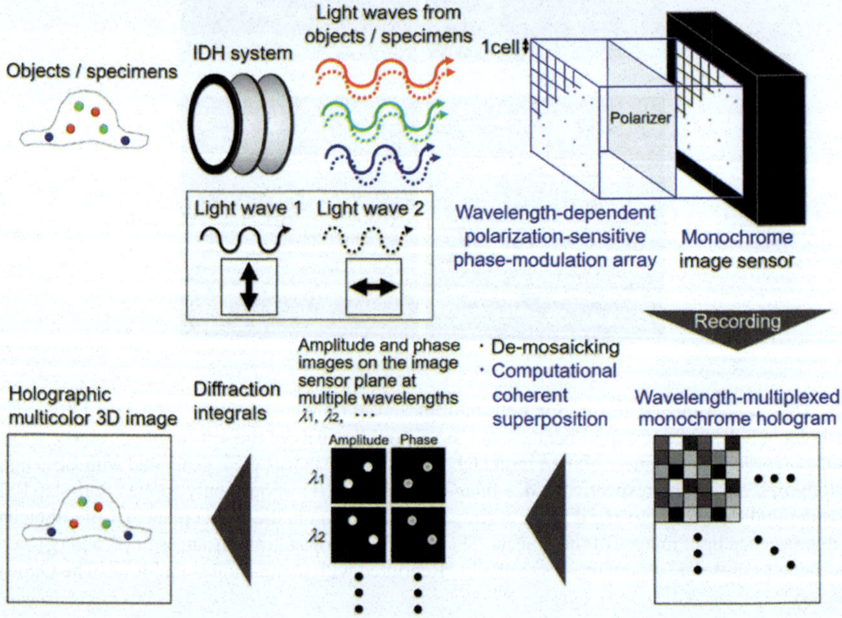

Fig. 18.9 Schematic of SS-CCS holography

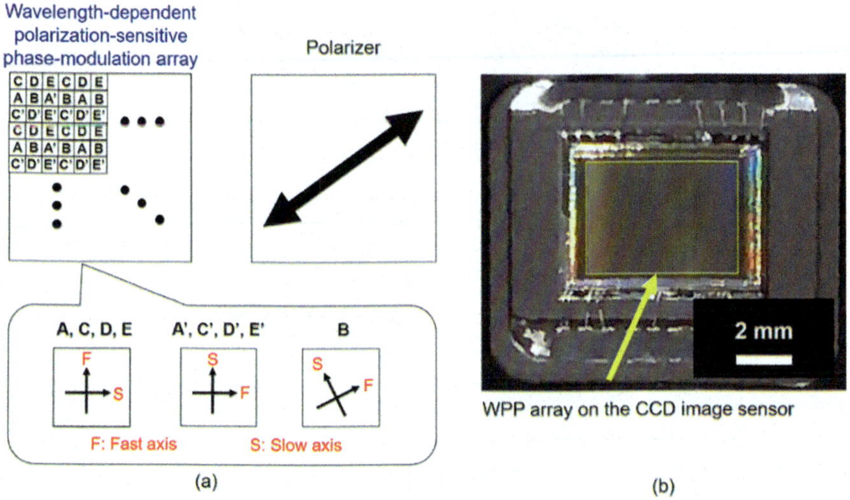

Fig. 18.10 Developed WPP array and image sensor. **a** Schematic of the designed WPP array and **b** photograph of the CCD image sensor developed with the WPP array. Adapted with permission from [43] ©AIP Publishing

the WPP array. Each WPP cell is composed of a photonic crystal, and a photonic crystal array is fabricated by the self-cloning technique [52]. The phase shifts of cells A, C, D, and E at a wavelength of 532 nm are 240°, 107°, 213°, and 320°, respectively. The wavelength dependence of the phase shift of the fabricated photonic crystal is used for the CCS algorithm. The developed image sensor is described in more detail in [43]. An SS-CCS holographic microscopy system, which comprised a fluorescence microscope, a CCS-IDH system, and the image sensor, was constructed to experimentally show its validity. The experimental conditions are described in detail in [43]. The experimental results shown in Fig. 18.11 indicate that fluorescence object waves in different wavelength bands are selectively extracted and that the 3D information in the respective wavelength bands is reconstructed successfully. Different types of fluorescence particle are identified from wavelength separations using the CCS scheme. The experimental results show that SS-CCS IDHM enables the color 3D imaging of fluorescence light from a single wavelength-multiplexed hologram. Improvements of the image quality and frame rate are ongoing, and the color 3D motion-picture recording of incoherent holograms with more than 70 fps and 4 megapixels has been performed [53].

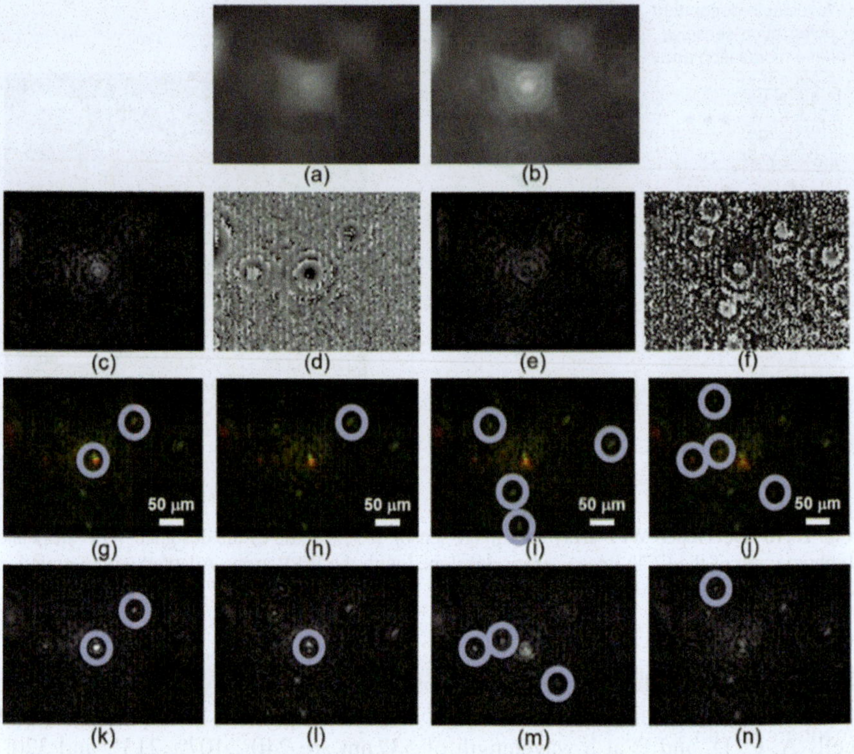

Fig. 18.11 Experimental results. **a** Recorded hologram and **b** wavelength-multiplexed hologram de-mosaicked from (**a**). **c** Intensity and **d** phase images of the object wave on the image sensor plane at a wavelength of 618 nm. **e** Intensity and **f** phase images on the image sensor plane at 545 nm. Color-synthesized images focused at depths of **g** 20.7 μm, **h** 23.7 μm, **i** 26.6 μm, and **j** 29.6 μm in the object plane. **k** 618 nm and **l** 545 nm components of (**g**). **m** 618 nm and **n** 545 nm components of (**j**). Blue circles highlight focused complex molecules. Adapted with permission from [43] ©AIP Publishing

18.2.3 Single-Shot Quantitative Phase Imaging with Single-Shot Phase-Shifting Digital Holography and Light-Emitting Diode

IDH generally adopts a self-interference interferometer to obtain a digital hologram. SSPS-IDH enables single-shot 3D imaging with natural light. However, it remains difficult to obtain a digital hologram of a transparent object with a self-interference interferometer. On the other hand, **self-reference DH (SRDH)** generates a reference wave from an object wave [54–58] and can be used for the **quantitative phase imaging (QPI)** of a transparent specimen with a commonly used light source such as a halogen lamp [55–57] or an LED [58]. The main difference of SRDH from

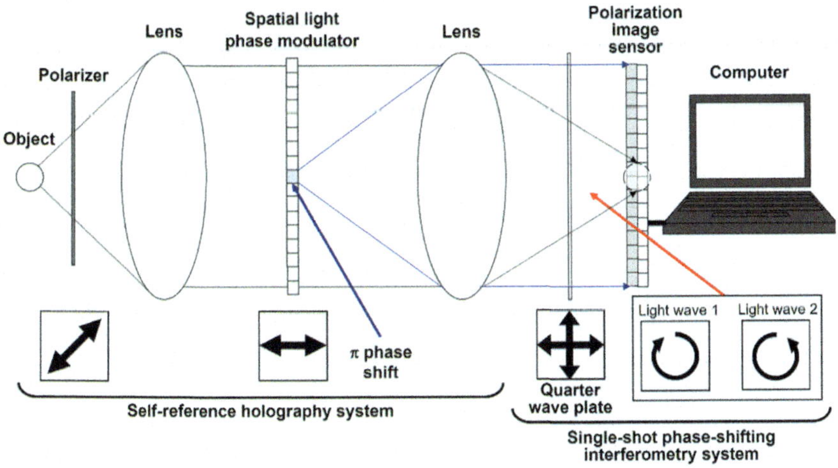

Fig. 18.12 Schematic of the SSPS-SRDH system

self-interference DH is in the generation of a reference wave whose wave vector is unique. The reference wave acts as a spatially or partially coherent light wave on the image sensor plane in SRDH. A self-reference holography system enables the measurement of the quantitative phase information of a transparent specimen with an incoherent light source and the generated reference wave, and it is utilized for QPM [54–58]. SRDH has been combined with SSPS to conduct single-shot QPI with an LED [59].

Figure 18.12 illustrates a schematic of the SSPS-SRDH system. An incoherent light source such as a halogen lamp or an LED is applicable, as demonstrated by Fourier phase microscopy with white light [56]. An object wave of a transparent specimen $O(x, y)$ passes through a polarizer to generate linear polarization from the random polarization of the object wave. After passing through the polarizer, the FT of the object wave FT[$O(x, y)$] is optically performed using a lens, where FT[] denotes the FT. The FT pattern of the magnified image is formed on the back focal plane of the lens, and a spatial light phase modulator is set on the plane. The transmittance axis of the polarizer and the working axis of the spatial light phase modulator have an angle of 45 °C between them. On the FT plane, the plane-wave component of the magnified image is collected on a spot, and the component is utilized as a reference wave. Its mathematical expression is as follows:

$$\begin{aligned} \mathrm{FT}[O(x, y)] \exp[j\delta(f_x, f_y)] = a(f_x, f_y)\mathrm{FT}[O(x, y)] \exp(j\delta_1) \\ +b(f_x, f_y)\mathrm{FT}[O(x, y)] \exp(j\delta_2), \end{aligned} \tag{18.4}$$

where

$$a(f_x, f_y) = \begin{cases} 1 & \text{when} \quad f_x^2 + f_y^2 > r^2, \\ 0 & \text{when} \quad f_x^2 + f_y^2 \leq r^2 \end{cases} \tag{18.5}$$

$$b(f_x, f_y) = \begin{cases} 1 & \text{when} \quad f_x^2 + f_y^2 > r^2, \\ 0 & \text{when} \quad f_x^2 + f_y^2 \leq r^2 \end{cases} \tag{18.6}$$

f_x and f_y are the horizontal and vertical axes in the FT plane, respectively, and r is the radius of $b(f_x, f_y)$. Moreover, δ_1 and δ_2 are the phase shifts of the object and generated reference waves, respectively. $b(f_x, f_y)$ is the aperture for the reference wave generated from the object wave, based on SRDH [54–58]. The spatial light phase modulator sets values of $\delta_1 = 0$ and $\delta_2 = \pi$ for the object and reference wave components, respectively, to set them as linearly polarized light waves with opposite directions. Another lens optically enables an inverse FT of these light waves. After that, a quarter-wave plate converts these linear polarizations to circularly polarized light waves with opposite handedness. As a result, the phase shifts between the two waves depend on the polarization directions. A **polarization-imaging camera** records an image $I(x, y)$, which is expressed as follows:

$$\begin{aligned} I(x, y) = &\left|\text{IFT}[a(f_x, f_y)\text{FT}[O(x, y)]]\exp[j(\delta_1 - \theta)]\right. \\ &\left.+\text{IFT}[b(f_x, f_y)\text{FT}[O(x, y)]]\exp[j(\delta_2 + \theta)]\right|^2 \\ = &\left|\text{IFT}[a(f_x, f_y)\text{FT}[O(x, y)]]\right|^2 + \left|\text{IFT}[b(f_x, f_y)\text{FT}[O(x, y)]]\right|^2 \\ &+ 2\left|\text{IFT}[a(f_x, f_y)\text{FT}[O(x, y)]]\right|\left|\text{IFT}[b(f_x, f_y)\text{FT}[O(x, y)]]\right| \cdot \\ &\cos\{\arg[O(x, t)] - [(\delta_1 - \delta_2) - 2\theta]\} \times \gamma(\Delta L) \end{aligned} \tag{18.7}$$

Here, IFT[] denotes the inverse FT, θ is the transmission axis of the micropolarizer in the polarization-imaging camera, $\gamma(\leq 1, \gamma(0) = 1)$ is a function related to the visibility of the interference fringes and the temporal coherency of light, and ΔL is the optical-path-length difference between the two light waves. Equation (18.7) indicates that intensity ratio is modulated by adjusting r. $\text{IFT}[b(f_x, f_y)\text{FT}[O(x, y)]]$ becomes a plane wave when $b(\xi, \eta)$ is the delta function. The spot size is based on the diffraction limit in the case where a coherent plane wave generated from a laser is subjected to FT. However, the spot size is enlarged on the FT plane using a spatially low-coherence light such as an LED. The intensity distribution of $\text{IFT}[b(f_x, f_y)\text{FT}[O(x, y)]]$ is too low to generate a digital hologram clearly when r is small. Therefore, r should be adjusted to obtain a suitable spot size of the light source on the FT plane. When the intensity ratio of $\text{IFT}[a(f_x, f_y)\text{FT}[O(x, y)]]$ to $\text{IFT}[b(f_x, f_y)\text{FT}[O(x, y)]]$ is small, r is not a small value, and the phase distribution of $\text{IFT}[b(f_x, f_y)\text{FT}[O(x, y)]]$ becomes a quasi-plane wave.

Next, note that the intensity of the second term on the right-hand side of Eq. (18.7) decreases as the optical-path-length difference increases between the two

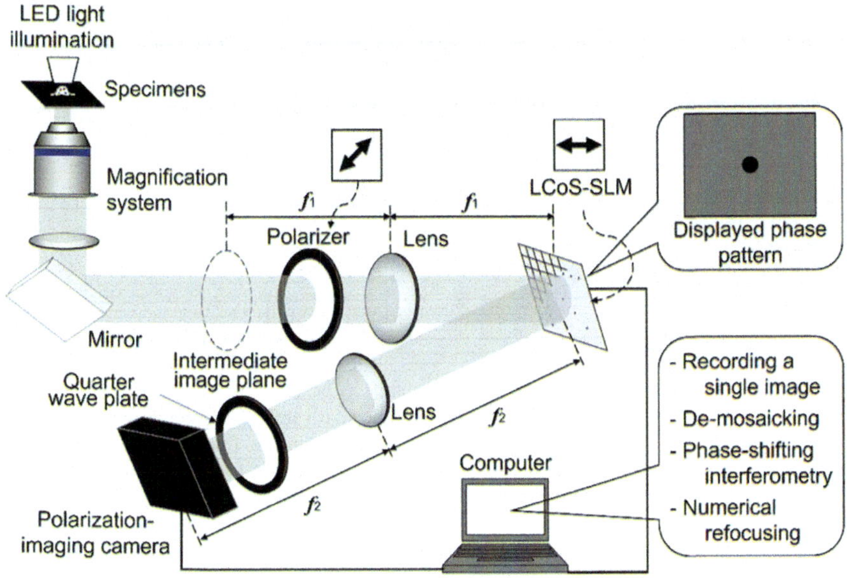

Fig. 18.13 Schematic of the constructed QPM system

waves in DH, as discussed in detail in [46]. The polarization-imaging camera has four transmission axes, $\theta = 0, \pi/4, \pi/2$, and $3\pi/4$, with a polarizer array that is composed of four types of linear micro-polarizers. Therefore, four phase-shifted incoherent digital holograms are simultaneously obtained.

The experimental results obtained using HeLa cells to demonstrate the QPI of transparent objects [58] is presented. The QPM system was constructed on the basis of SSPS-SRDH as shown in Fig. 18.13. A stage on which specimens were placed, a magnification system, and a mirror were the components of the commercially available inverted optical microscope (IX-73, Olympus). An oil-immersion microscope objective whose magnification and numerical aperture were 60 and 1.42, respectively, was included in the setup. A red LED with a nominal wavelength of 625 nm was used as the spatially and temporally low-coherence light source, which was mounted in a four-wavelength LED head (LED4D201, Thorlabs). A polarizer was set as illustrated in Fig. 18.13. A magnified image of the specimens was introduced to an SSPS-SRDH system through the output port of the microscope. Lenses whose focal lengths were 180 and 360 mm were selected to obtain two magnifications in the SSPS-SRDH system, and the total magnification of the QPM system was 120. An LCoS-SLM (X10468-01, Hamamatsu Photonics K.K.) was used and r was set as 300 μm (Fig. 18.14).

In summary, I have introduced IDH techniques and their optical systems for the acceleration of measurement. It has been reported that any variety of light including sunlight can be recorded as a digital hologram by IDH [47]. Many review articles on IDH have been published and can be freely downloaded [25–29, 60, 61]. These

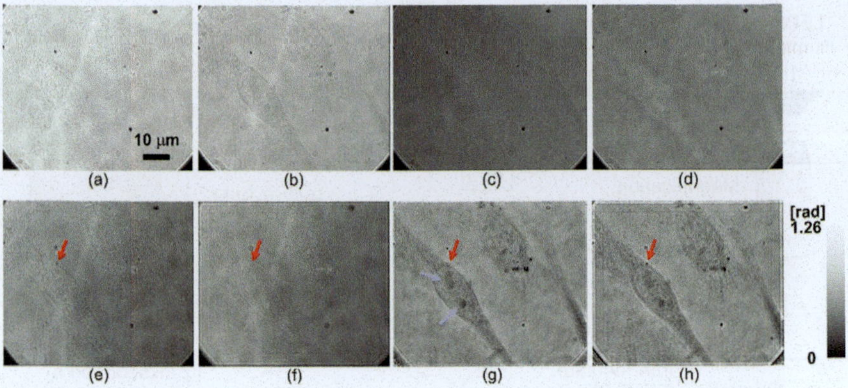

Fig. 18.14 QPI results for transparent objects. Multiple phase-shifted holograms with phase shifts of **a** 0, **b** $\pi/2$, **c** π, and $3\pi/2$ obtained from a recorded single image. Intensity distributions of the reconstructed images in which the numerical propagation distances for the magnified specimens are **e** 7 and **f** 30 mm. Quantitative phase images in which the numerical propagation distances are **g** 7 and **h** 30 mm. Nucleoli are focused in (**e**) and (**g**). A particle is focused in (**f**) and (**h**). Blue and red arrows indicate the nucleoli and particle, respectively. White (255) and black (0) in the phase images denote 1.26 and 0 rad, respectively. Reprinted with permission from [59] © The Optical Society

review articles will help readers study IDH in greater depth. Developments of a hardware architecture and algorithms for acceleration of image reconstruction have also been conducted in IDH [62–64]. In addition, a theoretical discussion of how the temporal coherency of light affects IDH is given in [46].

Fundings

The Mitsubishi Foundation (202111007); Precursory Research for Embryonic Science and Technology (PRESTO) (JPMJPR16P8); Cooperative Research Program of Network Joint Research Center for Materials and Devices (No. 20224020).

Acknowledgements The author in this chapter sincerely thanks the co-authors and co-workers of the articles discussed in this chapter.

References

1. LH Enloe, JA Murphy, and CB Rubinstein. Bstj briefs hologram transmission via television. *The Bell System Technical Journal*, **45**, 335–339, 1966.
2. Joseph W Goodman and RW Lawrence. Digital image formation from electronically detected holograms. *Applied Physics Letters*, **11**, 77–79, 1967.
3. Ting-Chung Poon and Adrianus Korpel. Optical transfer function of an acousto-optic heterodyning image processor. *Optics Letters*, **4**, 317–319, 1979.
4. Kazuyoshi Itoh, Takashi Inoue, Tetsuo Yoshida, and Yoshiki Ichioka. Interferometric supermultispectral imaging. *Applied Optics*, **29**, 1625–1630, 1990.

5. Laurent M Mugnier and Gabriel Y Sirat. On-axis conoscopic holography without a conjugate image. *Optics Letters*, **17**, 294–296, 1992.
6. Mitsuo Takeda, Wei Wang, Zhihui Duan, and Yoko Miyamoto. Coherence holography. *Optics Express*, **13**, 9629–9635, 2005.
7. Joseph Rosen and Gary Brooker. Digital spatially incoherent fresnel holography. *Optics Letters*, **32**, 912–914, 2007.
8. A Vijayakumar, Yuval Kashter, Roy Kelner, and Joseph Rosen. Coded aperture correlation holography–a new type of incoherent digital holograms. *Optics Express*, **24**, 12430–12441, 2016.
9. Jiachen Wu, Hua Zhang, Wenhui Zhang, Guofan Jin, Liangcai Cao, and George Barbastathis. Single-shot lensless imaging with fresnel zone aperture and incoherent illumination. *Light: Science & Applications*, **9**, 1–11, 2020.
10. Emmett N Leith and Juris Upatnieks. Reconstructed wavefronts and communication theory. *JOSA*, **52**, 1123–1130, 1962.
11. AW Lohmann. Reconstruction of vectorial wavefronts. *Applied Optics*, **4**, 1667–1668, 1965.
12. Pascal Picart, Eric Moisson, and Denis Mounier. Twin-sensitivity measurement by spatial multiplexing of digitally recorded holograms. *Applied Optics*, **42**, 1947–1957, 2003.
13. Jonas Kühn, Tristan Colomb, Frédéric Montfort, Florian Charrière, Yves Emery, Etienne Cuche, Pierre Marquet, and Christian Depeursinge. Real-time dual-wavelength digital holographic microscopy with a single hologram acquisition. *Optics Express*, **15**, 7231–7242, 2007.
14. Tatsuki Tahara, Toru Kaku, and Yasuhiko Arai. Digital holography based on multiwavelength spatial-bandwidth-extended capturing-technique using a reference arm (multi-spectra). *Optics Express*, **22**, 29594–29610, 2014.
15. Mitsuo Takeda, Hideki Ina, and Seiji Kobayashi. Fourier-transform method of fringe-pattern analysis for computer-based topography and interferometry. *JOSA*, **72**, 156–160, 1982.
16. Tatsuki Tahara, Takuya Gotohda, Takanori Akamatsu, Yasuhiko Arai, Tomoyoshi Shimobaba, Tomoyoshi Ito, and Takashi Kakue. High-speed image-reconstruction algorithm for a spatially multiplexed image and application to digital holography. *Optics Letters*, **43**, 2937–2940, 2018.
17. Tatsuki Tahara, Takanori Akamatsu, Yasuhiko Arai, Tomoyoshi Shimobaba, Tomoyoshi Ito, and Takashi Kakue. Algorithm for extracting multiple object waves without fourier transform from a single image recorded by spatial frequency-division multiplexing and its application to digital holography. *Optics Communications*, **402**, 462–467, 2017.
18. Tomoyoshi Shimobaba, Yoshikuni Sato, Junya Miura, Mai Takenouchi, and Tomoyoshi Ito. Real-time digital holographic microscopy using the graphic processing unit. *Optics Express*, **16**, 11776–11781, 2008.
19. Matteo Frigo and Steven G Johnson. Fftw: An adaptive software architecture for the fft. In *Proceedings of the 1998 IEEE International Conference on Acoustics, Speech and Signal Processing, ICASSP'98 (Cat. No. 98CH36181)*, volume 3, pages 1381–1384. IEEE, 1998.
20. Tomoyoshi Shimobaba, Takashi Kakue, and Tomoyoshi Ito. Review of fast algorithms and hardware implementations on computer holography. *IEEE Transactions on Industrial Informatics*, **12**, 1611–1622, 2015.
21. Takashige Sugie, Takanori Akamatsu, Takashi Nishitsuji, Ryuji Hirayama, Nobuyuki Masuda, Hirotaka Nakayama, Yasuyuki Ichihashi, Atsushi Shiraki, Minoru Oikawa, Naoki Takada, et al. High-performance parallel computing for next-generation holographic imaging. *Nature Electronics*, **1**, 254–259, 2018.
22. Banghe Zhu and Ken-ichi Ueda. Real-time wavefront measurement based on diffraction grating holography. *Optics Communications*, **225**, 1–6, 2003.
23. James E. Millerd, Neal J. Brock, John B. Hayes, Michael B. North-Morris, Matt Novak, and James C. Wyant. Pixelated phase-mask dynamic interferometer. In Katherine Creath and Joanna Schmit, editors, *Interferometry XII: Techniques and Analysis*, volume 5531, pages 304–314. International Society for Optics and Photonics, SPIE, 2004.
24. Yasuhiro Awatsuji, Masaki Sasada, and Toshihiro Kubota. Parallel quasi-phase-shifting digital holography. *Applied Physics Letters*, **85**, 1069–1071, 2004.

25. Tatsuki Tahara, Yaping Zhang, Joseph Rosen, Vijayakumar Anand, Liangcai Cao, Jiachen Wu, Takako Koujin, Atsushi Matsuda, Ayumi Ishii, Yuichi Kozawa, et al. Roadmap of incoherent digital holography. *Applied Physics B*, **128**, 193, 2022.
26. Joseph Rosen, A Vijayakumar, Manoj Kumar, Mani Ratnam Rai, Roy Kelner, Yuval Kashter, Angika Bulbul, and Saswata Mukherjee. Recent advances in self-interference incoherent digital holography. *Advances in Optics and Photonics*, **11**, 1–66, 2019.
27. Jung-Ping Liu, Tatsuki Tahara, Yoshio Hayasaki, and Ting-Chung Poon. Incoherent digital holography: a review. *Applied Sciences*, **8**, 143, 2018.
28. Joseph Rosen, Simon Alford, Vijayakumar Anand, Jonathan Art, Petr Bouchal, Zdeněk Bouchal, Munkh-Uchral Erdenebat, Lingling Huang, Ayumi Ishii, Saulius Juodkazis, et al. Roadmap on recent progress in finch technology. *Journal of Imaging*, **7**, 197, 2021.
29. Tatsuki Tahara. Review of incoherent digital holography: applications to multidimensional incoherent digital holographic microscopy and palm-sized digital holographic recorder–holosensor. *Front. Photonics*, **2**, 829139, 2022.
30. Jisoo Hong and Myung K Kim. Single-shot self-interference incoherent digital holography using off-axis configuration. *Optics Letters*, **38**, 5196–5199, 2013.
31. Xiangyu Quan, Osamu Matoba, and Yasuhiro Awatsuji. Single-shot incoherent digital holography using a dual-focusing lens with diffraction gratings. *Optics Letters*, **42**, 383–386, 2017.
32. Tatsuki Tahara, Takeya Kanno, Yasuhiko Arai, and Takeaki Ozawa. Single-shot phase-shifting incoherent digital holography. *Journal of Optics*, **19**, 065705, 2017.
33. Teruyoshi Nobukawa, Tetsuhiko Muroi, Yutaro Katano, Nobuhiro Kinoshita, and Norihiko Ishii. Single-shot phase-shifting incoherent digital holography with multiplexed checkerboard phase gratings. *Optics Letters*, **43**, 1698–1701, 2018.
34. KiHong Choi, Kyung-Il Joo, Tae-Hyun Lee, Hak-Rin Kim, Junkyu Yim, Hyeongkyu Do, and Sung-Wook Min. Compact self-interference incoherent digital holographic camera system with real-time operation. *Optics Express*, **27**, 4818–4833, 2019.
35. Dong Liang, Qiu Zhang, Jing Wang, and Jun Liu. Single-shot fresnel incoherent digital holography based on geometric phase lens. *Journal of Modern Optics*, **67**, 92–98, 2020.
36. Tatsuki Tahara, Tomoyoshi Ito, Yasuyuki Ichihashi, and Ryutaro Oi. Single-shot incoherent color digital holographic microscopy system with static polarization-sensitive optical elements. *Journal of Optics*, **22**, 105702, 2020.
37. Tatsuki Tahara, Ryota Mori, Shuhei Kikunaga, Yasuhiko Arai, and Yasuhiro Takaki. Dual-wavelength phase-shifting digital holography selectively extracting wavelength information from wavelength-multiplexed holograms. *Optics Letters*, **40**, 2810–2813, 2015.
38. Tatsuki Tahara, Ryota Mori, Yasuhiko Arai, and Yasuhiro Takaki. Four-step phase-shifting digital holography simultaneously sensing dual-wavelength information using a monochromatic image sensor. *Journal of Optics*, **17**, 125707, 2015.
39. Tatsuki Tahara, Reo Otani, Kaito Omae, Takuya Gotohda, Yasuhiko Arai, and Yasuhiro Takaki. Multiwavelength digital holography with wavelength-multiplexed holograms and arbitrary symmetric phase shifts. *Optics Express*, **25**, 11157–11172, 2017.
40. Tatsuki Tahara, Shuhei Kikunaga, and Yasuhiko Arai. Japanese patent, JP6308594, Sep. 2013.
41. Takayuki Hara, Tatsuki Tahara, Yasuyuki Ichihashi, Ryutaro Oi, and Tomoyoshi Ito. Multiwavelength-multiplexed phase-shifting incoherent color digital holography. *Optics Express*, **28**, 10078–10089, 2020.
42. Tatsuki Tahara, Tomoyoshi Ito, Yasuyuki Ichihashi, and Ryutaro Oi. Multiwavelength three-dimensional microscopy with spatially incoherent light, based on computational coherent superposition. *Optics Letters*, **45**, 2482–2485, 2020.
43. Tatsuki Tahara, Ayumi Ishii, Tomoyoshi Ito, Yasuyuki Ichihashi, and Ryutaro Oi. Single-shot wavelength-multiplexed digital holography for 3d fluorescent microscopy and other imaging modalities. *Applied Physics Letters*, **117**, 031102, 2020.
44. Tatsuki Tahara, Takako Koujin, Atsushi Matsuda, Ayumi Ishii, Tomoyoshi Ito, Yasuyuki Ichihashi, and Ryutaro Oi. Incoherent color digital holography with computational coherent superposition for fluorescence imaging. *Applied Optics*, **60**, A260–A267, 2021.

45. Tatsuki Tahara, Yuichi Kozawa, Ayumi Ishii, Koki Wakunami, Yasuyuki Ichihashi, and Ryutaro Oi. Two-step phase-shifting interferometry for self-interference digital holography. *Optics Letters*, **46**, 669–672, 2021.
46. Tatsuki Tahara. Multidimension-multiplexed full-phase-encoding holography. *Optics Express*, **30**, 21582–21598, 2022.
47. Myung K Kim. Full color natural light holographic camera. *Optics Express*, **21**, 9636–9642, 2013.
48. Tatsuki Tahara and Ryutaro Oi. Palm-sized single-shot phase-shifting incoherent digital holography system. *OSA Continuum*, **4**, 2372–2380, 2021.
49. Tatsuki Tahara and Yutaka Endo. Multiwavelength-selective phase-shifting digital holography without mechanical scanning. *Applied Optics*, **58**, G218–G225, 2019.
50. J Herriot Bruning, Donald R Herriott, Joseph E Gallagher, Daniel P Rosenfeld, Andrew D White, and Donald J Brangaccio. Digital wavefront measuring interferometer for testing optical surfaces and lenses. *Applied Optics*, **13**, 2693–2703, 1974.
51. Ichirou Yamaguchi and Tong Zhang. Phase-shifting digital holography. *Optics Letters*, **22**, 1268–1270, Aug 1997.
52. Takashi Sato, Takeshi Araki, Yoshihiro Sasaki, Toshihide Tsuru, Toshiyasu Tadokoro, and Shojiro Kawakami. Compact ellipsometer employing a static polarimeter module with arrayed polarizer and wave-plate elements. *Applied Optics*, **46**, 4963–4967, 2007.
53. Tatsuki Tahara, Takako Koujin, Atsushi Matsuda, Yuichi Kozawa, Yasuyuki Ichihashi, and Ryutaro Oi. 72 fps incoherent two-color digital motion-picture holography system for fluorescence cell imaging. In *Digital Holography and Three-Dimensional Imaging*, pages DTu6H–5. Optical Society of America, 2021.
54. Gabriel Popescu, Lauren P Deflores, Joshua C Vaughan, Kamran Badizadegan, Hidenao Iwai, Ramachandra R Dasari, and Michael S Feld. Fourier phase microscopy for investigation of biological structures and dynamics. *Optics Letters*, **29**, 2503–2505, 2004.
55. Zhuo Wang, Larry Millet, Mustafa Mir, Huafeng Ding, Sakulsuk Unarunotai, John Rogers, Martha U Gillette, and Gabriel Popescu. Spatial light interference microscopy (slim). *Optics Express*, **19**, 1016–1026, 2011.
56. Basanta Bhaduri, Krishnarao Tangella, and Gabriel Popescu. Fourier phase microscopy with white light. *Biomedical Optics Express*, **4**, 1434–1441, 2013.
57. Wu You, Wenlong Lu, and Xiaojun Liu. Single-shot wavelength-selective quantitative phase microscopy by partial aperture imaging and polarization-phase-division multiplexing. *Optics Express*, **28**, 34825–34834, 2020.
58. Tatsuki Tahara, Yuichi Kozawa, Atsushi Matsuda, and Ryutaro Oi. Quantitative phase imaging with single-path phase-shifting digital holography using a light-emitting diode. *OSA Continuum*, **4**, 2918–2927, 2021.
59. Tatsuki Tahara, Yuichi Kozawa, and Ryutaro Oi. Single-path single-shot phase-shifting digital holographic microscopy without a laser light source. *Optics Express*, **30**, 1182–1194, 2022.
60. J Hong and MK Kim. Overview of techniques applicable to self-interference incoherent digital holography. *Journal of the European Optical Society. Rapid publications*, **8**, 2013.
61. Tatsuki Tahara, Xiangyu Quan, Reo Otani, Yasuhiro Takaki, and Osamu Matoba. Digital holography and its multidimensional imaging applications: a review. *Microscopy*, **67**, 55–67, 2018.
62. Masahiro Tsuruta, Tomotaka Fukuyama, Tatsuki Tahara, and Yasuhiro Takaki. Fast image reconstruction technique for parallel phase-shifting digital holography. *Applied Sciences*, **11**, 11343, 2021.
63. Tomoyoshi Shimobaba, Tatsuki Tahara, Ikuo Hoshi, Harutaka Shiomi, Fan Wang, Takayuki Hara, Takashi Kakue, and Tomoyoshi Ito. Real-valued diffraction calculations for computational holography. *Applied Optics*, **61**, B96–B102, 2022.
64. Takayuki Hara, Takashi Kakue, Tomoyoshi Shimobaba, and Tomoyoshi Ito. Design and implementation of special-purpose computer for incoherent digital holography. *IEEE Access*, **10**, 76906–76912, 2022.

Part IV
FPGA-Based Acceleration for Computational Holography

Part IV consists of two chapters: hardware accelerators for computational holography typically use CPUs and GPUs, while Part IV presents examples of hologram and diffraction calculations implemented using FPGAs.

Chapter 19
FPGA Accelerator for Computer-Generated Hologram

Yota Yamamoto

Abstract A special-purpose computer for computer-generated hologram (CGH) has been built using an FPGA [1–13]. By constructing a special-purpose computation pipeline for CGH computations on FPGAs, faster computations than on CPUs can be attained. In this chapter, we present the fundamental principle of the special-purpose computer.

19.1 A Special-Purpose Computer for CGH Using FPGA

Efforts are being made to accelerate **CGH** computations with dedicated FPGA-based computing circuits. **Holographic reconstruction** (**HORN**) is a **special-purpose computer** for holography's high-speed computation [1–13]. HORN-1 [1], developed in 1993, comprises 26 IC chips on a universal board, with each IC being hand-wired (Fig. 19.1).

HORN-3 [3] was built with a programmable logic device (PLD), HORN-4 [4] was integrated using an FPGA, and the latest HORN-8 (Fig. 19.2), built using FPGAs, was 1,000 times faster than a conventional PC [8–10]. The construction of special-purpose computers using FPGAs has the benefit of removing the necessity for manual wiring between discrete ICs and the ability to redesign the circuit without rewiring. The reconfigurability of FPGAs allows the construction of circuits to be more effective. HORN-8 is an eight-FPGA dedicated board connected to a PC motherboard by PCI Express to communicate with a CPU.

The Xilinx Zynq series, which has an embedded CPU and FPGA on a single chip, has been built for the construction of CGH computers [11, 12]. Although Zynq is small, it is possible to implement HORN-8 similar to computing circuitry on this

Supplementary Information The online version contains supplementary material available at https://doi.org/10.1007/978-981-99-1938-3_19.

Y. Yamamoto (✉)
Faculty of Engineering, Tokyo University of Science, 6-3-1 Niijuku, Katsushika-ku, Tokyo 125-8585, Japan
e-mail: yy-yamamoto@rs.tus.ac.jp

T. Shimobaba and T. Ito (eds.), *Hardware Acceleration of Computational Holography*,
https://doi.org/10.1007/978-981-99-1938-3_19

Fig. 19.1 Special-purpose computer for holography (HORN-1)

Fig. 19.2 HORN-8 board

single chip. In the future, Zynq could be used for holographic head-mounted display applications. In the following sections, we present the construction of a special-purpose computer based on Zynq.

19.2 Algorithm for CGH Special-Purpose Computer

This section explains how to implement the CGH computation using Eq. (19.1) in FPGA:

$$I\left(x_a, y_a\right) = \sum_{j=0}^{M-1} A_j \cos\left[\frac{\pi}{\lambda z_j}\left\{\left(x_a - x_j\right)^2 + \left(y_a - y_j\right)^2\right\}\right], \tag{19.1}$$

where (x_a, y_a) represents the coordinate on the CGH plane, $\left(x_j, y_j, z_j\right)$ denotes the point cloud's coordinate, and A_j represents the point cloud's amplitude. Here, A_j is fixed to 1 to simplify the design. M denotes the point clouds' number and λ represents the reference light's wavelength.

When Eq. (19.1) is implemented in FPGAs, it is known that a recurrence algorithm can construct circuits with fewer arithmetic units [14]. However, for ease of understanding, Eq. (19.1) is implemented directly in FPGAs here.

Equation (19.1) includes division. In general, division in FPGAs necessitates more resources (LUTs) and lowers the operating frequency. Thus, we define a new variable as

$$\rho_j = \frac{\pi}{2\lambda z_j}, \tag{19.2}$$

ρ_j can be precomputed by a CPU. Using ρ_j, Eq. (19.1) is redefined as

$$I\left(x_\alpha, y_\alpha\right) = \sum_{j=0}^{M-1} \cos\left[\rho_j\left\{\left(x_a - x_j\right)^2 + \left(y_a - y_j\right)^2\right\}\right]. \tag{19.3}$$

The FPGA conducts Eq. (19.3) in a pipeline fashion. Furthermore, data parallelization for each pixel is achievable. High speed can be predicted using a special computation circuit. For more details, please refer to Chap. 7.

19.3 Overview of CGH Special-Purpose Computer Using Xilinx Zynq

Figure 19.3 shows a block diagram of a CGH-specific computer using Xilinx Zynq [15, 16]. In the Xilinx FPGA construction, we employed Vivado Design Suite (hereafter, Vivado) [17] as a construction tool. The design's basic flow is to define the circuit structure using register transfer level (RTL) description, compile the circuit using Vivado, and then download the circuit configuration data to the FPGA to create the special-purpose computer. Figure 19.3 shows that in Vivado, the intellectual property (IP) integrator [18] offers an aid to visually connect IP blocks.

In Fig. 19.3, the block labeled "ZYNQ UltraSCALE+" is the ARM CPU IP block, and the CGH computation block labeled "cgh_calculation_accelerator_v1_0" is implemented in the FPGA component. These blocks are connected through "AXI Interconnect," a block for AXI communication (see Chap. 7). The Zynq-based special-purpose computer employs the ARM CPU for pre- and post-computations (e.g., the computation of ρ_j in Eq. (19.3), object rotation processing, etc.) and the FPGA part's control.

Figure 19.4 reveals a block diagram of the CGH computation block (cgh_calculation_acceleator_v1_0) in the FPGA component. This block stimulates CGH computation through two parallelization methods (data and pipeline parallelization). Figure 19.4 shows that in the upper diagram, Eq. (19.3) is conducted using pipeline parallelization (see Chap. 7).

The shaded value in Fig. 19.4 illustrates the bit width; all operations inside the FPGA are in fixed-point numbers. Thus, the CGH coordinates and object point coordinates in Eq. (19.3) are normalized using the SLM's pixel pitch. CGH coordinates x_a and y_a are 12 bits, x_j, y_j, z_j of the point, cloud coordinates are 12 bits, and ρ is set to 32 bit width.

The bit width of the internal circuitry is measured by considering the digits' overflow. The notation [a: b] in the figure indicates that the signal line from the a-th to the b-th bit is handled. The adder's output bit width is increased by one bit from the maximum bit width of the input data. The output bit width of the multiplier denotes the sum of the bit widths of the two input data. The cosine function is processed using a **look-up table** (**LUT**). Bit 31–26 is the cos table's address; the LUT's output is empirically set to 6 bits. When computing the binary amplitude CGH, the most significant bit of the pipeline output is the computation finding.

Figure 19.5 shows the top-level block diagram of the CGH computation block (cgh_calculation_acceleator_v1_0). "cgh_c" is the unit that computes the CGH of a single pixel. "pcl_ram" is the memory for storing the point cloud. "controller" golds the point clouds' number and the CGH's size and regulates the whole circuit. The information for the point cloud is sent from the ARM CPU through the AXI connection. The transferred data are transiently stored in the FPGA memory and broadcast to the "cgh_c" units. Multiple "cgh_c" units compute several CGH pixels, which is data parallelization.

Zynq is equipped with a display controller. The CGH pixel data computed by the FPGA are written to the display memory through AXI, allowing CGH to be displayed on a display panel. Writing CGH pixels to display memory with an embedded CPU is slow; with direct memory access (DMA), enormous amounts of data can be read and written at a fast speed, permitting the development of very efficient dedicated computers.

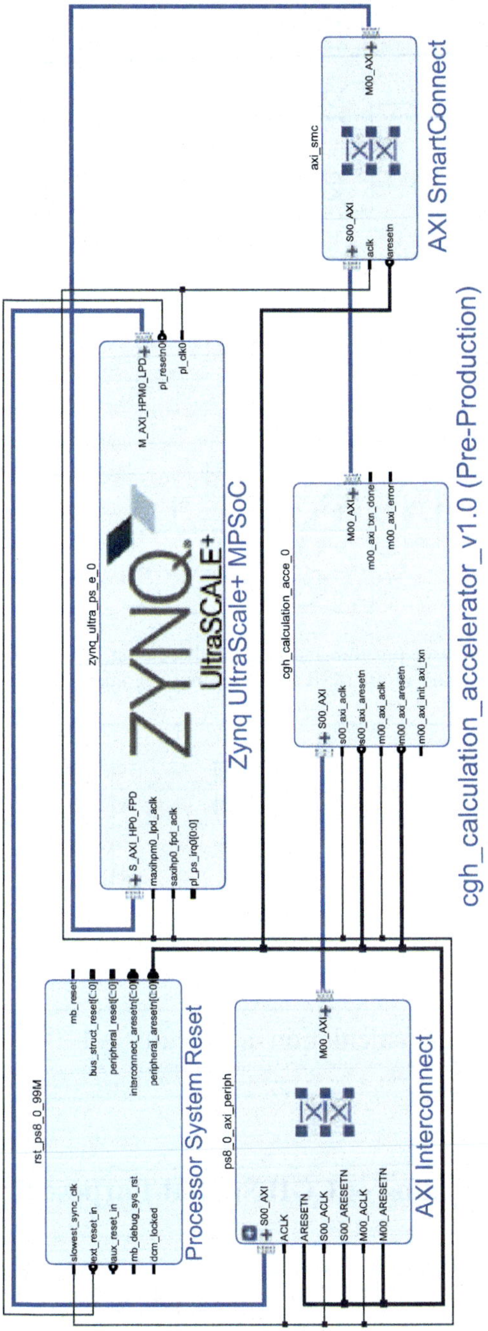

Fig. 19.3 Block diagram of the entire system

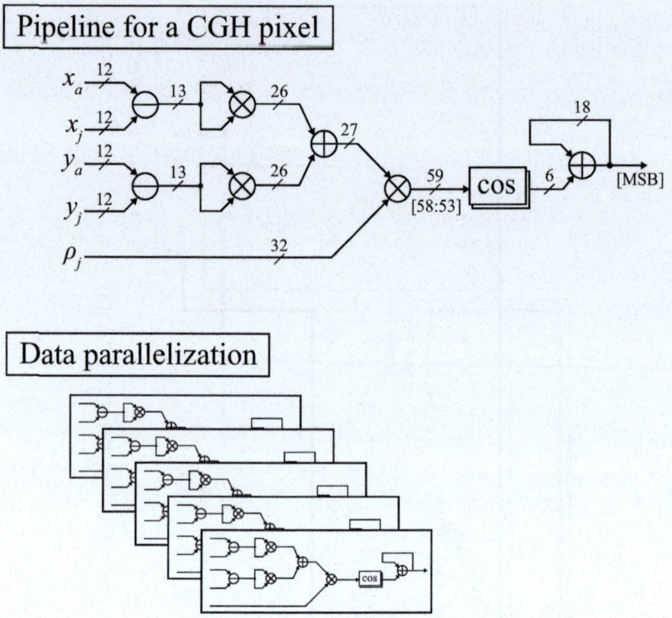

Fig. 19.4 Block diagram of pipeline and data parallelization for the CGH computation block (cgh_calculation_acceleator_v1_0). MSB stands for most significant bit

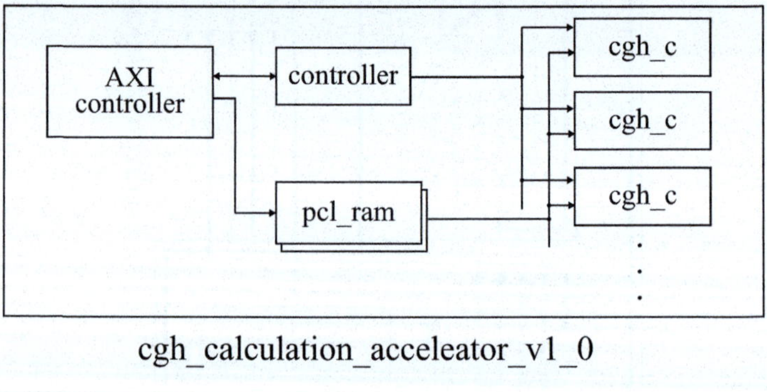

Fig. 19.5 Top-level block diagram

19.4 Implementation of CGH Special-Purpose Computer

Listing 19.1 reveals the source code for the "cgh_c" unit, which computes a single pixel on the CGH using Eq. (19.3) in a pipeline fashion. **SystemVerilog** was employed as the programming language. In Listing 19.1, the cosine computation is conducted using the **LUT** approach that stores the cosine function's computation finding in a read-only memory (ROM) ahead and obtains cosine values at high speed by referring to the ROM address. Here, we employed the LUT with 6-bit input and 6-bit output. Listing 19.2 shows the source code of the cos table, which is implemented in RAM64X8SW [19] (a small amount of memory) in the FPGA.

In the source code of "cgh_c," "clk" is a clock signal, which synchronizes to operate the circuit, "resetn" is an initialization signal for the circuit, "valid_in" is set to 1 when valid point cloud data are input, and "last_in" is set to 0 when the last point cloud data are input. The signals "x_a_in," "x_j_in," "y_a_in," "y_j_in," and "rho_j_in" correspond to x_a, x_j, y_a, y_j, and ρ_j in Eq. (19.3). The summing unit denotes the unit that conducts the summation computation. The "sum" unit for the accumulation in Eq. (19.3) accumulates the cosine table's output.

Listing 19.1 Source code of the "c_cgh" unit.

```
module cgh_c #(
    parameter int
        XY_I_WIDTH = 12,
        Z_DELTA_WIDTH = 32,
        Z_DELTA_FIXED_POINT = 32,
        THETA_WIDTH = 6,
        SUM_BUFFER_WIDTH = 18,
        INTENSITY_OUT_WIDTH = 8,
        SUB_XY_WIDTH = XY_I_WIDTH + 1,
        POW_2_XY_WIDTH = SUB_XY_WIDTH * 2,
        ADD_XY_WIDTH = POW_2_XY_WIDTH + 1,
        MULT_XYZ_WIDTH = ADD_XY_WIDTH + Z_DELTA_WIDTH,
        THETA_SHIFT = Z_DELTA_FIXED_POINT - THETA_WIDTH
)(
    input wire clk,
    input wire resetn,
    input wire valid_in,
    input wire last_in,
    input wire [XY_I_WIDTH-1:0] x_a_in,
    input wire [XY_I_WIDTH-1:0] x_j_in,
    input wire [XY_I_WIDTH-1:0] y_a_in,
    input wire [XY_I_WIDTH-1:0] y_j_in,
    input wire [Z_DELTA_WIDTH-1:0] rho_j_in,
    output wire intensity_valid_out,
    output wire [INTENSITY_OUT_WIDTH-1:0] intensity_out
);
    /* controll signal */
    logic valid[0:4], last[0:4];

    /* caluculation pipline reg */
    logic signed [SUB_XY_WIDTH-1:0] sub_x_ans, sub_y_ans;
    logic signed [POW_2_XY_WIDTH-1:0] pow2_x_ans, pow2_y_ans;
    logic signed [ADD_XY_WIDTH-1:0] add_x_y_ans;
    logic signed [MULT_XYZ_WIDTH-1:0] mult_xy_z_ans;
    logic signed [Z_DELTA_WIDTH-1:0] rho_j_deley [0:2];

    /* cos signal */
    logic signed [THETA_WIDTH-1:0] cos_out;
    logic signed [THETA_WIDTH-1:0] theta;
```

```

/* sum signal */
logic signed [SUM_BUFFER_WIDTH−1:0] cos_sum;
logic cos_sum_valid;

assign intensity_out = (cos_sum[SUM_BUFFER_WIDTH−1])?'d255:'d0;
assign intensity_valid_out = cos_sum_valid;
assign theta = mult_xy_z_ans[THETA_SHIFT+:THETA_WIDTH];

cos_table #(
    .THETA_WIDTH (THETA_WIDTH)
) cos_table_u (
    .clk (clk),
    .theta_in (theta),
    .cos_out (cos_out)
);

sum #(
    .IN_WIDTH (THETA_WIDTH),
    .SUM_BUFFER_WIDTH (SUM_BUFFER_WIDTH),
    .SUM_OUT_WIDTH (SUM_BUFFER_WIDTH)
) cos_sum_u (
    .clk (clk),
    .resetn (resetn),
    .valid_in (valid[4]),
    .last_in (last[4]),
    .val_in (cos_out),
    .sum_valid (cos_sum_valid),
    .sum_out (cos_sum)
);

always_ff @(posedge clk) begin
    if(!resetn) begin
        for (int i=0; i<5; i++) begin
            valid[i] <= 'b0;
            last[i] <= 'b0;
        end
    end else begin
        // 1st stage
        valid[0] <= valid_in;
        last[0] <= last_in;

        // 2nd stage
        valid[1] <= valid[0];
        last[1] <= last[0];

        // 3rd stage
        valid[2] <= valid[1];
        last[2] <= last[1];

        // 4th stage
        valid[3] <= valid[2];
        last[3] <= last[2];
```

```

        // 5th stage
        valid[4] <= valid[3];
        last[4] <= last[3];
      end
    end

    always_ff @(posedge clk) begin
      /* 1st stage */
      sub_x_ans <= $signed(x_j_in) − $signed(x_a_in);
      sub_y_ans <= $signed(y_j_in) − $signed(y_a_in);
      rho_j_deley[0] <= rho_j_in;

      /* 2nd stage */
      pow2_x_ans <= sub_x_ans ** 2;
      pow2_y_ans <= sub_y_ans ** 2;
      rho_j_deley[1] <= rho_j_deley[0];

      /* 3rd stage */
      add_x_y_ans <= pow2_x_ans + pow2_y_ans;
      rho_j_deley[2] <= rho_j_deley[1];

      /* 4th stage */
      mult_xy_z_ans <= add_x_y_ans * rho_j_deley[2];

      /* 5th stage */
      // cos table
    end
endmodule
```

Listing 19.2 Source code of the cosine table.

```
module cos_table#(
    parameter
      THETA_WIDTH = 6,
      TABLE_LEN = (1'b1 << THETA_WIDTH)
)(
    input wire clk,
    input wire [THETA_WIDTH−1:0] theta_in,
    output logic signed [THETA_WIDTH−1:0] cos_out
);
    logic signed [THETA_WIDTH−1:0] theta;

    always_ff @(posedge clk) begin
        theta <= theta_in;
    end

    RAM64X8SW #(
      .INIT_A('b00000000_00000000_11111111_11111111_\
11111111_11111110_00000000_00000000),
      .INIT_B('b00000000_00000000_11111111_11111111_\
11111111_11111110_00000000_00000000),
      .INIT_C('b00000000_00000000_11111111_11111111_\
```

```
11111111_11111110_00000000_00000000),
        .INIT_D('b11111111_11000000_11111000_00000000_\
00000000_00111110_00000111_11111111),
        .INIT_E('b11111100_00111000_11100111_10000000_\
00000011_11001110_00111000_01111111),
        .INIT_F('b11110011_00100100_10010110_01110000_\
00011100_11010010_01001001_10011111),
        .INIT_G('b11001010_10110110_01001101_01001110_\
11100101_01100100_11011010_10100111),
        .INIT_H('b00101111_11010010_10010111_11101001_\
00101111_11010010_10010111_11101001),
        .IS_WCLK_INVERTED('b0) // Optional inversion for WCLK
    ) RAM64X8SW_inst_cos (
        .O(cos_out), // 8-bit data output
        .A(theta), // 6-bit address input
        .D('b0), // 1-bit input: Write data input
        .WCLK(clk), // 1-bit input: Write clock input
        .WE('b0), // 1-bit input: Write enable input
        .WSEL('b000) // 3-bit write select
    );

endmodule
```

19.5 Control Program for CGH Special-Purpose Computer

Figure 19.3 shows the control program for the CGH special-purpose computer, referring to the program introduced in Chap. 8. In Fig. 19.3, the embedded CPU implements this program and computes ρ. The computed coordinate information for object points (x_j, y_j, ρ_j) is written to /dev/mem to send the data to the special-purpose computer on the FPGA side. Subsequently, the embedded CPU reads the control registers through /dev/mem and waits in a while loop for the computation to finish. Subsequently, the final CGH computation finding is generated from the FPGA side through /dev/mem.

Listing 19.3 The control program for the CGH special-purpose computer.

```
#include <stdio.h>
#include <stdint.h>
#include <stdlib.h>
#include <unistd.h>
#include <fcntl.h>
#include <sys/mman.h>
#include <math.h>

#define FPGA_ADDR_START 0x0A0000000
#define FPGA_ADDR_SIZE 0x000008000
#define FPGA_PIPELINE 810
#define PCD_MAX_LEN (1U<<16) // 2^16

```

```
struct point_cloud{
    float x;
    float y;
    float z;
};

struct point_cloud_fixed{
    int32_t x;
    int32_t y;
    int32_t rho;
};

struct point_cloud_fixed pcd_fixed[PCD_MAX_LEN];

extern struct point_cloud* load_point_cloud(const unsigned char* path, int* numpoint);
extern void unload_point_cloud(struct point_cloud* ptr);

int main(int argc, char *argv[])
{

    // open "/dev/mem"
    int fd = open("/dev/mem", O_RDWR | O_SYNC);
    if (fd < 1) {
        perror("failed to open devfile");
        return -1;
    }

    // map FPGA physical address into user space
    int32_t *uio = (int32_t *)mmap(NULL, FPGA_ADDR_SIZE, PROT_READ|
            PROT_WRITE, MAP_SHARED, fd, FPGA_ADDR_START);
    if (uio == MAP_FAILED) {
        perror("failed to mmap");
        close(fd);
        return -1;
  }

    /*
     * CGH calculation
     */
    int ret = 0; // for error
    {
        const double lambda = 0.000000532;
        const double pixel = 0.000008000;

        // cgh memory buffer for display
        unsigned char *cgh_buffer = (unsigned char *)calloc(1920*1080, sizeof(unsigned
                char));
        if (cgh_buffer == NULL) {
            printf("faild to make cgh buffer");
            ret = -1;
            goto FAILD_CALLOC_BUFFER;
        }

```

```
        // load point-cloud data from file
        struct point_cloud *pcd;
        int num_point;

        pcd = load_point_cloud(argv[1], &num_point);
        if (pcd == NULL) {
            printf("faild to load point-cloud data");
            ret = −1;
            goto FAILD_LOAD_POINT_CLOUD;
        }

        // convert fixed data
        for (int i = 0; i < num_point; i++) {
            pcd_fixed[i].x = (int32_t)pcd[i].x;
            pcd_fixed[i].y = (int32_t)pcd[i].y;
            pcd_fixed[i].rho = (int32_t)(pow(2.,32.)∗(pixel/(2.∗pcd[i].z)/lambda));
        }

        // verify that the FPGA is idle
      if((0x2&uio[0])==0x2){
            printf("FPGA is not idle");
            ret = −1;
            goto FAILD_VERIFY_IDLE;
        }

        // write point-cloud data from ARM CPU to fpga
        for (int i = 0; i < num_point; i++){
            uio[128] = pcd_fixed[i].x;
            uio[128] = pcd_fixed[i].y;
            uio[128] = pcd_fixed[i].rho;
      }

        // calculation start
        uio[0] = 0x1;

        for (int buffer_i = 0; buffer_i<1920∗1080; buffer_i=buffer_i+FPGA_PIPELINE){
        // wait for the pipeline to end
            while (0x2&uio[0]);
            // read CGH from FPGA to ARM CPU
            for (int i = 0; i < FPGA_PIPELINE; i++)
                cgh_buffer[buffer_i+i] = (unsigned char)uio[10+i];
        }
FAILD_VERIFY_IDLE:
        unload_point_cloud(pcd);
FAILD_LOAD_POINT_CLOUD:
        free(cgh_buffer);
    }

FAILD_CALLOC_BUFFER:
    // cleanup
    munmap((void∗)FPGA_ADDR_START, FPGA_ADDR_SIZE);
    close(fd);

```

```
    return ret;

}
```

Table 19.1 Comparison of computation time

Hardware	Calculation time [s]	Acceleration ratio
Zynq ZCU102	0.066	28.1
NVIDIA Jetson TX1	1.294	1.44
Intel Xeon	1.86	1.00

19.6 Comparison of Computation Times

The Zynq UltraScale+ MPSoC ZCU102 Evaluation Kit (ZCU102) offered by Xilinx was used for the implementation. A CPU (Intel Xeon CPU E5-2697 v2 2.70GHz) and integrated graphics processing unit (GPU) of NVIDIA Jetson TX1 were employed for comparison. Table 19.1 reveals the time needed to compute a CGH of 1,920 × 1,080 pixels from a point cloud of 6,500 points.

The Zynq ZCU102 has 810 pipelines ("cgh_c" units) and a 250 MHz operating frequency. This is 28 times the CPU's speed and 20 times the integrated GPU's speed.

19.7 Discussion

For simplicity of explanation, we have described the FPGA implementation of Eq. (19.3) in this chapter. However, by implementing a more hardware-appropriate computation algorithm, we can further stimulate CGH computation. For more details, please refer to the references [11, 13].

Funding This work was supported by JSPS KAKENHI Grant Number JP21K21294.

References

1. T. Ito, T. Yabe, M. Okazaki, and M. Yanagi: Special-purpose computer HORN-1 for reconstruction of virtual image in three dimensions. Comput. Phys. Commun. **82**, 104–110 (1994).
2. T. Ito, H. Eldeib, K. Yoshida, S. Takahashi, T. Yabe, and T. Kunugi: Special-purpose computer for holography HORN-2. Comput. Phys. Commun. **93**, 13–20 (1996).
3. T. Shimobaba, N. Masuda, T. Sugie, S. Hosono, S. Tsukui, and T. Ito: Special-purpose computer for holography HORN-3 with PLD technology. Comput. Phys. Commun. **130**, 75–82 (2000).

4. T. Shimobaba, S. Hishinuma, and T. Ito: Special-purpose computer for holography HORN-4 with recurrence algorithm. Comput. Phys. Commun. **148**, 160–170 (2002).
5. T. Ito, N. Masuda, K. Yoshimura, A. Shiraki, T. Shimobaba, and T. Sugie: Special-purpose computer HORN-5 for a real-time electroholography. Opt. Express **13**, 1923–1932 (2005).
6. Y. Ichihashi, H. Nakayama, T. Ito, N. Masuda, T. Shimobaba, A. Shiraki, and T. Sugie: HORN-6 special-purpose clustered computing system for electroholography. Opt. Express **17**, 13895–13903 (2009).
7. N. Okada, D. Hirai, Y. Ichihashi, A. Shiraki, T. Kakue, T. Shimababa, N. Masuda, and T. Ito: Special-purpose computer HORN-7 with FPGA technology for phase modulation type electro-holography. *Proc. IDW*, 1284-1287 (2012)
8. T. Sugie, T. Akamatsu, T. Nishitsuji, R. Hirayama, N. Masuda, H. Nakayama, Y. Ichihashi, A. Shiraki, M. Oikawa, N. Takada et al.: High-performance parallel computing for next-generation holographic imaging. Nat. Elec. **1**, 254–259 (2018).
9. T. Nishitsuji, Y. Yamamoto, T. Sugie, T. Akamatsu, R. Hirayama, H. Nakayama, T. Kakue, T. Shimobaba, and T. Ito: Special-purpose computer HORN-8 for phase-type electro-holography. Opt. Express **26**, 26722–26733 (2018).
10. Y. Yamamoto, H. Nakayama, N. Takada, T. Nishitsuji, T. Sugie, T. Kakue, T. Shimobaba, and T. Ito: Large-scale electroholography by HORN-8 from a point-cloud model with 400,000 points. Opt. Express **26**, 34259–34265 (2018).
11. Y. Yamamoto, N. Masuda, R. Hirayama, H. Nakayama, T. Kakue, T. Shimobaba, and T. Ito: Special-purpose computer for electroholography in embedded systems. OSA Continuum **2**, 1166–1173 (2019).
12. Y. Yamamoto, S. Namba, T. Kakue, T. Shimobaba, T. Ito, and N. Masuda: Special-purpose computer for digital holographic high-speed three-dimensional imaging. Opt. Engineering **59**, 054105 (2020).
13. Y.Yamamoto, T. Shimobaba, H.Nakayama, T.Kakue, N.Masuda, and T. Ito: System-on-a-chip-based special-purpose computer for phase electroholography. OSA Continuum **3**, 3407–3415 (2020).
14. T. Shimobaba, N. Masuda, and T. Ito: Simple and fast calculation algorithm for computer-generated hologram with wavefront recording plane. Opt. Lett. **34**, 3133–3135 (2009).
15. SoCs with Hardware and Software Programmability, Xilinx. https://japan.xilinx.com/products/silicon-devices/soc/zynq-7000.html
16. Zynq UltraScale+ MPSoC, Xilinx. https://www.xilinx.com/products/silicon-devices/soc/zynq-ultrascale-mpsoc.html
17. Vivado, Xilinx. https://www.xilinx.com/products/design-tools/vivado.html
18. ug995 vivado ip subsystems tutorial, Xilinx. https://www.xilinx.com/support/documents/sw_manuals/xilinx2022_1/ug995-vivado-ip-subsystems-tutorial.pdf
19. ug974 vivado ultrascale libraries, Xilinx. https://0x04.net/~mwk/xidocs/lib/ug974-vivado-ultrascale-libraries.pdf

Chapter 20
Special-Purpose Computer for Digital Holography

Nobuyuki Masuda and Ikuo Hoshi

Abstract Digital holography is a technique that digitally captures holograms using an image sensor. Because it can record the three-dimensional (3D) information of target objects, digital holography can be used for 3D measurement. This chapter introduces the implementation of a field programmable gate array (FPGA) for accelerating digital holography calculation.

20.1 Numerical Diffraction Calculation

The wave of diffracted light from a hologram is calculated using **Fresnel–Kirchhoff diffraction**:

$$\phi(x_i, y_i) = \frac{1}{i\lambda} \int_{-\infty}^{\infty} \int_{-\infty}^{\infty} I_\alpha \frac{\exp(ikr_{\alpha i})}{r_{\alpha i}} dx_\alpha dy_\alpha, \tag{20.1}$$

$$r_{\alpha i} = \sqrt{(x_\alpha - x_i)^2 + (y_\alpha - y_i)^2 + z_i^2}, \tag{20.2}$$

where $i = \sqrt{-1}$, $\phi(x_i, y_i)$ is the complex amplitude of the object light on a plane (x_i, y_i) at a distance z_i, λ is the wavelength of the reference light, $I_\alpha(x_\alpha, y_\alpha)$ denotes the intensity of the hologram, and k is the wavenumber of the reference light. Fur-

---

Supplementary Information The online version contains supplementary material available at https://doi.org/10.1007/978-981-99-1938-3_20.

N. Masuda (✉)
Department of Applied Electronics, Tokyo University of Science, 6-3-1 Niijuku, Katsushika-ku, Tokyo 125-8585, Japan
e-mail: n-masuda@rs.tus.ac.jp

I. Hoshi
Radio Research Institute, National Institute of Information and Communications Technology (NICT), 4-2-1 Nukuikitamachi, Koganei, Tokyo 184-8795, Japan
e-mail: hoshi@nict.go.jp

T. Shimobaba and T. Ito (eds.), *Hardware Acceleration of Computational Holography*,
https://doi.org/10.1007/978-981-99-1938-3_20

thermore, using the **Fresnel approximation** (assuming that the hologram size is sufficiently small compared with the distance of the z_i), we obtain

$$\phi\left(x_i, y_i\right)=\int_{-\infty}^{\infty} \int_{-\infty}^{\infty} I\left(x_\alpha, y_\alpha\right) g\left(x_i-x_\alpha, y_i-y_\alpha\right) d x_\alpha d y_\alpha, \tag{20.3}$$

where $g\left(x_i-x_\alpha, y_i-y_\alpha\right)$ is

$$g(x_i-x_\alpha, y_i-y_\alpha)=\frac{\exp(ikz_i)}{i\lambda z_i}\exp\left[\frac{ik}{2z_i}\{(x_i-x_\alpha)^2+(y_i-y_\alpha)^2\}\right]. \tag{20.4}$$

Equation (20.3) can be expressed in the form of a two-dimensional (2D) **convolution**:

$$\phi\left(x_i, y_i\right)=\mathcal{F}^{-1}\left[\mathcal{F}\left[I\left(x_\alpha, y_\alpha\right)\right] G(\mu, \nu)\right], \tag{20.5}$$

where $\mathcal{F}\left[\cdot\right]$ and $\mathcal{F}^{-1}\left[\cdot\right]$ denote the Fourier and its inverse transforms, respectively, and $G(\mu, \nu)$ is the Fourier transform of $g(x, y)$.

Considering a discretized version of Eq. (20.5), the Fourier transforms can be performed using fast Fourier transforms (FFTs) as well as the sampling periods ΔP in the spatial domain and $\Delta f = 1/(N\Delta P)$ in the frequency domain where the sampling number of $N \times N$, $G(\mu, \nu)$, can be rewritten as

$$\begin{aligned} G(n, m) &= \exp\left[ikz_i - i\pi\lambda z_i\left\{\left(\frac{n}{N\Delta P}\right)^2+\left(\frac{m}{N\Delta P}\right)^2\right\}\right] \\ &= \exp(ikz_i)\exp\left\{2\pi i\left(\frac{-\lambda z_i}{2N^2(\Delta P)^2}\right)(n^2+m^2)\right\}, \end{aligned} \tag{20.6}$$

where n and m denote the integer frequency coordinates.

In a case that the 3D reconstruction is reproduced by stacking the *intensity* images obtained at different propagation distances, the term $\exp(ikz_i)$ can be regarded as a constant and be omitted. Therefore, Eq. (20.6) can be rewritten as

$$G(n, m) = \exp\left\{2\pi i\left(\frac{-\lambda z_i}{2N^2(\Delta P)^2}\right)(n^2+m^2)\right\}. \tag{20.7}$$

20.2 Special-Purpose Computer System

In digital holography calculations, the computational load is particularly heavy during the FFT. The FFT is a simple calculation that repeats butterfly operations, making it suitable for hardware. In addition, the faster the camera, the smaller the number of pixels in the image sensor, so less memory is required during computation. These two facts work in favor of implementing a **special-purpose computer** for **digital holography** using a **field-programmable gate array (FPGA)** [1–4].

In this study, holograms of 128 × 128 pixels were used. In this case, the FPGA evaluation board used in this study can implement a special-purpose computer for digital holography using only internal memories. As a result, the overhead observed when using external memories can be eliminated, and very fast computation can be implemented. In this section, the special-purpose computer is described.

20.2.1 Calculation Circuit

The special-purpose computer was developed using the Virtex-7 FPGA VC707 Evaluation Kit (hereafter VC707) provided by Xilinx. This is an evaluation board with the high-performance FPGA chip Virtex-7 XC7VX485T-2FFG1761 C. Figure 20.1 and Table 20.1 show the appearance of VC707 and the specifications of the FPGA chip, respectively.

Fig. 20.1 Xilinx Virtex-7 FPGA VC707 Evaluation Kit

Table 20.1 Specifications of the FPGA chip

Evaluation kit	VC707
Family name	Virtex7
Device	XC7VX485T-2FFG1761C
Number of registers	60,720
Number of LUTs	30,360
Number of block RAMs	1,030

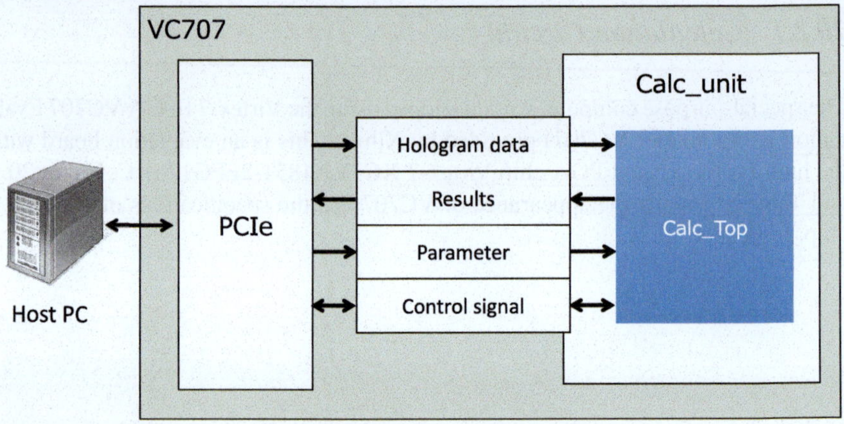

Fig. 20.2 Block diagram of the special-purpose computer system

20.2.2 Data Formats

Hologram data is represented in the 8-bit **fixed-point format**, but for the convenience of implementing the special-purpose computer, the hologram data is converted to a complex value with the real and imaginary parts of 16-bit fixed-point formats and 32 bits in total bit width. The calculated results have a total of 32 bits with the real and imaginary parts.

20.2.3 Overview of Special-Purpose Computer System

A block diagram of the special-purpose computer is shown in Fig. 20.2. This computing system uses **PCI Express** (hereinafter referred to as PCIe) for communication between the host PC and VC707. The host PC is a consumer computer with a PCIe slot. The host PC sends hologram data and parameters necessary for the calculation to VC707, controls the operations, receives the calculation results, and finally calculates the light intensity of complex amplitudes

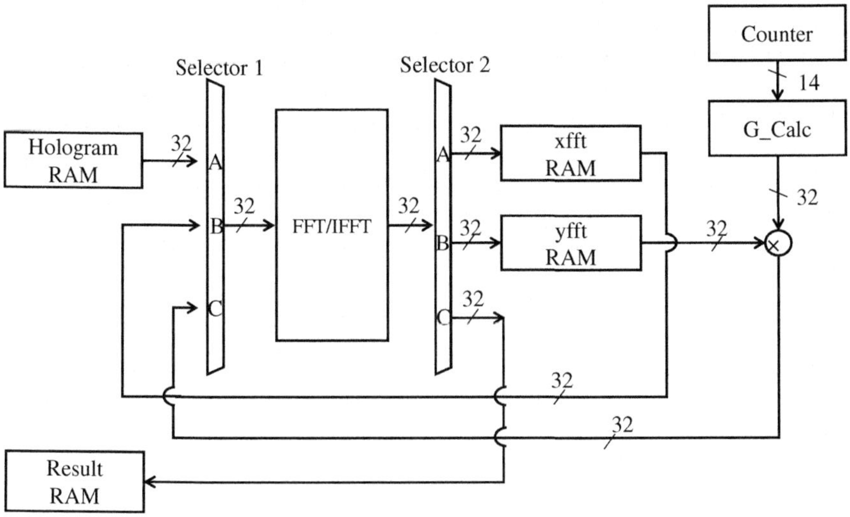

Fig. 20.3 Block diagram of Calc_Top that calculates the Fresnel diffraction of Eq. (20.5)

In the calculation circuit in the FPGA, the calculation circuit sequentially calculates one reconstructed image per input hologram but performs each step of the Fresnel diffraction in a pipeline fashion. In addition, the dual port random access memory (RAM) is implemented for temporary storage to separate the communication and calculation parts. A reconstructed image is saved in the save RAM and then sent to the host PC via PCIe. The communication circuit is designed to use burst transfers of PCIe to increase the communication performance.

The circuit can calculate reconstructed images at any wavelength and depth position by appropriately setting the parameters sent to the FPGA. The intensity image of the reconstructed data (complex values) calculated by the calculation circuit can be obtained on the host PC.

20.2.4 Overview of Calc_Top

Figure 20.3 shows a block diagram of Calc_Top that calculates the Fresnel diffraction of Eq. (20.5). The numbers on the diagonal lines in the figure represent the word length of the data, and the data are expressed in fixed-point formats.

First, the calculation flow in the FPGA chip is explained. Hologram data is saved in "Hologram RAM". As previously explained, each pixel of the hologram has a complex amplitude with 32 bits (16-bit real and 16-bit imaginary parts). Since the hologram has 128 $\times$ 128 pixels, 14 bits is sufficient for the address space because 7 bits (=128) $\times$ 7 bits (=128). In this circuit, the **intellectual property core (IP core)** of "Fast Fourier Transform v7.1" provided by Xilinx is used to perform one-

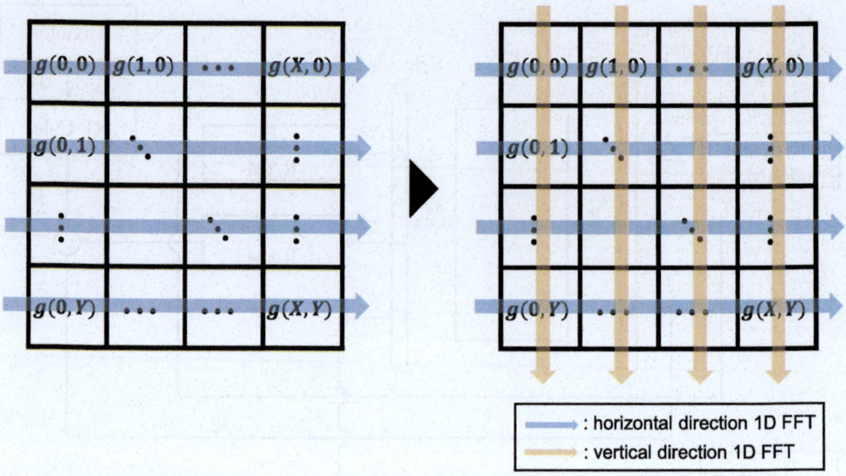

Fig. 20.4 Two-dimensional FFT using horizontal and vertical one-dimensional FFTs

dimensional (1D) FFT and inverse FFT (IFFT). For the Fresnel diffraction, it is necessary to perform a 2D FFT and IFFT. The 2D FFT transform can be calculated as 1D FFTs in each of the two directions, as shown below:

$$\begin{aligned}\Phi(n,m) &= \sum_{y=1}^{N_y}\sum_{x=1}^{N_x}\phi(x,y)\exp\left\{-2\pi i\left(\frac{nx}{N_x}+\frac{my}{N_y}\right)\right\}\\ &= \sum_{y=1}^{N_y}\left\{\sum_{x=1}^{N_x}\phi(x,y)\exp\left(\frac{-2\pi inx}{N_x}\right)\right\}\exp\left(\frac{-2\pi imy}{N_y}\right),\end{aligned} \tag{20.8}$$

where N_x and N_y represent the numbers of pixels in the horizontal and vertical directions, respectively. Figure 20.4 shows the 2D FFT using horizontal and vertical 1D FFTs. A 2D IFFT can be calculated in the same manner.

In calculations of FFT and IFFT, **overflow** can occur because the FFT accumulates complex values, as shown in Eq. (20.8). There are two methods to prevent overflow. One method is to ensure a sufficient data width to prevent overflow. However, this method requires more hardware resources. The other method is scaling to prevent the increase in data width by multiplying the calculation result with a constant less than one. The **scaling** technique can effectively save hardware resources. The IP cores described in this chapter can specify a scaling value and the timing of its multiplication. This is called a scaling schedule. The scaling schedule method was used for the special-purpose computer introduced in this chapter. The scaling schedule is set from the host PC.

Next, the reconstruction calculation method at different depths is explained. The distance z_i between the hologram and the reconstructed plane affects only $G(n, m)$ in Eq. (20.7). Therefore, the circuit only recalculates $G(n, m)$ at the given distance z_i and multiplies the recalculated $G(n, m)$ with the Fourier transform $\hat{I}(n, m)$ of a hologram. Repeating this process, the circuit outputs stacked reconstructions at different depths.

The module Calc_Top is controlled by a state machine with four states besides an idle state (IDLE). The roles of each state are as follows.

- STATE1: Read hologram data from Hologram RAM and perform 1D FFT (horizontal direction). Then, save the result to xfft RAM in the module Calc_Top. At this time, the scaling schedule is given to the FFT core. Repeat this for 128 rows. In this state, the initial distance of z_i is set.
- STATE2: The FFT result in STATE1 is vertically read from xfft RAM, 1D FFT is performed in the vertical data, and the results are saved in yfft RAM in the module Calc_Top. At this time, the scaling schedule is given to the FFT core. Repeat this for 128 columns. In this state, the interval dz between adjacent reconstructed planes is read.
- STATE3: The FFT calculation result in STATE2 is read from yfft RAM and multiplied by $G(n, m)$. Then, 1D IFFT is performed and the result is saved in xfft RAM in the module Calc_Top. At this time, the scaling schedule is given to the FFT core. Repeat this for 128 rows.
- STATE4: The FFT calculation result in STATE3 is vertically read from xfft RAM, 1D IFFT is performed on the vertical data, and the result is saved in the result RAM in the module Calc_Top. At this time, the scaling schedule is given to the FFT core. Repeat this for 128 columns.

Each state starts after the previous state processing is completed. Within each state, data for 128×128 pixels is continuously processed in a pipeline fashion. Listing 20.1 shows the very high-speed integrated circuit (VHSIC) **hardware description language (VHDL)** implementation of the module "Calc_top".

Listing 20.1 VHDL implementation of Calc_top

```
entity Calc_top is
    Port ( clk :in STD_LOGIC;
            rst :in STD_LOGIC;
            calc_start :in STD_LOGIC;
            fin_rst :in STD_logic;
            holo_ram_doutb :in STD_LOGIC_VECTOR (31 downto 0);
            --Upper16bit:Phi_Im / Lower16bit:Phi_re
            holo_ram_addrb :out STD_LOGIC_VECTOR (13 downto 0);
            calc_ram_web :out STD_LOGIC_vector(0 downto 0);
            calc_ram_addrb :out STD_LOGIC_VECTOR (13 downto 0);
            calc_ram_dinb :out STD_LOGIC_VECTOR (31 downto 0);
            --Upper16bit:Phi_Im / Lower16bit:Phi_re
            calc_fin :out STD_LOGIC;
            calc_stage :out STD_LOGIC_VECTOR(3 downto 0);
            fft_ovflo :out STD_LOGIC;
            ram2_addrb :out STD_LOGIC_VECTOR (13 downto 0);
```

```
            ram2_doutb :in STD_LOGIC_VECTOR (31 downto 0)
            );
end Calc_top;

architecture Behavioral of Calc_top is

    --cntrl signal
    signal s_cnt :std_logic_vector(13+1 downto 0):=(others=>'0');
    signal d_cnt :std_logic_vector( 13 downto 0):=(others=>'0');
    signal g_cnt :std_logic_vector( 13 downto 0):=(others=>'0');
    signal s_cnt2 :std_logic_vector( 13 downto 0):=(others=>'0');
    --fft
    signal fft_scale_sch_sig :STD_LOGIC_VECTOR(31 DOWNTO 0);
    --(yIFFT/xIFFT/yFFT/xFFT) Bit shifted and used in this order
    signal fft_start :std_logic;
    signal fft_fwd_inv :std_logic;
    signal fft_xn_re :std_logic_vector (15 downto 0);
    signal fft_xn_im :std_logic_vector (15 downto 0);
    signal fft_xk_re :std_logic_vector (15 downto 0);
    signal fft_xk_im :std_logic_vector (15 downto 0);
    signal fft_dv :std_logic;
    --xfft_output
    signal xfft_ram_addra : STD_LOGIC_VECTOR(13 DOWNTO 0);
    signal xfft_ram_dina : STD_LOGIC_VECTOR(31 DOWNTO 0);
    signal xfft_ram_addrb : STD_LOGIC_VECTOR(13 DOWNTO 0);
    signal xfft_ram_doutb : STD_LOGIC_VECTOR(31 DOWNTO 0);
    signal xfft_ram_wea : STD_LOGIC_VECTOR (0 downto 0);
    --yfft_output
    signal yfft_ram_addra : STD_LOGIC_VECTOR(13 DOWNTO 0);
    signal yfft_ram_dina : STD_LOGIC_VECTOR(31 DOWNTO 0);
    signal yfft_ram_addrb : STD_LOGIC_VECTOR(13 DOWNTO 0);
    signal yfft_ram_doutb : STD_LOGIC_VECTOR(31 DOWNTO 0);
    signal yfft_ram_wea : STD_LOGIC_VECTOR (0 downto 0);
    --G
    signal zparam_sig :STD_LOGIC_VECTOR(31 DOWNTO 0);
    signal G_re :std_logic_vector (15 downto 0);
    signal G_im :std_logic_vector (15 downto 0);
    --cmplx_mult
    signal cmplx_re :STD_LOGIC_VECTOR(15 downto 0);
    signal cmplx_im :STD_LOGIC_VECTOR(15 downto 0);

    COMPONENT Calc_G
        PORT ( clk : in STD_LOGIC;
        n : in STD_LOGIC_VECTOR (6 downto 0);
        m : in STD_LOGIC_VECTOR (6 downto 0);
        zparam : in STD_LOGIC_VECTOR (31 downto 0);
        G_re : out STD_LOGIC_VECTOR (15 downto 0);
        G_im : out STD_LOGIC_VECTOR (15 downto 0));
    END COMPONENT;

    COMPONENT IP_fft_core_128
      PORT (
```

```
            clk : IN STD_LOGIC;
            start : IN STD_LOGIC;
            xn_re : IN STD_LOGIC_VECTOR(15 DOWNTO 0);
            xn_im : IN STD_LOGIC_VECTOR(15 DOWNTO 0);
            fwd_inv : IN STD_LOGIC;
            fwd_inv_we : IN STD_LOGIC;
            scale_sch : IN STD_LOGIC_VECTOR(7 DOWNTO 0);
            scale_sch_we : IN STD_LOGIC;
            rfd : OUT STD_LOGIC;
            xn_index : OUT STD_LOGIC_VECTOR(6 DOWNTO 0);
            busy : OUT STD_LOGIC;
            edone : OUT STD_LOGIC;
            done : OUT STD_LOGIC;
            dv : OUT STD_LOGIC;
            xk_index : OUT STD_LOGIC_VECTOR(6 DOWNTO 0);
            xk_re : OUT STD_LOGIC_VECTOR(15 DOWNTO 0);
            xk_im : OUT STD_LOGIC_VECTOR(15 DOWNTO 0);
            ovflo : OUT STD_LOGIC
          );
        END COMPONENT;

        COMPONENT IP_RAM_32bit
          PORT (
            clka : IN STD_LOGIC;
            wea : IN STD_LOGIC_VECTOR(0 DOWNTO 0);
            addra : IN STD_LOGIC_VECTOR(13 DOWNTO 0);
            dina : IN STD_LOGIC_VECTOR(31 DOWNTO 0);
            clkb : IN STD_LOGIC;
            addrb : IN STD_LOGIC_VECTOR(13 DOWNTO 0);
            doutb : OUT STD_LOGIC_VECTOR(31 DOWNTO 0)
          );
        END COMPONENT;

        COMPONENT IP_cmplx_mult
            port (
            ar: in std_logic_vector(15 downto 0);
            ai: in std_logic_vector(15 downto 0);
            br: in std_logic_vector(15 downto 0);
            bi: in std_logic_vector(15 downto 0);
            clk: in std_logic;
            pr: out std_logic_vector(15 downto 0);
            pi: out std_logic_vector(15 downto 0));
        END COMPONENT;

        --state machine
        type CALC_STATE is (IDLE,STAGE1,STAGE2,STAGE3,STAGE4);
        signal NEXT_STAGE : CALC_STATE := IDLE;

begin

main:process(clk)
    begin
        if(clk'event and clk ='1')then
```

```
        case NEXT_STAGE is

         when IDLE =>
            --init
            calc_stage <= "0000";
            s_cnt <= (others=>'0');--for read
            d_cnt <= (others=>'0');
            g_cnt <= (others=>'0');--for write
            s_cnt2 <= (others=>'0');--for read (xifft)
            --holo data ram
            holo_ram_addrb <= (others=>'0');
            --calc result ram
            calc_ram_web <= "0";
            calc_ram_addrb <= (others=>'0');
            --paramater
            ram2_addrb <= "00000000000000";
            fft_scale_sch_sig <= ram2_doutb; --Determine the value of FFT scaling
            -----------------------------------------------------------
            --fft ram
            xfft_ram_wea <= "0";
            xfft_ram_addra <= (others=>'0');
            xfft_ram_addrb <= (others=>'0');
            yfft_ram_wea <= "0";
            yfft_ram_addra <= (others=>'0');
            yfft_ram_addrb <= (others=>'0');
            -----------------------------------------------------------
            if(fin_rst='1')then
               calc_fin <= '0';
            elsif(calc_start = '1')then
               NEXT_STAGE <= STAGE1;
            else
               NEXT_STAGE <= IDLE;
            end if;
---------------------------------------STAGE1 start
        when STAGE1 => --xfft
           --init
           calc_stage <= "0001";
           fft_fwd_inv <= '1'; --FFT
           --state count
           s_cnt <= s_cnt + '1';
           --xfft input
           holo_ram_addrb <= s_cnt(13 downto 0);
           fft_xn_re <= holo_ram_doutb(15 downto 0);
           --Starts with 2 clk delay from read due to the PCIe RAM.
           fft_xn_im <= holo_ram_doutb(31 downto 16);
           --xfft output
           xfft_ram_addra <= d_cnt;
```

```
                xfft_ram_dina <= fft_xk_im & fft_xk_re;
                --fft control
                if(s_cnt = "011111110000010")then
                --Stop just before inputting the last 128 hologram data (to de-assert the dv signal)
                 fft_start <= '0';
                elsif(s_cnt = "00000000000001")then
                --holo data Delayed start signal because it takes 2 clk to read RAM
                 fft_start <= '1';
                end if;
                --ram control
                if(fft_dv='1')then
                   xfft_ram_wea <= "1";
                    d_cnt <= d_cnt + '1';
                end if;
                if(s_cnt = "100000101010100")then --finish write ram
                    s_cnt <= (others=>'0'); --reset before next stage
                    d_cnt <= (others=>'0');
                    xfft_ram_wea <= "0";
                    xfft_ram_addra <= (others=>'0');
                    fft_scale_sch_sig(7 downto 0) <= fft_scale_sch_sig(15 downto 8); --Shift 8 bits to adjust the scaling value to the yFFT
                    NEXT_STAGE <= STAGE2; --Transition to yfft calculation
                end if;
--------------------------------------------STAGE1 end
--------------------------------------------STAGE2 start
            when STAGE2 => --yfft

                --init
                calc_stage <= "0010";
                --state cnt
                s_cnt <= s_cnt + '1';
                --yfft input
                xfft_ram_addrb <= s_cnt(6 downto 0) & s_cnt(13 downto 7); --vertical reading
                fft_xn_re <= xfft_ram_doutb(15 downto 0);
                fft_xn_im <= xfft_ram_doutb(31 downto 16);
                --yfft output
                yfft_ram_addra <= d_cnt;
                yfft_ram_dina <= fft_xk_im & fft_xk_re;
                --yfft control
                if(s_cnt = "011111110000001")then
                --Stop just before entering the last 128 pieces of data
                    fft_start <= '0';
                 elsif(s_cnt = "00000000000000")then
                    fft_start <= '1';
                end if;
                 --ram control
                if(fft_dv='1')then
```

```
                    yfft_ram_wea <= "1";
                    d_cnt <= d_cnt + '1';
                end if;
                if(s_cnt = "100000101010011")then --finish write ram
                  s_cnt <= (others=>'0'); --reset before next stage
                  d_cnt <= (others=>'0');
                  yfft_ram_wea <= "0";
                  yfft_ram_addra <= (others=>'0');
                  fft_scale_sch_sig(7 downto 0) <= fft_scale_sch_sig(23 downto 16); --Shift 8 bits to the right to adjust the scaling value to xIFFT
                  NEXT_STAGE <= STAGE3;--Transition to G-multiplication
                end if;
                --zparam control
                ram2_addrb <= "00000000000001";
                zparam_sig <= ram2_doutb; --Determine zparam value during stage2
-------------------------------------STAGE2 end
-------------------------------------STAGE3 start
            when STAGE3=> -- mult G and xIFFT
                --init
                calc_stage <= "0100";
                fft_fwd_inv <= '0'; --IFFT
               --state cnt
                s_cnt <= s_cnt + '1';
                s_cnt2 <= s_cnt2 + '1';
                --calc G
                g_cnt <= g_cnt + '1';
                --cmplx mult input
                yfft_ram_addrb <= s_cnt2(13 downto 0);
                if(s_cnt = "000000000001000")then --1clk before G calculation output
                 s_cnt2 <= (others=>'0');
                end if;
                --xifft input
                fft_xn_re <= cmplx_re;
                fft_xn_im <= cmplx_im;
                --xifft output
                xfft_ram_addra <= d_cnt;
                xfft_ram_dina <= fft_xk_im & fft_xk_re;
                --xifft control
                if(s_cnt = "011111110001110")then
                    fft_start <= '0'; ----xIFFT end
                elsif(s_cnt = "00000000001101")then --after 3clk from cmplx start
                    fft_start <= '1'; ----xIFFT start
               end if;
                --ram control
                if(fft_dv='1')then
                    xfft_ram_wea <= "1";
                    d_cnt <= d_cnt + '1';
               end if;
```

```
            if(s_cnt = "100000101100000")then --finish write ram
              --reset before next stage
              s_cnt <= (others=>'0');
              d_cnt <= (others=>'0');
              xfft_ram_wea <= "0";
              xfft_ram_addra <= (others=>'0');
              fft_scale_sch_sig(7 downto 0) <= fft_scale_sch_sig(31 downto 24); --Shift a bit to adjust the scaling value to yIFFT
              NEXT_STAGE <= STAGE4; --Transition to yIFFT calculation
            end if;
------------------------------STAGE3 end
------------------------------STAGE4 start
          when STAGE4 => --yIFFT
            --init
            calc_stage <= "1000";
            --state cnt
            s_cnt <= s_cnt + '1';
            --yifft input
            xfft_ram_addrb <= s_cnt(6 downto 0) & s_cnt(13 downto 7); --縦読み込み
            fft_xn_re <= xfft_ram_doutb(15 downto 0);
            fft_xn_im <= xfft_ram_doutb(31 downto 16);
            --yifft output
            calc_ram_addrb <= d_cnt;
            calc_ram_dinb <= fft_xk_im & fft_xk_re;
            --yifft control
            if(s_cnt = "011111110000001")then
            --Stop just before entering the last 128 pieces of data
              fft_start <= '0';
            elsif(s_cnt = "00000000000000")then
              fft_start <= '1';
            end if;
            --ram control
            if(fft_dv='1')then
              calc_ram_web <= "1";
              d_cnt <= d_cnt + '1';
            end if;
            if(s_cnt = "100000101010011")then --finish write ram
              s_cnt <= (others=>'0'); --reset before next stage
              calc_ram_addrb <= (others=>'0');
              calc_ram_web <= "0";
              calc_fin <= '1'; --termination flag
              NEXT_STAGE <= IDLE;--Transition to IDLE
            end if;
-----------------------------------STAGE4 end
          when others => null;

              end case ;
          end if;
      end process;
```

```

    --inst RAM
    xfft_ram : IP_RAM_32bit
      PORT MAP (
        clka => clk,
        wea => xfft_ram_wea,
        addra => xfft_ram_addra,
        dina => xfft_ram_dina,
        clkb => clk,
        addrb => xfft_ram_addrb,
        doutb => xfft_ram_doutb
      );
    yfft_ram : IP_RAM_32bit
      PORT MAP (
        clka => clk,
        wea => yfft_ram_wea,
        addra => yfft_ram_addra,
        dina => yfft_ram_dina,
        clkb => clk,
        addrb => yfft_ram_addrb,
        doutb => yfft_ram_doutb
      );

    --inst xfft
    inst_FFT : IP_fft_core_128
      PORT MAP (
        clk => clk,
        start => fft_start,
        xn_re => fft_xn_re,
        xn_im => fft_xn_im,
        fwd_inv => fft_fwd_inv,--FFT 1 / IFFT 0
        fwd_inv_we => fft_start,
        scale_sch => fft_scale_sch_sig(7 downto 0),
        scale_sch_we => fft_start,
        rfd => open,
        xn_index => open,
        busy => open,
        edone => open,
        done => open,
        dv => fft_dv,
        xk_index => open,
        xk_re => fft_xk_re,
        xk_im => fft_xk_im,
        ovflo => fft_ovflo
       );

    --inst G
    inst_Calc_G : Calc_G
        PORT MAP (
          clk => clk,
          n => g_cnt(6 downto 0),
          m => g_cnt(13 downto 7),
          zparam => zparam_sig,
```

```
            G_re => G_re,
            G_im => G_im
        );

    --cmplx mult (G*2DFFT) ---latency 4 clk
    inst_cmplx_mult : IP_cmplx_mult
        port map (
            ar => yfft_ram_doutb(15 downto 0), --yFFT_RE
            ai => yfft_ram_doutb(31 downto 16), --yFFT_IM
            br => G_re,
            bi => G_im,
            clk => clk,
            pr => cmplx_re,
            pi => cmplx_im
        );

end Behavioral;
```

20.2.5 Overview of Calc_G

Figure 20.5 shows a block diagram of Calc_G in Fig. 20.3. This module calculates Eq. (20.7). The equation is shown again below:

$$G(n, m) = \exp(2\pi i\theta) = \exp\left\{2\pi i\left(\frac{-\lambda z_i}{2N^2(\Delta P)^2}\right)(n^2 + m^2)\right\}. \qquad (20.9)$$

From Euler's formula $\exp(2\pi i\theta) = \cos 2\pi\theta + i \sin 2\pi\theta$, the real and imaginary parts of $G(n, m)$ are calculated using "cos ROM (read only memory)" and "sin ROM". sin ROM and cos ROM store the $\sin 2\pi\theta$ and $\cos 2\pi\theta$ values as tables, respectively. Because of the periodic nature of trigonometric functions, the integer part of θ is unnecessary and only its decimal part should be calculated. "zparam" in Fig. 20.5 is $\frac{-\lambda z_i}{2N^2(\Delta P)^2}$ and has only a decimal part of 32-bit width. Since zparam

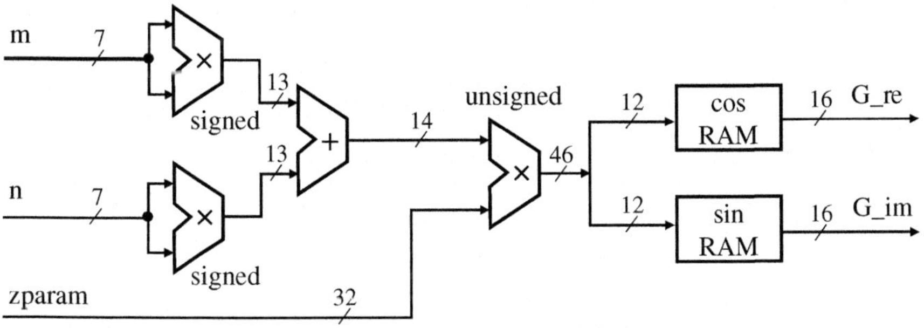

Fig. 20.5 Block diagram of Calc_G

requires complex calculations, zparam is calculated in advance on the host PC and transferred to the FPGA.

Next, the detailed operation of the circuit is described. The circuit receives n, m, and zparam, and then squares n and m, respectively. n and m are generated from the 14-bit binary counter in Fig. 20.3. The upper and lower 7 bits are assigned to m and n, respectively. In the case of 128×128, the result obtained by the discrete Fourier transform is the spectral intensity for -64 to 63 times the fundamental frequency. Therefore, to match the result of the Fourier transform, the square operation is a signed calculation in 2's complement. By this calculation, n and m can be set to $0, 1, 2, \ldots, 127$ instead of $0, 1, 2, \ldots, 63, -64, -63, \ldots, -1$. n^2 and m^2 are added and then multiplied with zparam. Finally, the cos and sin ROMs are used to obtain $\cos 2\pi\theta$ and $\sin 2\pi\theta$ values, respectively. The data width of the input is 12 bits, and that of the output is 16 bits. The $\cos 2\pi\theta$ and $\sin 2\pi\theta$ values can be calculated in one clock cycle. Listing 20.2 shows the VHDL implementation of the module "Calc_G".

Listing 20.2 VHDL implementation of Calc_G

```
entity Calc_G is
    port(
        clk : in std_logic;
        n : in std_logic_vector( 6 downto 0);
        m : in std_logic_vector( 6 downto 0);
        zparam : in std_logic_vector(31 downto 0);
        G_re : out std_logic_vector(15 downto 0);
        G_im : out std_logic_vector(15 downto 0)
    );
end Calc_G;

architecture Behavioral of Calc_G is

    -- Multiplier for n, m squared (signed 7bit x 7bit, latency
        = 3 ,ver11.2)
    component IP_mult_signed_7x7
        port (
            clk : in std_logic;
            a : in std_logic_vector( 6 downto 0);
            b : in std_logic_vector( 6 downto 0);
            p : out std_logic_vector(13 downto 0)
        );
    end component;

    signal n2, m2 : std_logic_vector(13 downto 0);
    signal n2_plus_m2 : std_logic_vector(13 downto 0);

    signal zparam_delay1 : std_logic_vector(31 downto 0);
    signal zparam_delay2 : std_logic_vector(31 downto 0);
    signal zparam_delay3 : std_logic_vector(31 downto 0);

    -- Multiplier to multiply zparam by (n^2 + m^2)
    -- (unsigned 32bit x 14bit, latency = 4,ver11.2)
    component IP_mult_unsigned32x14
        port (
```

```
            clk : in std_logic;
            a : in std_logic_vector(31 downto 0);
            b : in std_logic_vector(13 downto 0);
            p : out std_logic_vector(45 downto 0)
        );
    end component;

    signal mult_zparam_out : std_logic_vector(45 downto 0);
    signal theta : std_logic_vector(11 downto 0);

    -- cos ROM (addr 12bit, data 16bit,ver 7.3)
    component ipcosrom
        port (
            clka : in std_logic;
            addra : in std_logic_vector(11 downto 0);
            douta : out std_logic_vector(15 downto 0)
        );
    end component;

    -- sin ROM (addr 12bit, data 16bit)
    component ipsinrom
        port (
            clka : in std_logic;
            addra : in std_logic_vector(11 downto 0);
            douta : out std_logic_vector(15 downto 0)
        );
    end component;


begin

    -- n and m squared (clk 1-3)
    square_n : IP_mult_signed_7x7 port map (
        clk => clk,
        a => n,
        b => n,
        p => n2
    );

    square_m : IP_mult_signed_7x7 port map (
        clk => clk,
        a => m,
        b => m,
        p => m2
    );

    -- Calculate n^2 + m^2
    process (clk) begin
        if (clk'event and clk = '1') then
            n2_plus_m2 <= n2 + m2;--(clk 4)
        end if;
    end process;

```

```
-- Internal delay of zparam
process (clk) begin
    if (clk'event and clk = '1') then
        zparam_delay1 <= zparam;
        zparam_delay2 <= zparam_delay1;
        zparam_delay3 <= zparam_delay2;
        end if;
end process;


-- Multiply (n^2 + m^2) by zparam (clk 4-8)
--
mult_zparam : IP_mult_unsigned32x14 port map (
    clk => clk,
    a => zparam_delay3,
    b => n2_plus_m2,
    p => mult_zparam_out --clk 8
);


-- Rounding in theta (clk 9)
process (clk) begin
    if (clk'event and clk = '1') then
        theta <= mult_zparam_out(31 downto 20) + ("00000000000" &
                mult_zparam_out(19)); --(clk9)
    end if;
end process;


-- cos, sin ROM (clk 10)
cosrom : ipcosrom port map (
    clka => clk,
    addra => theta,
    douta => G_re
);

sinrom : ipsinrom port map (
    clka => clk,
    addra => theta,
    douta => G_im
);
end Behavioral;
```

20.2.6 Performance

In this section, to evaluate the performance of the special-purpose computer, calculations performed on the VC707 computer are compared with those of software alone, and the results are discussed. The communication interface between the FPGA evaluation board and the host PC was PCIe with a maximum bandwidth of 4 GB/s.

Table 20.2 Software environment

Motherboard	msi X995 SLI PLUS
CPU	Intel Core i7-5820K 3.30 GHz
Memory	16 GB (8 GB 2)
Operating system	CentOS Linux 7.1.1503
FPGA design software	ISE 14.4

Table 20.3 Summary of the hardware resource usage. Numbers in parentheses indicate hardware resource utilization

FPGA	
FPGA chip	Virtex7
Register	6,459 (1%)
LUT	5,325 (1%)
Block RAM	99 (9%)
Max frequency	277.362 MHz
Frequency	250 MHz

Table 20.2 shows the software environment for the FPGA evaluation board. Table 20.3 gives summary of the hardware resource usage for the FPGA chip.

From Table 20.3, it can be seen that there is room in circuit resources for the registers and LUTs. Therefore, the calculation can be accelerated by parallelizing the calculation circuit since the harware resources are still available.

20.2.7 Calculation Speed

The calculation speeds of the software alone and the computer system were evaluated. Each calculation time is shown in Table 20.4. The total processing time in the special-purpose computer, including communication time, was 31 ms, and the pure calculation time of Eq. (20.5) (excluding communication time) was only 2 ms. The software alone used FFTTW3.3.4 with six CPU threads, which is an open-source FFT library. Table 20.4 shows that the special-purpose computer was able to

Table 20.4 Calculation times of software alone and special-purpose computer

Processor	Calculation time [ms]	Acceleration ratio
Software alone	51.0	1.0
Special-purpose computer (total processing time)	31.0	1.65
Special-purpose computer (only internal processing time)	2.0	25.5

calculate Eq. (20.5) 1.65 times faster than the software alone when communication time was included and 25.5 times faster when communication time was excluded.

20.2.8 Reconstructed Images

Figure 20.6 shows the simulated hologram of two point objects. One point object is at (0.0[m], 0.0[m], 0.10[m]) and other is at (0.0[m], 0.0[m], 0.09[m]). Figures 20.7 and 20.8 show the reconstructed images calculated by the special-purpose computer. Figure 20.7 shows the reconstructed image at z = 0.10[m]. Figure 20.8 shows the image at z = 0.09[m]. The special-purpose computer was able to obtain correct reconstructions.

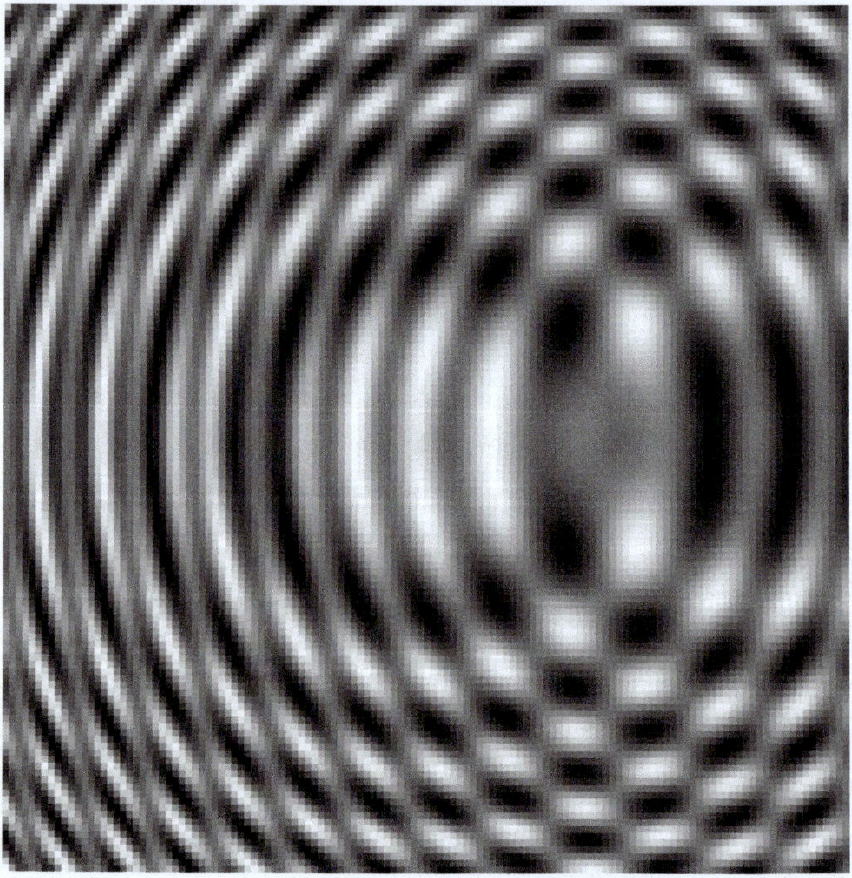

Fig. 20.6 Simulated hologram of two point objects

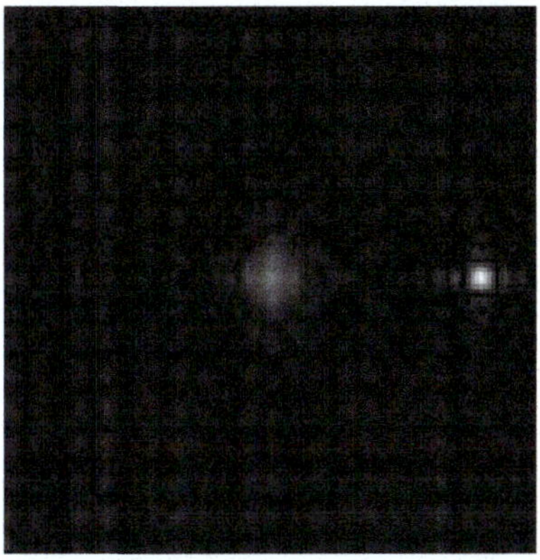

Fig. 20.7 Reconstructed image at z = 0.9[m]

Fig. 20.8 Reconstructed image at z = 1.0[m]

References

1. N. Masuda, T. Ito, K. Kayama, H. Kono, S. Satake, T. Kunugi, and K. Sato, "Special purpose computer for digital holographic particle tracking velocimetry," Opt. Express **14**, 587–592 (2006).
2. Y. Abe, N. Masuda, H. Wakabayashi, Y. Kazo, T. Ito, S. Satake, T. Kunugi, and K. Sato, "Special purpose computer system for flow visualization using holography technology," Opt. Express **16**, 7686–7692 (2008).
3. N. Masuda, T. Sugie, T. Ito, S. Tanaka, Y. Hamada, S. Satake, T. Kunugi, and K. Sato, "Special purpose computer system with highly parallel pipelines for flow visualization using holography technology," Comput. Phys. Commun. **181**, 1986–1989 (2010).
4. C.-J. Cheng, W.-J. Hwang, C.-T. Chen, and X.-J. Lai, "Efficient FPGA-based Fresnel transform architecture for digital holography," J. Disp. Technol. **10**, 272–281 (2013).

References

Index

A
Accurate PAS, 259
Advanced extensible interface (AXI), 106
Affine matrix, 216
Aliasing-free cone, 264
Amplitude hologram, 26, 29, 201
Analytical method, 215
Angular spectrum, 9
Angular spectrum method, 8, 131, 281
Atomic, 267

B
Barrier, 148
Binary CGH, 228
Binomial theorem, 27
Bjøntegaard-Delta peak signal-to-noise ratio (BD-PSNR), 278
Block, 73
Burst transfer, 61

C
Cache blocking, 63
Cache line, 63
Cache memory, 62, 157
Central processing unit (CPU), 47
Circular convolution, 120
Coefficient shrinking, 256
Coherence, 5
Compensate phase, 196
Complex amplitude, 5
Complex hologram, 202
Compressed sensing (CS), 241
Compressible, 242
Compressive holography, 242
Computational coherent superposition (CCS), 312
Computer-generated hologram (CGH), 26, 137, 169, 227, 329
Compute unified device architecture (CUDA), 67, 126, 267
Conjugate wave, 18
Convolution, 9, 119, 344
Convolution theorem, 9
Core, 57
CuFFT, 126

D
Data parallelization, 102
Dennis Gabor, 17
Depth camera, 193
Depth of field, 283
Device code, 72
Diffraction, 7
Digital holography, 31, 345
Digital signal processor (DSP), 98
Direction cosine, 4, 10
Double buffering, 233
Dynamic random access memory (DRAM), 49

F
Fast iterative shrinkage-thresholding algorithm (FISTA), 247
Fast-math operation, 170

T. Shimobaba and T. Ito (eds.), *Hardware Acceleration of Computational Holography*, https://doi.org/10.1007/978-981-99-1938-3

FFT shift, 120
FFTW, 123
Field-programmable gate array (FPGA), 97, 345
Fixed-point format, 346
Floating-point arithmetic, 104
Floor operator, 266
Fourier transform, 8
Fraunhofer diffraction, 14
Fraunhofer hologram, 21
Fresnel approximation, 27, 228, 344
Fresnel diffraction, 10, 119
Fresnel hologram, 21
Fresnel incoherent correlation holography (FINCH), 313
Fresnel–Kirchhoff diffraction, 343
Full-color hologram, 303
Full-field propagation, 281
Full widths at half maximum (FWHMs), 314

G
Gabor hologram, 244
Gather-scatter, 61
Gaussian random matrix, 243
Gaussian window, 258
Global memory, 231
GPU cluster, 228
Gradient projection, 248
Graphic processing unit (GPU), 67, 169, 227,, 242
Grating equation, 257

H
Hamming window, 258
Hann window, 258
Hardware description language (HDL), 98, 349
Hogels, 257
Hologram, 18
Holographic reconstruction (HORN), 329
Holographic stereogram, 257
Holography, 17
Host code, 72
Hyper-threading technology, 57

I
Ill-posed inverse problem, 245
Image hologram, 21
Impulse response, 9
Inclination factor, 8
Incoherence, 243
Incoherent digital holography (IDH), 309
In-line holography, 20
Integer type, 162
Integral photography (IP), 186
Intellectual property core (IP core), 347
Interference, 6
Interpolation, 213

J
Jacobian, 210
JPEG Pleno Holography, 273

K
Kernel, 73
Kernel functions, 72
Kinoform, 29

L
L1 cache, 160
Latency, 53, 101, 154
Layer image, 194
Layer method, 194
Lens array, 186
Light field, 257
Light-ray information, 185
Linear convolution, 120, 194
Liquid crystal phase modulator, 313
Logic synthesis, 98
Look-up table (LUT), 155, 170, 176, 332, 334
Lozenge cell, 264

M
Many-core CPU, 57
MATLAB, 211, 281
Memory caching, 263
Message passing interface (MPI), 231
Multi-core CPU, 57
Multidimension-multiplexed full-phase-encoding holography (MPH), 312
Multithread, 123, 146
Multiwavelength-multiplexed phase-shifting incoherent color digital holography (MP-ICDH), 313
Mutual coherence, 243

N
Node, 228
Non-locality, 273

No Operation (NOP), 51
Nyquist–Shannon sampling theorem, 241

O
Objective quality assessment, 277
Objective VQA, 277
Object wave, 18
Off-axis holography, 20, 34
OpenCL, 83
Open Computing Language, 83
OpenMP, 123, 145
Orthographic view reconstruction, 286
Overflow, 348

P
Parallel reduction, 79
Paraxial approximation, 11
PC cluster, 228
PCI Express, 346
Perspective reconstruction, 284
Phase-added stereogram (PAS), 257
Phase-only hologram, 29, 201
Phase-shifting digital holography, 38
Phase space, 285
Piezo actuator, 313
Pipeline, 230
Pipeline bubble, 51
Pipeline delay, 50
Pipeline depth, 50
Pipeline parallelization, 102
Pipeline processing, 48
Pipeline stage, 50
Pipeline stall, 51
Plan, 123
Plane wave, 5
Plane wave expansion method, 8
Plenoptic function, 257
Plenoptic imaging, 274
Point cloud, 25, 258
Point-cloud method, 201
Point spread function (PSF), 258
Polarization-imaging camera, 320
Polygon-based hologram, 207
Pragma directive, 146
Programmable shader, 69
Proximal operator, 248
Python, 261

Q
Quantitative phase imaging (QPI), 318

R
Random phase, 196
Rank, 231
Ray-tracing method, 171
Recurrence formula, 27, 151
Reduction, 78, 149
Register, 48
Restricted isometry property (RIP), 243
RGB-D, 193
Rotation matrix, 210
Rounding operator, 261

S
Sampling theorem, 36
Scalar approximation, 3
Scalar computer, 55
Scaling, 348
Sectional image, 196
Self-reference DH (SRDH), 310, 318
Shading language, 69
Shared memory, 231
Shifted and scaled diffraction, 15
Short-time Fourier transform (STFT), 257
Single instruction multiple data (SIMD), 54, 145
Single-shot full-color off-axis DH, 304
Single-shot phase-shifting digital holography, 41
Single-shot phase-shifting incoherent digital holography (SSPS-IDH), 310
Single-shot phase-shifting (SSPS), 309
Soft thresholding, 248
Sommerfeld diffraction, 8
Sparse, 241
Sparse CGH, 255
Sparseness significance ranking measure, 279
Sparsity, 253, 263
Spatial coherence, 5
Spatial frequency, 36
Spatial frequency-division multiplexing, 304
Special-purpose computer, 329, 345
Speckle noise, 274
Spherical wave, 5
Split-LUT (S-LUT), 176
Superscalar computer, 55
Synchronize, 78
SystemVerilog, 99, 334

T
Taylor approximation, 260

Taylor expansion, 10, 27
Temporal coherence, 5
Thread, 73
3D affine transformation, 221
Throughput, 53, 101
Total variation (TV), 245
Transfer function, 9
Twin-image problem, 34

V
Vector processor, 55
Versatile similarity metric (VSM), 279
VHDL, 99, 349
Visual quality, 272
Visual quality assessment (VQA), 272

W
Wavefront recording plane (WRP), 257
Wave vector, 4, 35
Window function, 258
Wraparound effect, 120

Z
Zero padding, 120, 194, 212
Zeroth-order diffraction, 18
Zone plate, 258

The manufacturer's authorised representative in the EU is Springer Nature Customer Service Centre GmbH, Europaplatz 3, 69115 Heidelberg, Germany. If you have any concerns regarding our products, please contact ProductSafety@springernature.com

Printed and bound by CPI Group (UK) Ltd, Croydon, CR0 4YY
01/10/2026
02225900-0013